TRANSISTOR ENGINEERING

McGRAW-HILL SERIES IN SOLID-STATE ENGINEERING

ALVIN B. PHILLIPS, *Consulting Editor*

Phillips—TRANSISTOR ENGINEERING
Uzunoglu—SEMICONDUCTOR NETWORK ANALYSIS AND DESIGN

TRANSISTOR ENGINEERING

and Introduction to Integrated Semiconductor Circuits

ALVIN B. PHILLIPS

Manager—Integrated Circuits Operations
Semiconductor Products Division
Motorola, Inc., Phoenix, Arizona

McGRAW-HILL BOOK COMPANY, INC.

New York Toronto London

TRANSISTOR ENGINEERING

 Library of Congress Catalog Card Number: 61-18729

49786

To My Wife Sally

and Sons Richard, Gregory, and Robert

PREFACE

This textbook has been written to satisfy the needs of both the student and the graduate engineer. The duality of purpose arises from the fact that the technological growth of transistors has been so rapid and intense since their conception in 1948 that neither the colleges nor the electronics industry has been properly equipped to keep up with the required fundamental understanding of transistors. Although most colleges have introduced semiconductors into their engineering curriculum, the student soon discovers that the treatment is either too generalized or has been already outdated. In all probability, this is attributable to the fact that most available textbooks emphasize the circuit and applications point of view of transistors and provide only a cursory treatment of their theory and design. Furthermore, since these college courses have been offered only in *recent* years, the many engineers, physicists, chemists, metallurgists, *et al.* already in industry find themselves partially handicapped by a lack of fundamental understanding of transistor devices, even though all are well trained in their respective technical fields.

These problems were particularly acute in 1956, when the idea for this book was born. The author was associated with the Semiconductor Products Department of the General Electric Company and had prepared and taught a course on the theory and design of junction transistors aimed at bridging the gap between transistor theory and actual engineering practice. Since its original inception, the material has been presented to over 150 engineers at General Electric. Most recently the author has been associated with the Motorola Semiconductor Products Division and has taught a considerably up-dated version of the course—which is the substance of this textbook—to over 100 technical personnel. These five years of teaching experience, coupled with the constructive comments of the students, have aided immensely in correcting and maturing the manuscript.

The purpose of the text is to equip the reader with a basic understanding of the principles of transistor theory and design and the relationship of all measured electrical characteristics to the physical properties of transistor material. It is not a book on semiconductor process technology. Rather, it points out that regardless of what process is used to manufacture a particular transistor, the theoretical principles are universally applicable. Thus, any transistor design may be *synthesized* from designated electrical characteristics, or vice versa; any transistor electrical characteristic may be *analyzed* from a designated transistor design. For the student, this provides a firm base for his entry into the field of solid-state engineering; for the semiconductor scientist and engineer, it provides the means of handling the design, development, and evaluation problems of the transistor; for the transistor circuit designer, the text will deepen his understanding of the device, enabling him to more properly optimize the application as well as to select the most suitable type from the various transistors in production today. Most important of all, the reader will realize that all the theory and principles presented in this book are equally applicable to the design and understanding of integrated semiconductor circuits.

A logical "building-block" approach has been used in establishing the pattern of this text; that is, the content of each chapter has a continuous dependence on the material in the chapters preceding it. The book begins with simplified arguments, and each chapter progressively introduces new factors, until the complex whole is understood at the end. Chapter 1 acquaints the reader with the many and varied transistors in use today. Chapter 2 provides a qualitative treatment of solid-state physics and introduces the concept of energy bands. Chapters 3 and 4 explain what semiconductors are and give a detailed description of carrier phenomena and the measurable properties of semiconductor materials. Chapters 5 and 6 introduce the theory of the p-n junction and cover, in detail, its pertinent electrical characteristics, recognizing that it is a fundamental part of the transistor. Chapters 7, 8, and 9 introduce transistors first from a qualitative and then from a quantitative point of view; that is, the generalized theoretical description of transistor mechanisms is followed by an explicit analysis of the important d-c characteristics. Chapter 10 extends the low-level theory to explain the d-c properties of transistors at high currents. With the d-c behavior as a reference, Chaps. 11 and 12 bring in the low-frequency characteristics of transistors, including the explanation of the small-signal h parameters. Chapters 13 and 14 extend the theory to the high-frequency performance of transistors. Chapters 15 and 16 treat the device as an amplifier and a switch, respectively, and illustrate the requirements for optimum performance. Finally, Chap. 17 demonstrates the application of the theoretical principles to the various transistor types, pointing out the limitations on design.

Throughout the text, the theory is modified to account for all the transistor types known today. Wherever possible, the requirements for optimum design and/or performance are discussed. Furthermore, almost all the design formulas are mathematically derived from basic fundamentals and are reduced to practical approximations. Additionally, the most recently accumulated data on semiconductor materials are contained throughout in the form of curves or tables wherever appropriate. Finally, the problems at the end of each chapter have been carefully selected to offer both the design approach and the necessary understanding of the chapter. The author has found it essential that the problems be worked out, since they provide the illustrative numerical examples of the theory.

It goes without saying that this book would not have been possible without the many contributions of research and development workers in the semiconductor industry. Included in the references are many prominent technical papers which are unquestionable classics in the field. In the organization and preparation of the material, the author wishes to give particular credit to W. Shockley, J. Early, and R. L. Pritchard, whose technical contributions to transistor theory are salient within the profession. Other individuals deserving important recognition are R. Beaufoy and J. J. Sparkes, J. J. Ebers and J. L. Moll, R. N. Hall, H. Krömer, and W. M. Webster. The author also would like to thank the management of the Semiconductor Products Department of General Electric for their cooperation in permitting him to start the course that led to this book. Thanks are also extended to his colleagues both at General Electric and Motorola for the many helpful technical discussions and the comments received. Special thanks go to Dr. C. Lester Hogan, Vice-president and General Manager, Semiconductor Products Division, Motorola, Inc., for his encouragement in the completion of the book, and to Dr. Daniel E. Noble, Executive Vice-president, Motorola, Inc., for the many stimulating discussions concerning the impact of solid-state engineering on modern electronics. Finally, the author is very grateful for the untiring efforts of Diane Smith, who typed the final manuscript, and to his secretary, Lee Belt, who assisted him during the proofreading and production stages of the book.

A. B. Phillips

CONTENTS

LIST OF SYMBOLS

A	Area
A_e	Emitter area
A_s	Surface recombination area
a	Impurity grade constant
α	Grounded-base current gain
α^*	Collector multiplication ratio
α_i	Carrier ionization rate
α_o	Low-frequency grounded-base current gain
α_R	Inverse current gain
BV_{CBO}	Collector-base breakdown voltage
BV_{CEO}	Collector-emitter breakdown voltage (open base)
BV_{CES}	Collector-emitter breakdown voltage (shorted base)
BV_{EBO}	Emitter-base breakdown voltage
b	Ratio of electron mobility to hole mobility
β	Grounded-emitter current gain
β_o	Low-frequency grounded-emitter current gain
β^*	Base-transport factor
β_o^*	Low-frequency base-transport factor
C_{Dc}	Collector diffusion capacitance
C_{De}	Emitter diffusion capacitance
C_T	Transition capacitance per unit area
C_{Tc}	Collector transition capacitance
C_{Te}	Emitter transition capacitance
C_c	Collector capacitance
C_n	Electron capture probability
C_p	Hole capture probability
c	Carrier concentration

γ	Emitter efficiency
γ_o	Low-frequency emitter efficiency
γ_R	Inverse collector efficiency
D	Diffusion constant
D_n	Electron diffusion constant
D_{nb}	Electron diffusion constant in base
D_{nc}	Electron diffusion constant in collector
D_{ne}	Electron diffusion constant in emitter
D_p	Hole diffusion constant
D_{pb}	Hole diffusion constant in base
D_{pc}	Hole diffusion constant in collector
D_{pe}	Hole diffusion constant in emitter
E	Energy
E_A	Acceptor-impurity energy
E_C	Conduction-band energy
E_D	Donor-impurity energy
E_F	Fermi-level energy
E_G	Band-gap energy
E_{GO}	Band-gap energy at absolute zero
E_R	Recombination-center energy level
E_V	Valence-band energy
E_i	Impurity ionization energy
E_{iA}	Acceptor-impurity band-gap energy
E_{iD}	Donor-impurity band-gap energy
e	Carrier generation rate
e_g	Generator voltage
$\mathcal{E}$	Electric field
$\mathcal{E}_M$	Maximum electric field
$\mathcal{E}_{MB}$	Maximum electric field at avalanche breakdown
ϵ_o	Permittivity of free space
f	Frequency
f_T	Current-gain-bandwidth frequency
f_b	Base-cutoff frequency
f_b'	Graded base-cutoff frequency
$f_{\max}$	Maximum frequency of oscillation
$f_{\alpha b}$	Alpha-cutoff frequency
$f_{\alpha e}$	Beta-cutoff frequency
G	Power gain
G_b	Grounded-base power gain
G_e	Grounded-emitter power gain

g_c Collector conductance
g_{oe} Grounded-emitter matched output conductance
g_L Load conductance

h_{FE} Grounded-emitter d-c current gain

h_{ib}	Short-circuit input impedance	grounded-base h parameters
h_{fb}	Short-circuit forward current transfer	
h_{ob}	Open-circuit output admittance	
h_{rb}	Open-circuit reverse-voltage feedback	
h_{ie}	Short-circuit input impedance	grounded-emitter h parameters
h_{fe}	Short-circuit forward current transfer	
h_{oe}	Open-circuit output admittance	
h_{re}	Open-circuit reverse-voltage feedback	

θ Phase angle

I Current
I_B Base current
I_{B1} Turn-on base current
I_{B2} Turn-off base current
I_{BX} Excess base current
I_C Collector current
I_{CBO} Collector-base reverse current
I_{CEO} Collector-emitter reverse current (open base)
I_{CES} Collector-emitter reverse current (shorted base)
I_D Volume diffusion current
I_E Emitter current
I_{EBO} Emitter-base reverse current
I_F Forward-biased current
I_G Charge generation current
I_R Reverse-biased current
I_{RE} Emitter recombination current
I_S Surface-leakage current
I_n Electron current
I_{nC} Collector electron current
I_{nE} Emitter electron current
I_p Hole current
I_{pC} Collector hole current
I_{pE} Emitter hole current
I_{sB} Surface-recombination base current
I_{vB} Volume-recombination base current
i_b A-C base current
i_c A-C collector current
i_e A-C emitter current

i_{nc}	A-C collector electron current
i_{pc}	A-C collector hole current
i_{ne}	A-C emitter electron current
i_{pe}	A-C emitter hole current
J	Current density
J_n	Electron current density
J_{nC}	Collector electron current density
J_{nE}	Emitter electron current density
J_p	Hole current density
J_{pC}	Collector hole current density
J_{pE}	Emitter hole current density
K_R	Sheet resistance ratio factor
K_θ	Excess phase-shift factor
k	Boltzmann's constant
κ	Dielectric constant
L	Diffusion length
L_n	Electron diffusion length
L_{nb}	Electron diffusion length in base
L_{nc}	Electron diffusion length in collector
L_{ne}	Electron diffusion length in emitter
L_p	Hole diffusion length
L_{pb}	Hole diffusion length in base
L_{pc}	Hole diffusion length in collector
L_{pe}	Hole diffusion length in emitter
l	Length
λ	Wavelength
M	Avalanche-current multiplication factor
m	Mass of electron
m_{eff}	Effective carrier mass
m_n	Effective electron mass
m_p	Effective hole mass
μ	Mobility
μ_I	Impurity mobility
μ_L	Lattice mobility
μ_{Ln}	Electron lattice mobility
μ_{Lp}	Hole lattice mobility
μ_{bc}	Base voltage-feedback factor
μ_{ec}	Emitter voltage-feedback factor
μ_n	Electron mobility

μ_{nb}	Electron mobility in base
μ_{nc}	Electron mobility in collector
μ_{ne}	Electron mobility in emitter
μ_p	Hole mobility
μ_{pb}	Hole mobility in base
μ_{pc}	Hole mobility in collector
μ_{pe}	Hole mobility in emitter
N	Impurity concentration
N_A	Acceptor impurity concentration
N_B	Base impurity concentration
N_B'	Base impurity concentration at emitter junction
N_{BC}	Background impurity concentration
N_C	Density of states in conduction band
N_D	Donor impurity concentration
N_E	Emitter impurity concentration
N_I	Ionized impurity concentration
N_R	Recombination-center concentration
N_V	Density of states in valence band
N_o	Surface impurity concentration
n	Electron concentration or density
n_D	Density of electrons bound to donors
n_i	Intrinsic carrier concentration
n_n	Majority electron concentration in n region
n_{nb}	Majority electron concentration in base
n_{nc}	Majority electron concentration in collector
n_{ne}	Majority electron concentration in emitter
n_p	Minority electron concentration in p region
n_{pb}	Minority electron concentration in base
n_{pc}	Minority electron concentration in collector
n_{pe}	Minority electron concentration in emitter
ν	Wave number
ν	Carrier injection ratio
p	Hole concentration or density
p_A	Density of holes bound to acceptors
p_C	Base hole concentration at collector
p_E	Base hole concentration at emitter
p_c	A-C base hole concentration at collector
p_e	A-C base hole concentration at emitter
p_n	Minority hole concentration in n region
p_{nb}	Minority hole concentration in base
p_{nc}	Minority hole concentration in collector

p_{ne} Minority hole concentration in emitter
p_p Majority hole concentration in p region
p_{pb} Majority hole concentration in base
p_{pc} Majority hole concentration in collector
p_{pe} Majority hole concentration in emitter

Q_B Stored base charge
Q_{BS} Stored base charge in saturation
Q_{CS} Stored collector charge in saturation
Q_{OFF} Emitter delay charge
Q_{Tc} Collector transition-capacitance charge
Q_{Te} Emitter transition-capacitance charge
Q_X Excess stored base charge
q Charge of electron

R Resistance
R Carrier-recombination rate
R_{BB} Base sheet resistance
R_{CC} Collector sheet resistance
R_{EE} Emitter sheet resistance
R_L Load resistance
R_n Electron recombination rate
R_p Hole recombination rate
r Radius
r_B' Base resistance
r_B'' Effective base resistance
r_{SC} Collector series resistance
r_{SE} Emitter series resistance
r_c A-C collector resistance
r_e A-C emitter resistance
r_g Generator resistance
r_{ie} Grounded-emitter matched input resistance

ρ Fixed charge density
ρ Resistivity
ρ_b Base resistivity
ρ_b' Effective base resistivity
ρ_c Collector resistivity
ρ_e Emitter resistivity
ρ_i Intrinsic resistivity
ρ_n Resistivity of n region
ρ_p Resistivity of p region

S Stripe dimension
s Surface recombination velocity

σ Conductivity

T Absolute temperature
t Time
t_{OFF} Turn-off time
t_{ON} Turn-on time
t_b Base transit time
t_b' Graded-base transit time
t_{dr} Diode recovery time
t_d Delay time
t_f Fall time
t_r Rise time
t_s Storage time

τ Lifetime
τ_B Base time constant
τ_C Collector time constant
τ_{CS} Saturated collector time constant
τ_E Emitter time constant
τ_{ER} Reverse emitter time constant
τ_F Fall time constant
τ_R Rise time constant
τ_S Storage time constant
τ_b Base lifetime
τ_{eff} Effective lifetime
τ_n Electron lifetime
τ_{nb} Electron lifetime in base
τ_{nc} Electron lifetime in collector
τ_{ne} Electron lifetime in emitter
τ_{nr} Electron lifetime in highly p-type region
τ_o Low-level lifetime
τ_∞ High-level lifetime
τ_p Hole lifetime
τ_{pb} Hole lifetime in base
τ_{pc} Hole lifetime in collector
τ_{pe} Hole lifetime in emitter
τ_{pr} Hole lifetime in highly n-type region
τ_{sb} Base surface lifetime
τ_v Volume lifetime

V Voltage or potential
V_B Avalanche-breakdown voltage
V_{BE} Base-emitter forward bias voltage
$V_{BE(\mathrm{OFF})}$ Emitter reverse bias voltage

V_C Collector voltage
V_{CB} Collector-base voltage
V_{CC} Collector supply voltage
V_{CE} Collector-emitter voltage
$V_{CE(\mathrm{SAT})}$ Grounded-emitter saturation voltage
V_E Emitter voltage
V_{EBO} Emitter floating potential
V_F Forward bias voltage
V_G Band-gap potential
V_{PT} Punchthrough voltage
V_R Reverse bias voltage
V_T Junction contact potential
V_j Junction voltage drop
V_o Applied junction voltage
V_s Series voltage drop
v Velocity
v_D Average drift velocity
v_{Dn} Electron drift velocity
v_{Dp} Hole drift velocity
v_c A-C collector voltage
v_e A-C emitter voltage
v_{sc} Scattering-limited velocity

ϕ Fermi-level potential
ϕ_n Fermi-level potential in n region
ϕ_p Fermi-level potential in p region
ϕ_n' Quasi-Fermi level in n region
ϕ_p' Quasi-Fermi level in p region

W Effective base width
W_e Epitaxial layer thickness
W_o Physical base width
W_1 Amplitude of a-c base-width variation

x Distance
x_C Collector junction position
x_E Emitter junction position
x_j Junction depth from surface
x_m Depletion-layer thickness
x_{mB} Depletion-layer thickness at avalanche breakdown
x_n Depletion-layer thickness in n region
x_p Depletion-layer thickness in p region
x_1 Depletion-layer thickness on base side of graded collector junction

y Admittance

ψ Electrostatic potential
ψ_n Electrostatic potential in n region
ψ_p Electrostatic potential in p region

Z Impedance
Z Injection factor

ω Angular frequency
ω_T Angular current-gain-bandwidth frequency
ω_b Angular base-cutoff frequency
ω_c Angular collector-delay cutoff frequency
ω_d Angular depletion-layer-transit cutoff frequency
ω_e Angular emitter-delay cutoff frequency
$\omega_{\alpha b}$ Angular alpha-cutoff frequency
$\omega_{\alpha e}$ Angular beta-cutoff frequency

TABLE OF CONSTANTS

Boltzmann's constant $k = 8.63 \times 10^{-5}$ ev/°K
Electron charge $q = 1.6 \times 10^{-19}$ coulomb
$q = 4.8 \times 10^{-10}$ esu
Free space permittivity $\epsilon_0 = 8.85 \times 10^{-14}$ farad/cm
Planck's constant $h = 6.63 \times 10^{-27}$ erg-sec
Electron mass $m = 9.11 \times 10^{-28}$ g
1 ev $= 1.6 \times 10^{-12}$ erg
$kT = 0.026$ ev ($T = 300$°K)
$kT/q = 0.026$ volt ($T = 300$°K)
1 mil $= 10^{-3}$ in.
1 $\mu = 10^{-4}$ cm
1 $\mu = 10^{4}$ A
1 in. $= 2.54$ cm
1 mil$^2 = 6.45 \times 10^{-6}$ cm^2

1

Transistor Processes and Types

It may be said that the discovery of the transistor was one of the great significant technological developments of our modern era, exclusive of the harnessing of nuclear energy. For the electronics industry, it represents the most radically different and unique invention since that of the vacuum tube at the beginning of the twentieth century. Since its inception in 1948, the transistor has not only created a new field of endeavor for engineers and scientists, but has already effected profound changes in the design philosophy and characteristics of electronic systems and equipments. The unusual features of transistors have revolutionized the hearing-aid industry, introduced vest-pocket radios, made portable television sets possible, stimulated more intensive computer development, and, last but not least, provided the military with the opportunity to attain greater efficiency and reliability for our defense systems and weapons. There is no doubt that the consequences of these dramatic technological changes will affect the lives of all of us today and for years to come.

1-1. Comparison with Vacuum Tubes. Basically speaking, the transistor, like the vacuum tube, is an electronic amplifying device. The fundamental distinction, however, rests with the fact that in the vacuum tube, current flow arises from the motion of charge carriers within a *vacuum*, whereas in the transistor, current flow arises from the motion of charge carriers within a *crystalline solid*. Further, in the vacuum tube the carriers must be generated by thermionic emission from a hot cathode whereas in the transistor the carriers are structually inherent to the solid material, and exist to the extent that they contribute to the electrical conductivity of the solid. In other words, we come to the first primary advantage that transistors have over vacuum tubes, namely, no filaments are necessary. This advantage manifests itself in the complete elimination of the A battery or equivalent filament power in transistorized equipment. For communications and computer applications in particular this represents an enormous saving in power consumption and cooling equipment.

A second advantage that transistors have over vacuum tubes is in the area of component reliability. Although both devices require an absolute minimum of contamination within the envelope to assure maximum operating life, the vacuum tube suffers from the drawback that eventually the filament becomes exhausted and loses its emission properties. In the case of the transistor, "emission" is simply a matter of circuit transfer of charge carriers and consequently would be self-sustaining ad infinitum. This is not to imply that a transistor cannot possibly fail in operation. Contamination, hermetic-seal failure, overheating, etc., can contribute to a degradation of pertinent transistor characteristics. Nevertheless, the reliability of today's transistors has been shown to be vastly superior to that of vacuum tubes. Statistical studies of the reliability of transistors which have been life-tested for millions of unit-hours indicated a maximum failure rate of about 0.001 per cent per thousand hours.

Thirdly, another feature which makes the transistor so extraordinary is its efficiency as an amplifier for typical small-signal applications. Transistors require biasing conditions which involve power consumptions far less than those of vacuum tubes. The disadvantage of vacuum tubes in this regard is that a moderately higher voltage is necessary, since, in principle, a strong electric field is required to accelerate the electrons to the plate. On the other hand, for transistors, voltages in the order of 1 to 10 volts are all that is necessary for proper operation. For example, transistor audio amplifiers can offer more than 40 db power gain at bias dissipations as low as 2 to 5 milliwatts.[1,*]

From the point of view of frequency and power, transistors have already made tremendous strides over vacuum tubes. Devices have been designed with maximum frequency of oscillation in the kilomegacycle region. Power transistors are being produced which are capable of delivering 10 watts of r-f power in the 160-megacycle region. High-frequency switching transistors turn on and off with such fantastic speeds that they have made possible the development of computers operating at clock rates of 100 megacycles per second.

Size is a fourth advantage of transistors over tubes. In Fig. 1-1 is an interesting comparison of the dimensions of a typical junction transistor and a 6AU6 miniature vacuum tube. This volume advantage is definitely appealing to all electronic equipment manufacturers who are concerned with miniaturization. An excellent example would be the design of military aircraft electronic equipments in which space and weight are at a

* References, indicated in the text by superscript figures, are listed at the end of the chapter.

premium. In addition to size, the transistor's inherent ruggedness makes it a strong contender with the vacuum tube. Constructed wholly of solid materials, transistors have been designed to withstand tremendous shocks, centrifugal forces, and vibrations, whereas tubes have been limited somewhat by their delicate grid structures.

In all, one might conclude that the vacuum tube would eventually be completely displaced by the transistor. Undoubtedly, in the majority of applications, as we have seen, this change has taken place. However, there is a very long way to go before transistors can displace tubes in every respect. As of 1961, the tube industry has been in existence for almost 55 years (as compared to the transistor industry's 13) and has

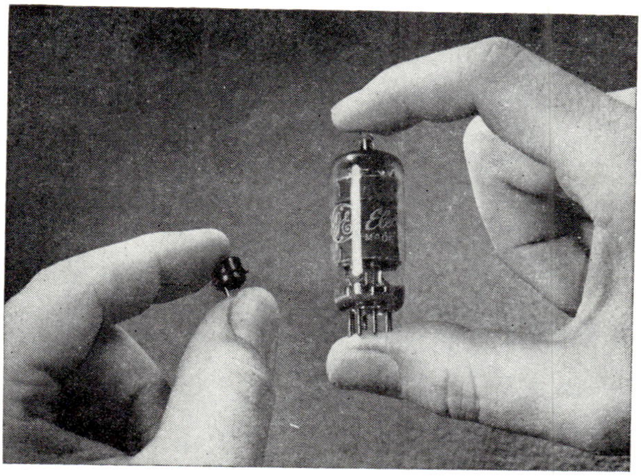

FIG. 1-1. Size comparison of junction transistor and typical miniature vacuum tube

developed a superb technology which assures the user of a good product both technically and economically. Nevertheless, the transistor has taken its place beside the vacuum tube, and as in all things, either device will be used in the application that warrants it.

1-2. History of Transistor Development. The history of transistors is actually the history of semiconductor research.[2] Our knowledge of semiconductors did not come about overnight, but resulted from a great many years of experimental exploration in physics and chemistry. Perhaps the first significant contribution to our understanding of solid-state phenomena and semiconductors was in 1833 when Faraday discovered that silver sulfide has a negative temperature coefficient of resistance which distinguishes it from metals. In the course of the next 50 years, the only significant additions to the knowledge of semiconductors were the observations of contact rectification and photoconduction. It was not until 1879 that E. H. Hall discovered that a current-carrying conductor would develop a transverse voltage when placed in a magnetic field perpendicular to the direction of current flow.[3] This now-famous Hall effect enabled workers to determine both the concentration and mobility of carriers in materials and thereby provided a means of differentiating semiconductors from metals on the basis of conductivity behavior. Although it could not be explained at the time, the Hall measurement also indicated the existence of positively charged carriers (holes).

During the twenties, our increased understanding of quantum mechanics added greatly to the explanation of electron behavior in solids. Diffraction experiments by Davison and Germer and Schrödinger's work in 1924 yielded the wave-nature concept of electrons. In 1928, Sommerfeld published his findings on the application of Fermi-Dirac statistics to explain electrical conduction.[4] In 1931, A. H. Wilson presented the first real theory of semiconductors using quantum mechanics as the basis of analysis.[5] This fundamental work introduced the concept of energy bands in solids and conduction by lattice-site deficiencies (holes), and formed the basis of our understanding semiconductors today.

Before World War II, activity in the device area was along the lines of copper-oxide and selenium rectifiers, which tended to be more a matter of art than of science. Of course, one may recall the many hours of sheer listening enjoyment derived from the simple galena detector which utilized the famous "catwhisker." At the time, rectification was not well understood, and it was not until 1939 that Schottky and Mott presented the first space-charge dipole theory. Just prior to the war, silicon detectors appeared on the scene and demonstrated their usefulness as high-frequency diodes in radar systems. The superiority of silicon as such soon launched the purification technology that we are familiar with today. In fact, at the Bell Telephone Laboratories, Scaff and Theuerer were successful in growing the first silicon p-n junction, which exhibited extraordinary rectification properties.

After World War II, developments came rapidly. Increased emphasis was placed on semiconductor research, and improved metallurgical techniques in crystal processing facilitated matters greatly. Out of an attempt to fit a theory to the many observed rectification phenomena, J. Bardeen offered his theory of surface states for metal-to-semiconductor contacts.[6] In 1948, this was followed by further experimentation by W. Shockley, who was concerned with field effects on the conductivity of semiconductor surface films. Also in 1948, using an electrolyte to bias a semiconductor surface, J. Bardeen and W. H. Brattain performed the work that created the first point-contact transistor and thereby opened up a new era of progress in electronics.[7] Soon after, in 1949, there came a work of comparable importance, that of Shockley's development of the theory of p-n junctions and junction-type transistors.[8] The establishment of a sound theory for semiconductor-to-semiconductor junctions stimulated rapid progress in the development of suitable technology for fabricating transistors of several types, including grown-junction transistors. At General Electric, R. N. Hall and W. C. Dunlap devised a method of preparing p-n junctions by alloy diffusion.[9] This is the basic process in the manufacture of the alloy-junction transistors of today.

By 1951, the potential impact of transistors on the field of electronics became apparent to many people. Recognizing the tremendous advantages that transistors would offer in the design of military electronic equipment, the United States Defense Department embarked upon a large program to accelerate the research, development, and production of semiconductors. Under the joint sponsorship of the military services, rather substantial contracts were let throughout the industry. The results have been quite rewarding for both the government and the electronics industry in that transistors have now become a major field of endeavor. In fact, so rapid was the progress since 1948, that in 1956 Drs. Bardeen, Brattain, and Shockley won the Nobel Prize in physics for their work in developing the transistor. Since then, contributions that furthered the theoretical understanding of transistors and offered tremendous improvements in process technology mounted at an extremely rapid rate. Of salient interest was the announcement in 1956 by researchers at Bell Telephone Laboratories of diffused-base *mesa* transistors.[10,11] This basic work made possible the reasonably low-cost production of the very-high-frequency high-reliability transistors which are in widespread use today. Added to this achievement was the introduction of the planar-epitaxial process in 1960, which made possible the truly universal transistor having optimum frequency, d-c, and reliability characteristics.

1-3. Fundamental Transistor Structure. As a possible definition for an active circuit element such as the transistor or vacuum tube, Pearson and Brattain suggested the following: "Two phase boundaries close enough together so that deviations from equilibrium occurring at either will influence the other and with electrical connections to each of the three regions bounded by the two phase boundaries."[2]

In the case of the transistor, the general definition does explain its fundamental operation, which is the scope of the entire text to follow. Thus, the transistor consists of two closely spaced boundaries, or more specifically, *junctions*, as is shown in Fig. 1-2. The narrow active region between the junctions is referred to as the *base*, and the two remaining regions are called the *emitter* and *collector*, respectively. In most transistors, the structure of one of the outer regions is sufficiently different from the other to designate it as the emitter.

The significant aspect of the transistor structure is that all the regions in the transistor are of solid materials. All three regions may comprise a homogenous crystalline semiconductor solid such as germanium or silicon, with the exception that each of the regions differs from the others primarily in the percentage concentration of certain intentionally-introduced impurities. Some of the elements used for impurities are indium, gallium, antimony, and aluminum. It is the change in impurity characteristic in going from one region to another that establishes the so-called junctions. In certain cases, it is possible for only the base region to be a crystalline solid, whereas both the emitter and collector are metals. In these cases, the junctions exist at the contact of the metal to the semiconductor. This is called a metal-to-semiconductor junction. Regardless of how the transistor is composed, i.e., whether the junctions consist of semiconductor-to-semiconductor or metal-to-semiconductor contacts, the aforementioned definition is applicable as long as the junctions are very closely spaced (in the order of 0.001 in. or less) and suitable electrical contact is made to each region.

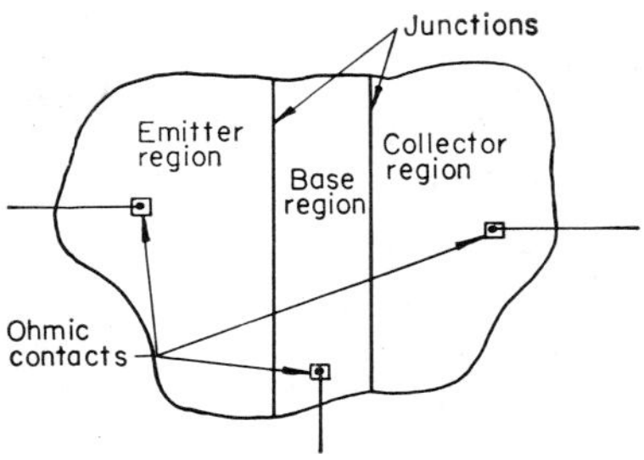

FIG. 1-2. Elements of the transistor as an active circuit device.

As will be studied later on in greater detail, the active aspect of the transistor, namely its amplification properties, is dependent upon the application of appropriate voltage potentials across the junctions by means of the electrical contacts to the individual regions. At the emitter-to-base junction the potential is such to permit excess carriers (current) to flow easily into the base region, wherein they move across to the collector-to-base junction. The potential at the latter junction is opposite to that at the emitter, such as to minimize the equilibrium flow of current from collector to base. However, the collector potential does not retard the flow of carriers in the base which arrive there from the emitter. These carriers therefore appear in the collector region as output current. It will be shown later on that, for these conditions, we obtain the transfer of a common current from a low-resistance junction (emitter) to and out of a high-resistance junction (collector). As a result of this effect, we can obtain signal amplification or power gain from the transistor. Thus, in accordance with the definition, we see that the carrier deviation from the equilibrium value at the emitter junction influences the current flow at the collector junction. It will also be shown later that voltage variations at the collector junction about the equilibrium value will influence conditions at the emitter junction.

Although the hypothetical structure of Fig. 1-2 is fundamental to transistor operation, there are an infinite number of combinations and ways of achieving the result physically. This brings out the essential point of this section, that the electrical characteristics of transistors are wholly structure dependent. This covers a host of items pertinent to transistor design, such as the nature of the junctions discussed in this section and many others which will be examined in detail later. For example, the choice and quality of the semiconductor material is important; also, the distribution and types of impurities in the various regions become vital factors. From this it becomes evident that the design of a particular transistor structure to achieve certain electrical properties would be related to the manner in which the various solid materials are

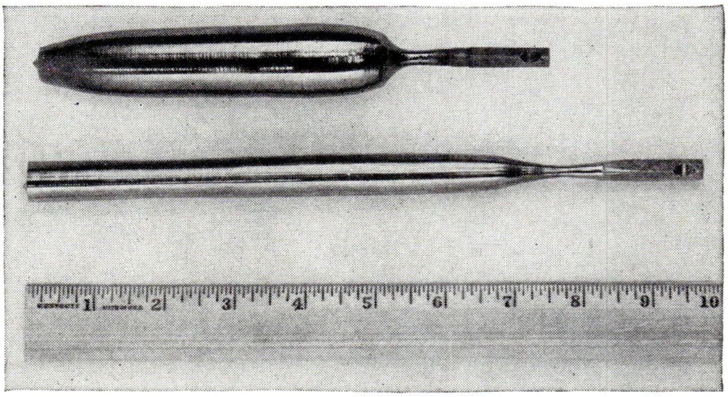

FIG. 1-3. Production-size single crystals of germanium and silicon. (*Courtesy of Motorola, Inc.*)

fabricated and processed. In the next sections, we shall discuss the various process methods that are being used to produce junction-transistor structures. It is not the intention here to cover all the ramifications of transistor process technology, but rather to provide the necessary background for transistor theory and design.

1-4. Semiconductor Crystal Growing. Regardless of what final manufacturing process is employed, the heart of any transistor is the tiny section of single-crystal semiconductor material, usually either germanium or silicon. These small sections are actually obtained from a large single crystal which is very carefully grown in specially designed furnaces. Typical production-type crystals of germanium and silicon used in the manufacture of transistors are shown in Fig. 1-3. Crystals of this size, when properly sawed and diced into pellets, will provide sufficient material to fabricate thousands of junction transistors.

Crystal growing is commonly done by the widely adopted Czochralski technique, although there are other preferred methods. The growing, or "pulling," as it is often called, is done in a special furnace apparatus such as the one shown in Fig. 1-4. The entire process is carried out in an inert-gas atmosphere within a relatively contaminant-free quartz enclosure. Extreme cleanliness is absolutely essential here, since the smallest amount of undesirable foreign impurities can readily degrade the quality of the crystal being pulled.

In the Czochralski method, the highly purified semiconductor material is placed in a crucible which is mounted at the bottom of the furnace.

FIG. 1-4. Crystal-growing furnaces. (*Courtesy of Motorola, Inc.*)

In the case of germanium, which melts at about 960°C, the crucible material is usually made of graphite. However, silicon, which melts at about 1420°C, will react strongly with carbon at that temperature, and therefore high-quality quartz is used for the crucible when silicon crystals are grown. Heat is applied to the crucible by means of radio-frequency induction or resistance heating in order to melt the material and bring it to a temperature just a few degrees above the melting point. At the upper portion of the furnace is a pulling rod containing a single-crystal seed of the semiconductor mounted in a chuck. The seed is usually precisely cut at a specific crystallographic plane, in order that the crystal to be pulled will grow at a desired orientation. The seed is lowered into the surface of the melt and then slowly withdrawn. Under conditions of accurate temperature control, the melt will freeze out onto the seed in the form of a single crystal of the same orientation. The seed is pulled continuously at a constant rate until all the melt in the crucible is exhausted. Although the entire process is quite critical, very uniform crystals can be pulled repeatedly.

Actually, extremely pure crystals are not desirable for transistor manufacture, since the electrical conductivity would be too low. It is the practice in semiconductor technology, therefore, to purify the material to the highest degree attainable, and then to add a known quantity of desired impurity during the crystal growing to obtain the required conductivity. This procedure is referred to as *doping*. When donor-type impurities such as phosphorus, arsenic, and antimony are added, the crystal is called "n type." A crystal doped with acceptor-type impurities such as aluminum, gallium, and indium is called "p type." The significance of this terminology will be treated in detail in Chap. 3. Since we are dealing with impurity dopings which range from one part in ten thousand to as little as one part in ten million, it is essential that there be adequate mixing during crystal growth in order to obtain good uniformity. This is accomplished by rotating the seed during the pull, so as to stir the melt just below the freezing line.

As will be seen in subsequent sections, for certain methods of transistor manufacture it is necessary to add both donor and acceptor impurities to the semiconductor melt, in order to have both types in the final crystal. The resulting crystal is either n or p type, depending on which impurity is predominant.

1-5. Basic Methods of Junction Formation. As was defined earlier, a semiconductor junction is formed when one can establish within the semiconductor crystal a change from a donor-type impurity concentration to an acceptor-type impurity concentration or vice versa. Generally speaking, junction formation is accomplished by introducing an impurity of one type into a crystal already doped with an impurity of the opposite type. The most popular methods of introducing desired impurities into the semiconductor are *segregation* from an impurity-doped melt or alloy and solid-state *diffusion* from an impurity concentration source at the boundary of the semiconductor. Since segregation and diffusion are two of the most fundamental mechanisms in modern transistor technology, it is essential that these be treated at this point.

Segregation makes use of the property that most impurities are more soluble in the liquid or molten semiconductor than in the solid. Therefore, when a crystal freezes out from a melt containing a certain impurity concentration, the concentration of that impurity in the solidified crystal will be less than originally in the liquid. The ratio of the concentration in the solid to the concentration in the liquid is called the *segregation* or *distribution coefficient.* Each of the various impurities has a unique set of segregation coefficients for germanium and silicon. Experimentally established values are given in Table 1-1.[12]

TABLE 1-1. IMPURITY SEGREGATION COEFFICIENTS

Impurity	In germanium	In silicon
Boron (B)	17.4	0.8
Aluminum (Al)	0.1	0.002
Gallium (Ga)	0.1	0.008
Indium (In)	0.001	0.0004
Phosphorus (P)	0.08	0.35
Arsenic (As)	0.02	0.3
Antimony (Sb)	0.003	0.023

As an illustrative example of the significance of these coefficients, suppose one were to alloy pure aluminum in both germanium and silicon. Then the resultant aluminum concentration in the resolidified germanium would be 50 times greater than the concentration in silicon. It is interesting to note that when both donor and acceptor impurities are in a common solution, they will segregate independently, according to their respective coefficients.

Diffusion is probably the most widely used method of forming junctions in semiconductors. This is especially true for very-high-frequency transistor processes. In this method, the desired impurities are introduced either at the surface of the semiconductor from a gaseous ambient or from an abrupt change in impurity concentration already contained within the solid semiconductor. In the former method, the semiconductors to be diffused are placed in a high-temperature furnace through which an inert gas containing a small amount of the desired impurity is flowing. The surface concentration of the impurity is determined by its vapor pressure, which is controlled by the source temperature. The depth of penetration of the diffusing impurity (diffusant) is determined by the time and temperature of the furnace cycle and by the diffusion constant D for the impurity. Figures 1-5 and 1-6 show useful curves of the experimental diffusion constants as a function of temperature for various impurities in germanium and silicon.[12] These curves provide information as to the relative rate of diffusion of various impurities. For example, in germanium, indium is a very slow diffusant, more than two orders of magnitude slower than antimony at a given temperature. The reader may find a very complete discussion of diffusion technology in the reference by Smits.[13]

1-6. Methods of Transistor Manufacture. The objective in transistor-device processing is to obtain a specified distribution of impurities within the semiconductor such that the final result consists of two closely spaced parallel junctions. Furthermore, each region should be of the highest quality consistent with the impurity level there. This involves the well-known semiconductor parameters, such as mobility and lifetime, which will be discussed in Chap. 4.

At the present time, there are three major groups of processes that have been commercially accepted for manufacturing junction transistors. This author wishes to introduce his own terminology to classify the processes along more proper categorical lines. These processes are

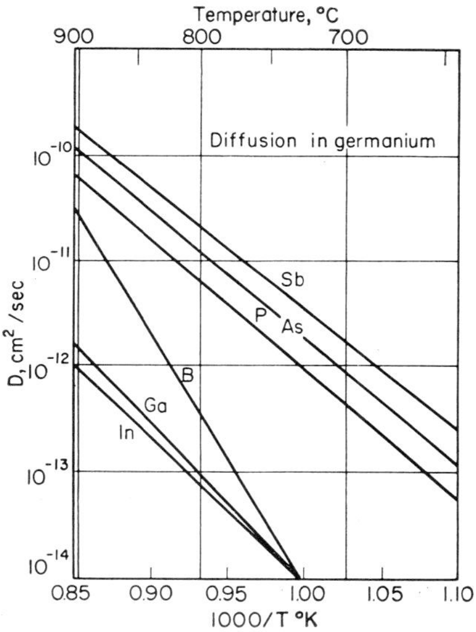

FIG. 1-5. Diffusion coefficients of impurities in germanium.

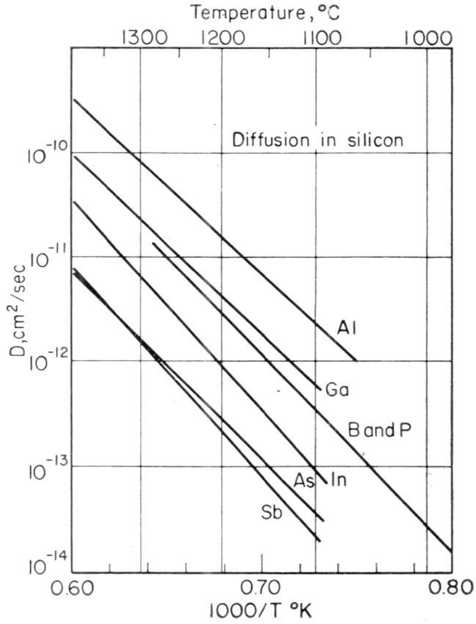

FIG. 1-6. Diffusion coefficients of impurities in silicon.

called (1) the grown-junction process, (2) the double-ended impurity-contact process, and (3) the single-ended impurity-contact process. Under each of these categories there are several variations of the fundamental principle which is suggested by the name. These are listed in Table 1-2.

Essentially, all three methods accomplish the objective of attaining proper impurity control in each of the transistor regions, but they differ primarily in the requirements for the initial semiconductor crystal. The highlights of the actual processes will be examined briefly in the next sections. In Chap. 17 the subject will be treated in greater detail in order to bring out the design limitations of these process types.

It should be noted that both the point-contact and surface-barrier

TABLE 1-2. TRANSISTOR MANUFACTURING PROCESSES

Grown-junction	Double-ended impurity-contact	Single-ended impurity-contact
Double-doped	Alloy	Diffused-base mesa or planar†
Rate grown	Diffused-base alloy*	Diffused-emitter-base mesa or planar†
Meltback		Alloy-diffused
Meltback-diffused		
Grown-diffused		

* Includes both drift and MADT types.
† Also includes epitaxial types.

transistor processes were omitted from Table 1-2 because of the fact that the junctions are formed from metal-to-semiconductor contacts, whereas all the others are homogenous semiconductor-to-semiconductor junctions. As a matter of definition, transistors that are made from processes that yield semiconductor-to-semiconductor junctions will be called *junction transistors*. There is by far a greater number of junction transistors being produced and used commercially than metal-to-semiconductor types. Because of this and the fact that the theory has been more fully developed for junction transistors, the scope of this text will be confined to these types only.

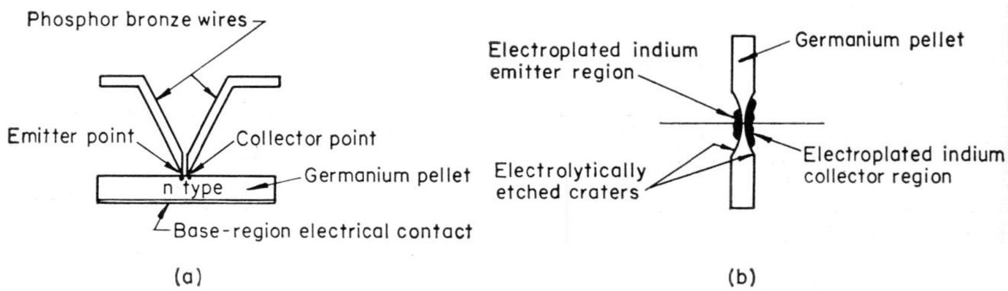

FIG. 1-7. Metal-to-semiconductor transistor types. (*a*) Point-contact transistor; (*b*) surface-barrier transistor.

For completeness, however, we will briefly describe the point-contact and surface-barrier types. Although the point-contact transistor was the first to be developed and fabricated, it rapidly gave way to the junction types. As shown in Fig. 1-7*a*, the point-contact transistor consists of two finely pointed whisker wires closely spaced on the surface of an n-type germanium pellet. The wires are attached by a process called *forming*, which is a variation of electric welding. This transistor was not conducive to uniform manufacture because of the very critical problem of forming and spacing the point contacts; this resulted in somewhat uncontrollable electrical characteristics. Point-contact transistors have current gains greater than unity and are characterized with negative-resistance regions, which made them useful as switches.

The surface-barrier types, on the other hand, lend themselves to reasonable manufacturing processes. These transistors, introduced by the Philco Corporation,[14] are manufactured by a combination of electrolytic etching and plating. A semiconductor pellet is electrolytically etched with jet streams playing on opposite sides until the pellet is thinned down to the desired thickness. The polarity is then reversed and the same solution acts to electroplate on metal emitter and collector regions. This technique is capable of producing extremely thin base regions, which in turn yield high-frequency response. A cross-sectional drawing of a typical germanium surface-barrier transistor is shown in Fig. 1-7*b*. The thin base region, however, limits the useful application of these transistors to low power dissipations because of the very low voltage rating and high thermal resistance of this type of structure. Contemporary junction-transistor designs are bettering the frequency performance of the surface-barrier transistor with much greater power-handling capacity as well.

Additional information on both the point-contact and surface-barrier types, as well as the theory of rectification of metal-to-semiconductor contacts, may be found in the literature.[15]

1-7. Grown-junction Processes. The grown-junction processes have as a common denominator the fact that the final transistor structure is in the form of a solid crystal bar in which there is established a thin base region. There are two basic methods of producing the desired junction structure. One involves the control of both the concentration and type of impurity that segregates from the melt into the solid during the initial crystal growth. For a relatively short interval of time, a thin section of the growing crystal will become of opposite type to that of the remainder, thereby creating a base region in a plane transverse to the growth line. The other method involves the redistribution of controlled amounts of impurities already contained in a bar which is sawed out from a previously grown crystal. The former grown-junction process includes double-doping, rate-growing, and grown-diffusion; the latter includes meltback and meltback-diffused.

a. Double-doped Process. This grown-junction process makes use of the principle of changing the impurity doping concentration in the melt at appropriate intervals during the crystal-growing cycle.[16] This technique is generally used for growing n-p-n structures in either germanium or silicon. At the outset, the crystal is grown n type with a conductivity corresponding to the collector region. At approximately the halfway point, a doping pill containing acceptor impurities is dropped into the melt and effectively stirred by the crystal rotation. Although the crystal is still growing, it abruptly changes to p type and is permitted to continue growing p type for another 0.5 to 1 mil. Then the melt is doped again with a donor pill and the crystal reverts to n type of relatively higher conductivity corresponding to the emitter region. The one disadvantage of the double-doped process is that only one base region can be grown during a crystal pull. However, this is offset by the fact that the conductivity of each region can be independently controlled, offering a limited degree of transistor designability. After the crystal pull, a wafer containing the p region is sawed out and then diced into n-p-n bars. The remaining segments of the double-doped crystal are not usable, but may be reclaimed by further purification.

b. Rate-grown Transistors. Rate-growing is a grown-junction technique developed by R. N. Hall of the General Electric Research Laboratory[17] and is based on the way the segregation coefficients of different impurities will vary with rate of crystal growth. This method has been successfully applied to growing n-p-n germanium bars using antimony as the donor and gallium as the acceptor. In the rate-growing process, the crystal is pulled from a melt containing both impurities. At the normal pulling temperature, both antimony and gallium are segregating from the melt; however, there is a predominance of antimony and the crystal grows n type. It was found, however, that if the temperature is raised suddenly to decrease the growth rate, the segregation of antimony drops almost to zero, whereas the gallium remains practically unaffected by the temperature change. Consequently, for the short interval that the furnace temperature is increased, the gallium predominates over the antimony and the crystal grows p type. When the temperature is returned to normal, the crystal goes back to n type. The unique feature of rate growing is that the temperature cycling may be repeated at fixed intervals during the crystal-growing period, yielding as many base regions as temperature cycles. Such a crystal is shown in Fig. 1-8. The ridge lines correspond to the p-type base regions, formed when the furnace temperature is increased. This crystal, after waferizing and dicing, would yield approximately 2,000 rate-grown n-p-n transistor bars. Rate-grown transistors are characterized by fairly abrupt emitter junctions and graded collector junctions. The latter contributes to higher collector voltage ratings and lower output capacities.

c. Meltback Transistors. The principle involved in this method is identical to that previously described for rate-growing. It differs in that the junctions are formed in individual bars directly rather than in the crystal. For germanium transistors, bars are cut from crystals that are doped with a predetermined ratio of antimony-to-gallium impurities. The antimony is in excess of the gallium, making the bar uniformly n type.

FIG. 1-8. Rate-grown germanium crystal. (*Courtesy of General Electric Co.*)

If one end of the bar is melted in a special furnace and permitted to freeze again, the impurities will segregate as shown in Fig. 1-9*a*. Initially, at the liquid-solid interface the gallium concentration predominates. However, as the bar cools the antimony will segregate more rapidly than the gallium because of the increase in growth rate. The net effect is that a p region is formed where the gallium is in excess, producing an n-p-n germanium transistor bar. After meltback, the bar acquires a teardrop shape at one end and the base region is right at the frozen melt line.

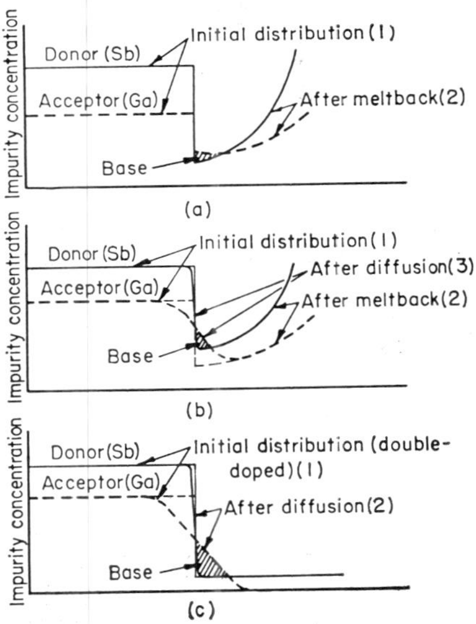

FIG. 1-9. Typical impurity profiles for transistors made from grown-junction processes. (*a*) Germanium meltback transistor; (*b*) silicon meltback-diffused transistor; (*c*) silicon grown-diffused transistor.

d. Meltback-diffused Transistors. This method is applicable to n-p-n silicon or p-n-p germanium types. In silicon, for example, the bars are doped with antimony and gallium and are melted back in the same manner as in the meltback process just described. However, after meltback no base region is formed because antimony segregates more rapidly than gallium in silicon.

The result is the formation of an n-type region of considerably lower impurity concentration just at the melt side of the interface. This is shown in Fig. 1-9*b*. The bar is then subjected to a long-time high-temperature thermal cycle which causes the large concentration of impurities in the unmelted region to *diffuse* into the region of lower concentration. Since gallium diffuses much faster than antimony in silicon, the final distribution after diffusion will be such that a p-type base region is formed on the melt side of the interface. Because the diffusion takes place over a period of many hours, precise timing is not required, and consequently the process is capable of producing very thin base regions with good reproducibility.

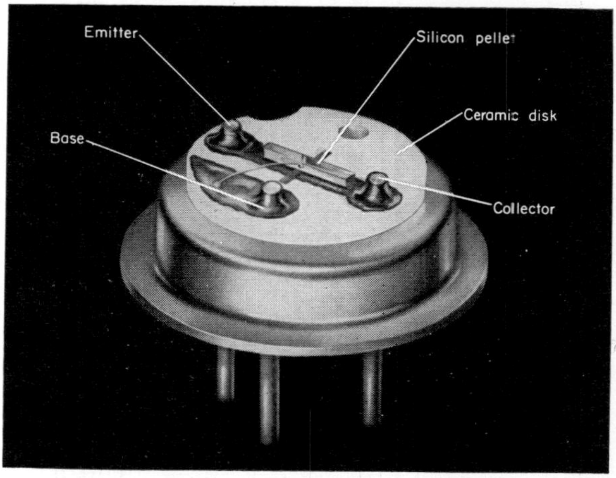

FIG. 1-10. An n-p-n silicon transistor made by the grown-diffused process. A silicon bar is fixed to a metalized ceramic disk. (*Courtesy of General Electric Co.*)

e. Grown-diffused Transistors. The grown-diffused process, developed by Texas Instruments, Inc., is fundamentally similar to the meltback-diffused process, but employs the double-doping method of growing the initial crystal. The crystal is grown n type, lightly doped with antimony, before the doping pill is added to the melt. The doping pill contains a mixture of antimony and gallium, with the former impurity in excess. The result is that the crystal being pulled abruptly changes to a more heavily doped n-type composition. As shown in Fig. 1-9*c*, this second region contains a moderate level of p-type gallium. Because of the very high temperature at the solid-liquid interface, the gallium impurities diffuse into the region of lower impurity concentration, thereby forming a thin base region of moderately high p-type conductivity. After diffusion, a wafer containing the diffused base region is sawed from the crystal and diced into n-p-n bars. A typical silicon transistor fabricated from a grown-diffused bar is shown in Fig. 1-10.

The grown-junction processes just described all offer the advantage of achieving good control over the geometry and thickness of the transistor base regions. This is important in obtaining uniform electrical characteristics in the final devices. However, because of the bar geometry, frequency response is somewhat limited by the smallest cross-sectional area that can be practically handled during manufacture. Furthermore, the inherent frequency performance is also degraded by the contact wire to the base region which, of necessity, overlaps into the adjacent regions of the bar.

1-8. Double-ended Impurity-contact Processes. In the double-ended impurity-contact process for making junction transistors, the transistor structure is obtained by placing the appropriate impurities in contact with both sides of a homogenous pellet of the semiconductor crystal and applying heat to form the junctions. Depending upon the temperature and materials involved, the impurities will alloy with and/or diffuse into the parent crystal. With other processing factors equal, careful control of time and temperature will produce uniform base regions. The alloying and diffusion methods are discussed below.

FIG. 1-11. A p-n-p junction transistor made by the alloy process. (*Courtesy of Motorola, Inc.*)

a. Alloy Transistors. This process by far accounts for the largest number of junction transistors in use today. This impurity-contact process, originated by Hall and Dunlap,[9] became readily adapted to germanium devices using indium metal as the acceptor impurity. Figure 1-11 is a photograph of an uncapped typical germanium-alloy transistor. During the manufacture, round dots of indium are fixed to opposite sides of a small square pellet of n-type germanium. The pellet dimensions may be ⅛ in. square and about 0.010 in. thick. The collector dot is usually made larger than the emitter to improve the current-gain characteristics. The assembly is brought up to a temperature of approximately 600°C, whereupon the indium alloys into the germanium until a saturated liquid solution of both materials is formed on both sides. The assembly is then permitted to cool down and the germanium commences to freeze out of solution back onto the parent crystal. The recrystallized layers are identical in structure with the parent crystal, but become heavily doped with indium, making them highly p type. If the pellet is cut at a specific crystal orientation, the liquid alloy will penetrate the germanium in a flat plane, resulting in planar-parallel junctions. Figure 1-12*a* is a typical cross section. The final electrical characteristics of the alloy transistor are determined in part by the area of the dots and their depth of penetration. As will be seen, the alloy process lends itself to the production of transistors having excellent low-frequency-gain characteristics and switching characteristics and high current ratings for power applications. Furthermore, both p-n-p and n-p-n types in either germanium or silicon may be made by this process.

b. Diffused-base Alloy Transistors. As a means of increasing the frequency response of alloy transistors, the diffused-base alloy process family was developed. This process is fundamentally similar to the alloy process, with the exception that an additional impurity is diffused into one

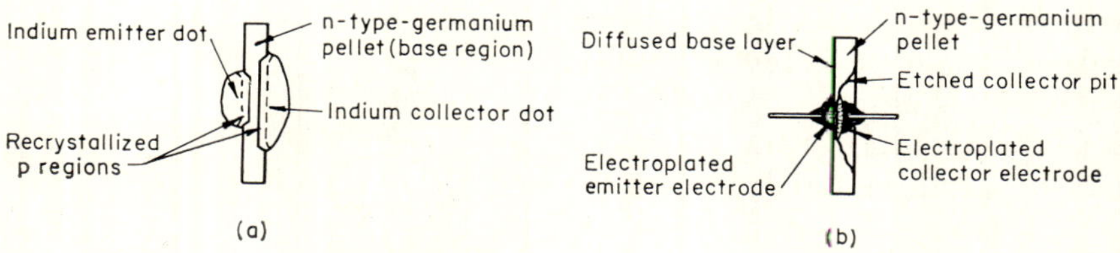

FIG. 1-12. Cross sections of transistor structures made by the double-sided impurity-contact processes. (*a*) Alloy process; (*b*) diffused-base alloy process (MADT).

side of the base pellet prior to the alloying of the emitter and collector materials. Because of this base diffusion, one gains a very significant improvement in frequency response, since very thin base regions can be formed. At this time, there are available two basic transistor types that utilize this process principle, viz., the drift transistor and the microalloy diffused transistor.

The drift transistor, which was introduced commercially by RCA,[18,19] is so named because of the presence of the accelerating drift field in the base layer. This comes as a result of the diffused base-impurity distribution. The drift field provides a manifold increase in frequency performance.

The microalloy diffused transistor, commercially described by Philco as the MADT,[20] is a diffused-base alloy transistor which is manufactured by electrochemical techniques similar to that described previously for the surface-barrier transistor. In this process, an impurity is diffused into a semiconductor pellet to form the base layer. Pits are then etched into both sides of the pellet in order to reduce the base to a very thin membrane, approximately 0.0004 in. or less in thickness. Electrode contacts are then electroplated into the pits and microalloyed (shallow penetration) to form the junctions, resulting in the cross section shown in Fig. 1-12*b*. The MADT is a very-high-frequency transistor having excellent amplifier and switching characteristics. However, because of its basic structure, it must be restricted to low-voltage applications and low-power-dissipation operation.

1-9. Single-ended Impurity-contact Processes. All these processes have in common the fact that one starts with a semiconductor wafer corresponding to the collector and that the junctions are formed by the subsequent addition of selected impurities into one side of the wafer. These impurities are introduced by either diffusion or alloying or combinations of both. The one unique feature of these processes is that the use of vacuum-evaporation and photoetch techniques permits the formation of several hundred transistor junction structures with contacts on one

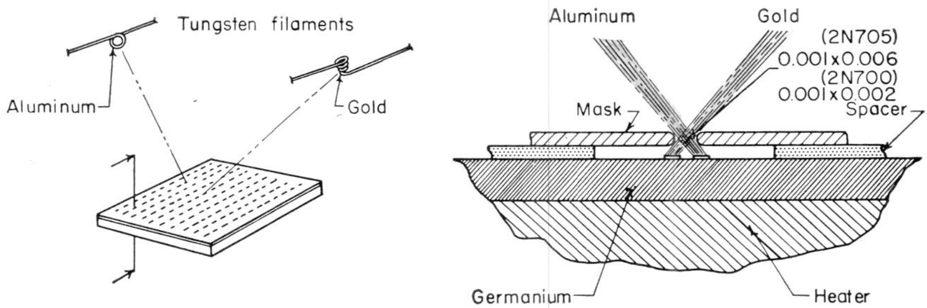

FIG. 1-13. Schematic of cross-evaporation technique used in forming stripes in germanium diffused-base mesa process. (*Courtesy of Motorola, Inc.*)

wafer. This wafer is subsequently cut into individual dice which are then ready for assembly. Presently, there are three variations of the single-ended impurity-contact process, viz., the diffused-base mesa, the diffused-emitter-base mesa, and the alloy-diffused transistor.

a. Diffused-base Mesa. In this process, the first junction is formed by the gaseous diffusion of an impurity opposite in type to that already contained in the semiconductor wafer. Using a p-n-p germanium mesa transistor as an example,[21] we would have a diffused n-type base region established on one side of a p-type wafer. Such a wafer may have a diameter of 0.5 in. and a thickness of 0.005 in.; the thickness of the diffused base layer would be in the order of 0.00004 in. The emitter junction is formed on the base layer by the vacuum evaporation of a p-type impurity such as aluminum through a metal mask containing an array of rectangular slots. Since the masks are prepared by a photoetch process, one can achieve rectangular stripes of aluminum having dimensions of the order of 0.001 in. wide and 0.006 in. long. Furthermore, as many as 400 stripes may be evaporated onto a wafer at one time. This emitter-stripe evaporation is commonly done by the cross-evaporation method illustrated in Fig. 1-13, in order that the adjacent base-contact stripe may be formed at the same time. The base-contact stripe is usually evaporated gold containing a small percentage of n-type impurity. As can be seen from the figure, by cross-evaporating the aluminum emitter and the gold base through the same slot, one achieves a very close spacing between the stripes, which is essential for good high-frequency performance. Thus, the use of the evaporation principle and the precision mask permits the formation of extremely small stripe areas with excellent accuracy and reproducibility. By simply subjecting the resulting wafer to a short high-temperature cycle, all the stripes alloy and form junctions that have very shallow penetration into the diffused base layer. The last step in the wafer process involves the selective etching of the *mesa* (Spanish word meaning table), which eliminates any edge shorts across the collector junction and reduces the active area of the

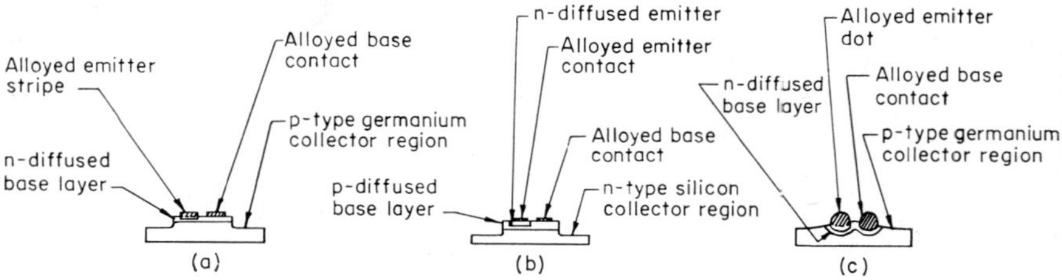

FIG. 1-14. Cross sections of transistor structures made by the single-sided impurity-contact processes. (*a*) p-n-p germanium diffused-base mesa transistor; (*b*) n-p-n silicon diffused-emitter-base mesa transistor; (*c*) p-n-p germanium alloy-diffused transistor.

collector junction to a small rectangle enclosing the pair of stripes. The resulting junction structure is clearly illustrated in Fig. 1-14*a*. The completed wafer is then scribed into dice which are then ready for assembly. A photograph of a highly magnified p-n-p germanium diffused-base mesa transistor is shown in Fig. 1-15*a*. Electrical connections to each of the stripes are made by the thermocompression bonding of high-purity gold wire having a diameter of 0.0005 in.

b. Diffused-emitter-base Mesa. This single-sided transistor process is similar in principle to the diffused-base mesa previously described, with the exception that the emitter junction is formed by diffusion rather than by evaporation and alloying. A cross section of a typical n-p-n silicon diffused-emitter-base transistor-junction structure is shown in Fig. 1-14*b*. In this process, the base layer is formed by diffusing a p-type impurity such as boron into an n-type wafer of silicon. After the base diffusion, a thin layer of silicon oxide is thermally grown on the surface of the base layer. The oxide has the property of blocking the diffusion of certain impurities such as phosphorus, which will be used to diffuse the n-type emitter into the p-type base layer.[22] By covering the silicon oxide with a thin coat of photosensitive, etch-resistant material (e.g., Kodak Photo Resist), one may employ photographic techniques to selectively remove the resist and leave an array of rectangular openings across the wafer. The resulting uncovered rectangular stripes of oxide are then etched, the resist is removed, and there remains an accurate silicon oxide diffusion mask. The wafer is then subjected to the diffusion of the emitter impurity, i.e., phosphorus, which will diffuse into the base layer only

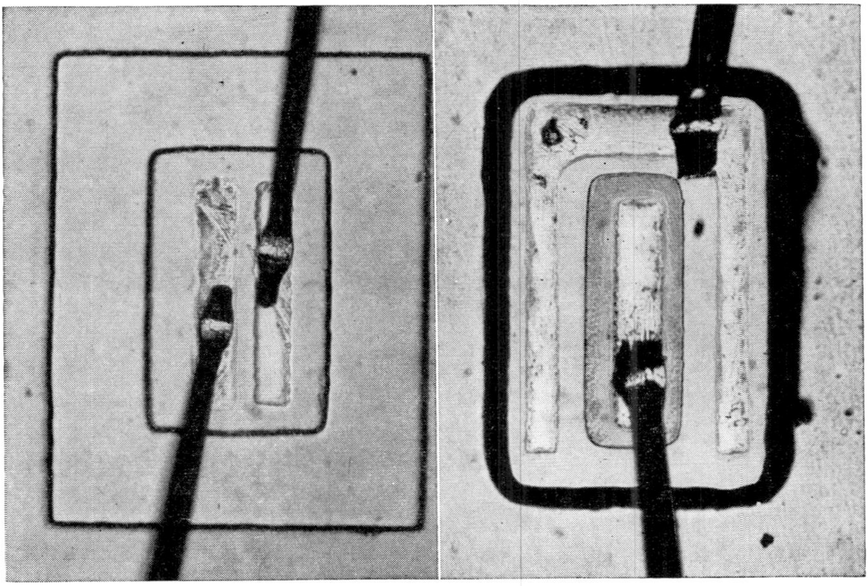

(*a*) (*b*)

FIG. 1-15. High-frequency mesa transistor. The metallic stripes are approximately 0.001 in. wide. (*a*) p-n-p germanium diffused-base mesa transistor; (*b*) n-p-n silicon diffused-emitter-base mesa transistor. (*Courtesy of Motorola, Inc.*)

where the oxide has been removed. Thus, diffused-emitter junctions are established in the form of precise rectangles. Actually, any shape or pattern can be obtained simply by using the appropriately designed photographic plate. Before the mesas are etched, ohmic contacts to the emitter and base regions are applied by means of either mask evaporation or the photoetch method. Either technique involves a critical alignment of the mask or plate to the diffused silicon wafer. A photograph of a typical n-p-n silicon mesa transistor is shown in Fig. 1-15*b*.

c. Alloy-diffused Transistor. Philips Research of Holland was the first to introduce this method commercially. This technique utilized the same principles involved in forming the base layer in the grown-diffusion process in that the base is formed by an impurity which diffuses out from an alloyed emitter. In the Philips process for making p-n-p alloy-diffused transistors, pairs of n-type doped lead shot are alloyed into a p-type germanium wafer. To one dot, which will become the emitter, there is added a large concentration of a p-type impurity such as aluminum. During the high-temperature alloying, the n-type impurity (antimony) diffuses rapidly from the liquid-solid interfaces of both dots and forms an n-type base layer which joins together in the narrow space between the alloy dots. Upon cooling, the aluminum, having a very high segregation coefficient, forms an alloyed emitter junction. The other alloy dot remains n type, forming an ohmic contact to the diffused base layer. The resultant structure is illustrated in Fig. 1-14*c*. Although this process involves relatively noncritical process steps, it is restricted to relatively larger junction configurations because of the dot-handling problem.

Because of the extremely thin diffused base layers and the very small junction areas that are achieved by each of the three aforementioned single-sided impurity-contact processes, these transistor types offer a superior combination of very-high-frequency and d-c characteristics. Mesa transistors make excellent signal and power amplifiers, oscillators, and high-speed switching transistors. Furthermore, since the collector regions of these structures can be mounted directly to heat sinks, these devices have high power-dissipation ratings, good operating efficiency, and high reliability. Lastly, because large numbers of wafers can be processed at a time, with several hundred potential devices represented by each wafer, these transistor types are very economical to manufacture.

1-10. The Epitaxial-planar Process. As had been indicated in previous sections, the two basic processes that yield the highest-frequency transistor types are the microalloy-diffused-transistor and the mesa-transistor processes. With particular regard to switching applications, the mesa transistor's power-dissipation capability was offset by the MADT's switching characteristics. Both of these factors are results of the design of the collector regions of these transistors. In 1960, however, the epitaxial technique was introduced into the mesa process and yielded the epitaxial-mesa transistor, which not only retained its power-dissipation capability, but also equaled the switching characteristics of the MADT.

The epitaxial process[23] provides a means of growing a very thin high-purity single-crystal layer of semiconductor material on a very heavily doped crystal wafer of the same type. For either germanium or silicon, this may be done by the hydrogen reduction at high temperature of $GeCl_4$ or $SiCl_4$, leaving the pure semiconductor atom to grow as a crystalline extension of the wafer substrate.[24] By incorporating the epitaxial wafer into the mesa-transistor process in place of the conventional homogenous wafer, one obtains a very significant improvement in frequency-response, d-c, and switching characteristics. In the mesa transistor, the collector region is designed as a compromise between switching performance and voltage; in the epitaxial device, because of the thin high-resistivity layer, no design compromise is necessary. The principles involved here will be more clearly understood as the design theory is uncovered in the chapters to follow.

Another major improvement in transistor technology that occurred in 1960 was the development of the silicon planar process by Fairchild.[25] The planar method is primarily a means of improving the reliability of diffused-emitter-base transistors. In this process, the base-collector junction area is determined by an oxide diffusion mask which is photoetched just prior to the base diffusion. The emitter is then diffused in a

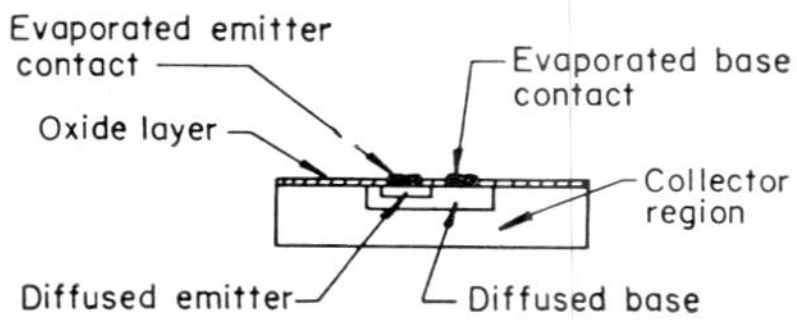

FIG. 1-16. Cross section of n-p-n silicon planar transistor.

manner identical to that described in the previous section. After both diffusions, a final silicon oxide is thermally grown over the surface of the entire wafer, thereby passivating all the surface junctions against the environment. The resulting structure is as shown in Fig. 1-16. The essential feature of the planar process is that no mesa etching is required; with the edge of the collector junction in the plane of the oxide-protected wafer surface, very low leakage currents result. Furthermore, additional improvement in noise figure and low-current-gain characteristics are obtained as a result of the gettering action of the oxide on undesirable impurities in the semiconductor material.

The salient features of the epitaxial and the planar processes are combined with those of the silicon diffused-emitter base to make the epitaxial-planar transistor. The process-flow diagram for this device is given in Fig. 1-17. Although the various processes in the sequence are somewhat sophisticated and critical, modern transistor technology has reached a point wherein the epitaxial-planar transistor is very economical to produce. It appears that this process is the first major step toward achieving the universal transistor, offering the ultimate in switching speed, very-high-frequency amplification, d-c characteristics, voltage ratings, power dissipation, operating temperature, and reliability.

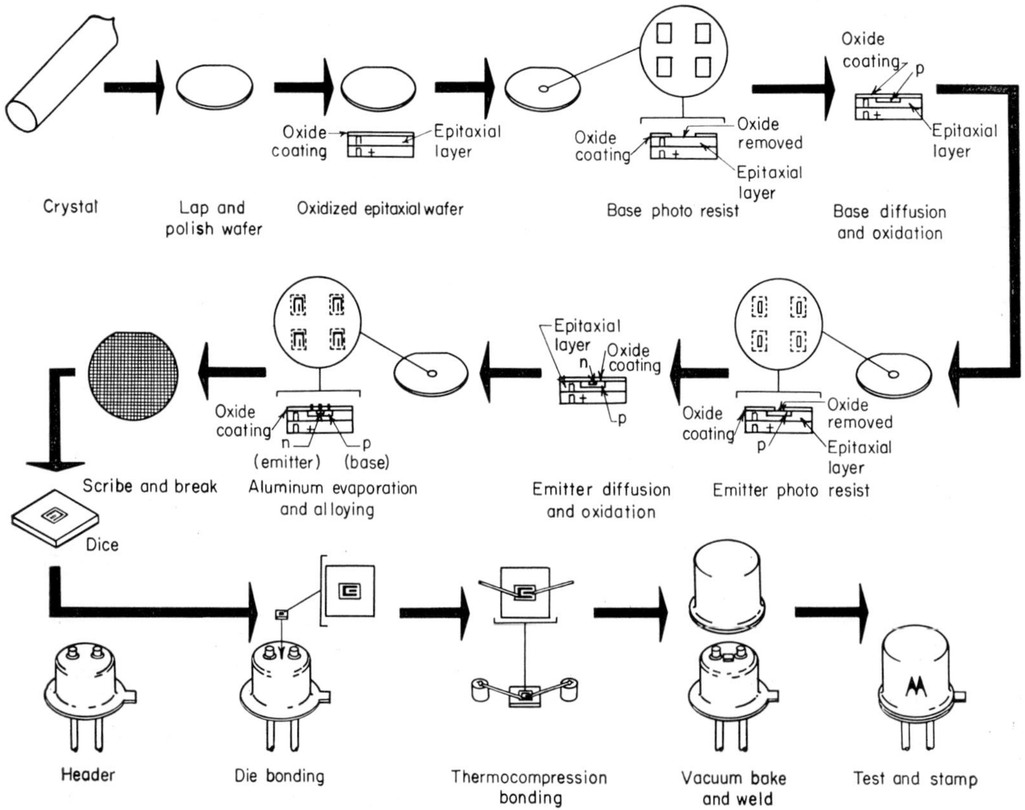

FIG. 1-17. n-p-n silicon epitaxial-planar transistor fabrication. (*Courtesy of Motorola, Inc.*)

1-11. Scope of Transistor Technology. The various processes for manufacturing junction transistors that were presented in the preceding sections cover only the means of preparing the basic junction structures. No attempt was made to treat all the process procedures related to making electrical connections to the emitter, base, and collector regions, nor was there any discussion of the problems associated with the surface preparation, packaging, and encapsulation of the final transistor assembly. These are problems which, in the last analysis, are important to junction-transistor design, but are really a matter of additional process technology subject to the discretion of the individual manufacturers. At the present time, each manufacturer has his own particular ideas as to what the over-all physical design shall be. The many variations of the alloy transistor in production today show this. Ultimately, it is expected that the transistor industry will adopt standard designs and processes.

Nevertheless, the process descriptions given in this chapter should provide a broad understanding of what is involved in the manufacture of junction transistors. It must not be inferred that these descriptions represent the full limit of contemporary research and development. There are many ramifications and variations of these techniques which were not illustrated here, but are well covered in the literature. For example, in several of the processes, n-p-n or p-n-p transistors in either germanium or silicon can be obtained simply by selecting the correct combination of impurities. The essential point to note, however, is that regardless of what process is employed to make the transistor, the theory to be presented in the subsequent chapters will be applicable. In other words, as long as the specific properties and distributions of the impurities in the semiconductor regions are known, it is possible to predict with the theory what the electrical characteristics of the transistor will be. From the device-design point of view, the theory and design formulas can be used to determine the impurity structure that is necessary to satisfy a given set of electrical specifications for the transistor.

REFERENCES

1. Fink, D. G.: Transistors vs. Vacuum Tubes, *Proc. IRE*, vol. 44, p. 481, April, 1956.
2. Pearson, G. L., and W. H. Brattain: History of Semiconductor Research, *Proc. IRE*, vol. 43, pp. 1794–1806, December, 1955.
3. Hall, E. H.: On a New Action of the Magnet on Electric Currents, *Am. J. Math.*, vol. 2, November, 1879.
4. Sommerfeld, A.: Zur Electronentheorie der Metalle auf Grund der Fermischen Statistik, *Z. Physik*, vol. 47, February, 1928.
5. Wilson, A. H.: The Theory of Electronic Semiconductors, *Proc. Roy. Soc. (London)*, vol. A133, October, 1931.
6. Bardeen, J.: Surface States and Rectification at a Metal to Semiconductor Contact, *Phys. Rev.*, vol. 71, pp. 717–727, May 15, 1947.

7. Bardeen, J., and W. H. Brattain: The Transistor, A Semiconductor Triode, *Phys. Rev.*, vol. 74, pp. 230–231, July 15, 1948.
8. Shockley, W.: The Theory of PN Junctions in Semiconductors and PN Junction Transistors, *Bell System Tech. J.*, vol. 28, pp. 435–489, July, 1949.
9. Hall, R. N., and W. C. Dunlap: PN Junctions Prepared by Impurity Diffusion, *Phys. Rev.*, vol. 80, p. 467, Nov. 1, 1950.
10. Lee, C. A.: A High Frequency Diffused Base Germanium Transistor, *Bell System Tech. J.*, vol. 35, pp. 23–24, January, 1956.
11. Tanenbaum, M., and D. E. Thomas, Diffused Emitter and Base Silicon Transistors, *Bell System Tech. J.*, vol. 35, pp. 1–22, January, 1956.
12. Hall, R. N.: Fabrication Techniques for High-frequency Transistors, G. E. Research Laboratory Report 58-RL-1874, January, 1958.
13. Smits, F. M.: Formation of Junction Structures by Solid-State Diffusion, *Proc. IRE,* vol. 46, pp. 1049–1061, June, 1958.
14. Philco Research Division Technical Staff: The Surface-barrier Transistor, parts I to V, *Proc. IRE,* vol. 41, pp. 1702–1720, December, 1953.
15. See, for example, Hunter, L. P.: "Handbook of Semiconductor Electronics," McGraw-Hill Book Company, Inc., New York, 1956.
16. Teak, G. K., M. Sparks, and E. Buehler: Growth of Germanium Single Crystal Containing PN Junctions, *Phys. Rev.*, Vol. 81, p. 637, Feb. 15, 1951.
17. Hall, R. N.: PN Junctions Produced by Rate Growth Variation, *Phys. Rev.*, vol. 88, p. 139, Oct. 1, 1952.
18. Krömer, H.: The Drift Transistor, *Naturwissenschaften,* vol. 40, p. 578, 1954.
19. Kestenbaum, A. L., and N. H. Ditrick: Design, Construction, and High Frequency Performance of Drift Transistors, *RCA Rev.*, vol. 18, pp. 12–23, March, 1957.
20. Thornton, C. G., and J. B. Angell: Technology of Micro-alloy Diffused Transistors, *Proc. IRE,* vol. 46, pp. 1166–1176, June, 1958.
21. Knowles, C. H.: New Transistor Design—The "Mesa," *Electronic Ind.*, August, 1958.
22. Aschner, J. F., C. A. Bittman, W. F. J. Hare, and J. J. Kleimack: A Double Diffused Silicon High-frequency Switching Transistor Produced by Oxide Masking Techniques, *J. Electrochem. Soc.*, vol. 106, June, 1959.
23. Christensen, H., and H. C. Theurer: Unpublished work, Bell Telephone Laboratories.
24. Russell, G.: Epitaxial Growth of Germanium and Silicon Single Crystal Films from the Vapor Phase, Professional Group on Electron Devices, Washington, D.C., October, 1960.
25. Hoerni, J. A.: Planar Silicon Diodes and Transistors, Professional Group on Electron Devices, Washington, D.C., October, 1960.

2

Atomic Structure and Energy-band Theory

2-1. Introduction. The conduction processes observed in semiconductors such as the transistor involve the mass movement of "particles" whose mode of behavior may be analyzed by classical methods. As long as the existence and nature of these particles is accepted, then a rather complete device theory can be formulated. Although this may be satisfactory, it leaves much to be desired if no attempt is made to describe the origin of this particle behavior from fundamental energy considerations. Such a study would provide a better understanding of semiconductor mechanisms.

It is the purpose of this chapter, therefore, to present an explanation of the nature of electron energy bands in crystals, based on atomic structure and the electromechanical properties of electrons. The next chapter will show how these energy bands give rise to the particles, viz., holes and electrons. A thorough treatment of this subject would necessitate a postgraduate course in solid-state physics and quantum mechanics and would cloud the text with rigorous mathematical equations and abstract concepts which in themselves have no obvious physical meaning. Nevertheless, an excellent presentation of the subject can be made using simplified models and analogies as a basis; exact analysis would be only of academic interest. It is for these reasons that the treatment of so formidable a subject in this chapter is largely qualitative.

2-2. Planck's Quantum Hypothesis. Our understanding of the atom and its structure historically embraces the ideas and experiments of many great scientists. The present knowledge, however, is based largely on the work of Max Planck, who in 1900 introduced his quantum theory to account for physical phenomena that were unexplainable at the time.

Planck was concerned with the nature of radiation emitted from heated blackbodies.[1] * It is observed from experimentation that the distribu-

* References, indicated in the text by superscript figures, are listed at the end of the chapter.

tion of energy as a function of wavelength is of the form shown in Fig. 2-1.

At that time, no complete explanation of the shape of the radiation curves was available. Wilhelm Wien was able to theoretically account for the energies at the shorter wavelengths by applying thermodynamic considerations. Similarly, Rayleigh and Jeans came up with a formula which worked very well for large values of λ. Neither theory held for the entire blackbody spectrum.

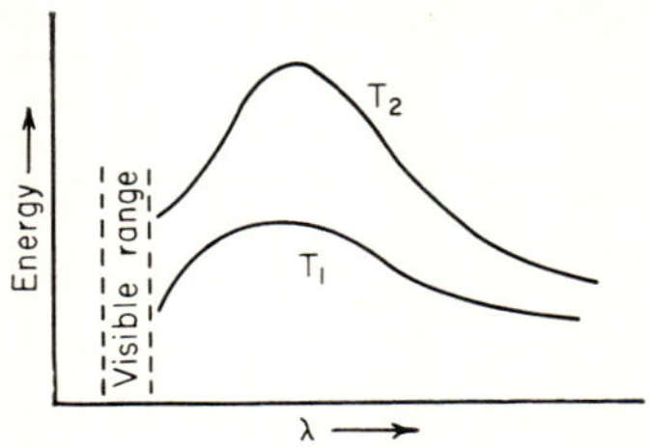

FIG. 2-1. Distribution of radiaated energy.

Planck approached the problem by introducing the idea that a radiating body is composed of many harmonic oscillators, each capable of responding to a particular frequency. Furthermore, he postulated that the energy distribution of the oscillators was not continuous, but occurred in multiples of some discrete energy which he called a quantum. The magnitude of a quantum of energy, which can be absorbed or emitted by any oscillator of frequency f, is given by the following equation:

$$E = hf \tag{2-1}$$

where E = energy, ergs
f = frequency
h = Planck's constant, 6.63×10^{-27} erg-sec

From this hypothesis, Planck was able to formulate an equation which satisfactorily explained the entire radiation spectrum.

It should be noted that this granular or quantum concept of energy is radically different from that maintained by the classical Newton mechanics, whereby the energy of a body can change from one value to another through every conceivable value. Planck's assumptions represent an empirical reasoning done in order to bring theory into agreement with experiment. The basic concept, interpreted from Eq. (2-1), is as follows: Any physical system is capable of having only a discrete set of possible energy values; the difference between any two values is absorbed or emitted as radiation equal to $f = \Delta E/h$. The next sections will treat the application of this concept to the energies of electrons in the atom.

2-3. The Planetary Atom. It is now commonly accepted that an atom consists of a small nucleus composed of positive electrical charges called *protons* and electrically neutral particles called *neutrons*. The nucleus is surrounded by a number (equal to the number of protons) of negatively charged particles, namely, *electrons*. This number is referred to as the atomic number of the element. The neutrons are recognized as fundamental units of mass and, together with the protons, determine the atomic weight of the element. Figure 2-2 illustrates the planetary configuration of an atom of oxygen. The nucleus is composed of eight protons and eight neutrons and the total positive charge is neutralized by eight electrons. Thus, the atomic number for oxygen is 8 and its atomic weight is 16.0.

In this model of the atom, the electrons travel about the nucleus in circular orbits, wherein the centripetal force of the revolving electron is balanced by the Coulomb attraction of the nucleus charge.[2] Consider, for example, the hydrogen atom, which consists of one electron revolving about a single proton nucleus. The hydrogen nucleus contains no neutrons and its atomic number is 1. The force exerted upon the electron would be equal to $-q \times q/\kappa r^2$ or $-q^2/\kappa r^2$ where r is the radius of orbit, q is the electron charge (4.80×10^{-10} esu), and κ is the dielectric constant, equal to unity for free space. If the electron is revolving with linear velocity v, the acting centripetal force is mv^2/r, where m is the mass of the electron, equal to 9.11×10^{-28} g. If the magnitudes of the two forces are equated,

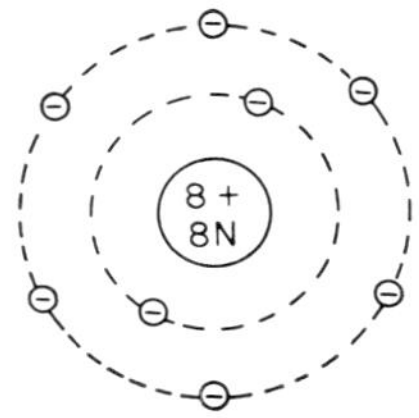

FIG. 2-2. Oxygen atom.

$$\frac{q^2}{\kappa r^2} = \frac{mv^2}{r} \tag{2-2}$$

the radius of orbit is found to be

$$r = \frac{q^2}{\kappa m v^2} \tag{2-3}$$

Additionally, the energy of the electron is made up of its potential energy, $-q^2/\kappa r$, and its kinetic energy, $\frac{1}{2}mv^2$. Thus the total energy of the electron is

$$E = \frac{-q^2}{\kappa r} + \tfrac{1}{2}mv^2 = \frac{-q^2}{2\kappa r} \tag{2-4}$$

The foregoing equations could be satisfied by any value of the radius r, and if no other limitations were imposed, the electron could have any conceivable value of energy. If this were the case, on the basis of Planck's Eq. (2-1), the spectrum of hydrogen would comprise all frequencies. Spectrographic analysis of hydrogen reveals that this is contrary to fact, indicating that Eq. (2-4) is not completely correct.

2-4. Hydrogen Spectra. It is evident from the spectrum of a burning hydrogen flame that the electron could not transcend a continuous range of energies, since only four discrete spectral lines are prominently visible. The wavelengths of these lines are 6,563, 4,861, 4,340, and 4,102 A[2]. Although no theoretical basis was offered at the time, in 1885 Johann

Balmer found that these spectral values formed a series which satisfied the following empirical relationship:

$$\lambda = 3{,}646 \frac{N^2}{N^2 - 4} \tag{2-5}$$

where λ is the wavelength in angstroms and N is an integer having values greater than 2. If N is given successive values of 3, 4, 5, and 6, then the four previously mentioned wavelengths are obtained.

In addition to these, other spectral lines have been obtained beyond the visible region. Lyman made measurements in the ultraviolet, and Paschen established a series in the infrared region. Both of these series were found to satisfy equations similar to Balmer's. All were merged into the single equation

$$\nu = R\left(\frac{1}{M^2} - \frac{1}{N^2}\right) \tag{2-6}$$

where ν is the so-called wave number, that is, the number of waves per centimeter and R is a constant, called the Rydberg number, equal to 109,737 cm^{-1}. M and N are integers and are given values, depending on the series, as follows:

Series	Spectrum	M	N
Lyman	Ultraviolet	1	$N = 2, 3, 4, \ldots$
Balmer	Visible	2	$N = 3, 4, 5, \ldots$
Paschen	Infrared	3	$N = 4, 5, 6, \ldots$

It is interesting to note that other equations similar to Eq. (2-6) were found to be in good agreement with spectrographic data for other elements. In fact, the Rydberg constant came to be empirically universal.

2-5. The Bohr Atom. In 1913, Niels Bohr applied Planck's quantum hypothesis to the planetary picture of the atom and was thus able to offer the first physical explanation of the discrete line spectra. Bohr postulated that the revolving electron could occupy any one of a definite number of orbits without radiating; each orbit represents a definite energy level. Energy is radiated or absorbed only by jumps between orbits of different energy levels.

For the hydrogen atom, the normal state corresponds to the electron residing in its innermost orbit. When the atom receives energy, the electron jumps to a larger orbit; when the atom loses energy by radiation, the electron falls back to definite inner orbits. To obtain agreement with experiment, Bohr assumed that the angular momentum of the electron would have to be an exact multiple of $h/2\pi$, where h is Planck's constant.[2]

Mathematically, this is written as

$$mvr = n\frac{h}{2\pi} \tag{2-7}$$

where n is an integer, referred to as a quantum number. In 1916, Sommerfield extended the Bohr concept from circular to elliptic orbits for the revolving electron. In a circular orbit, the angle of rotation is the only variable; in an elliptic orbit, both the angle of rotation and radius vary during one revolution. In any case, the conservation of angular momentum mvr is maintained. In other words, for an elliptic orbit, the closer the electron comes to the nucleus, the faster it will travel, and conversely.

For the circular orbit, by combining Eqs. (2-7) and (2-3), eliminating v between them, and solving for the radius, we obtain

$$r = \frac{n^2h^2\kappa}{4\pi^2mq^2} \tag{2-8}$$

where r becomes the radii of allowed orbits for the Bohr hydrogen atom. The energy of the electron at each orbit is found by substituting Eq. (2-8) into Eq. (2-4) and obtaining

$$E = \frac{-2\pi^2mq^4}{n^2h^2\kappa^2} \tag{2-9}$$

Thus, the only allowed energy levels exist for integral values of n, i.e., $n = 1, 2, 3, \ldots, \infty$. The lowest energy state corresponds to $n = 1$, while $n = \infty$ indicates that the electron is free and far from the attraction of the nucleus; that is, the atom is ionized.

As indicated previously, the energy radiated by an electron jumping from one level to another is equal to hf. Application of Eq. (2-9) shows that the total energy change is given by

$$\Delta E = \frac{2\pi^2mq^4}{h^2\kappa^2}\left(\frac{1}{n_1{}^2} - \frac{1}{n_2{}^2}\right) = hf \tag{2-10}$$

where the n's correspond to the respective energy states. Since the frequency of radiation f is equal to νc, where c is the velocity of light in centimeters per second, Eq. (2-10) may be solved for the wave number.

$$\nu = \frac{2\pi^2mq^4}{h^3c\kappa^2}\left(\frac{1}{M^2} - \frac{1}{N^2}\right) \tag{2-11}$$

This relationship gives the wave number of the spectral line emitted by this energy transfer, and is identical in form to Eq. (2-6), which was derived empirically. If numerical values are applied to the constants, it will be found that the coefficient will be in excellent agreement with the Rydberg constant, $R = 109{,}737\ \text{cm}^{-1}$.

Figure 2-3 shows the energy levels and spectral series for the hydrogen atom for values of n up to $n = 6$. From the figure, it is seen that the Lyman series corresponds to jumps to the lowest level $n = 1$; the Balmer series represents jumps to the second level $n = 2$; the Paschen series represents jumps to the third level. Recently, other series which are related to jumps to the higher orbits have been observed in the extreme infrared.

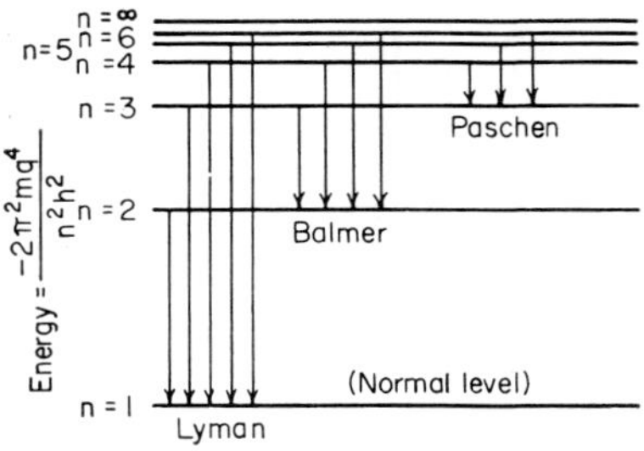

FIG. 2-3. Energy levels for hydrogen.

2-6. Quantum Wave Mechanics. The Bohr theory of the atom, referred to as the old quantum theory, has been superseded in recent years by a new theory which still preserves the concept of discrete energy states, but renders an entirely abstract interpretation for the nature of matter. This new theory is called the new quantum theory of wave mechanics, and has as its basis the work of de Broglie, who in 1924 suggested that matter might be considered to have a wave nature similar to that of light energy.[3] For the electron particle, de Broglie postulated that the "matter waves" would have a wavelength of

$$\lambda = \frac{h}{mv} \tag{2-12}$$

where m and v are the mass and velocity of the electron. It is seen that this relationship is consistent with Bohr's quantized momentum, i.e., $mvr = h/2\pi$, where $\lambda = 2\pi r$. This suggests that the circumference of any electron orbit corresponds to an integral number of wavelengths. The wave hypothesis was later verified by the work of Germer and Davisson, who discovered that electrons can be diffracted in crystals as though they were waves of the wavelength predicted by de Broglie. These developments soon led to the mathematical wave functions, introduced by Schrödinger, which opened up a new era of understanding of atomic physics.

Schrödinger hypothesized that the laws of motion of the electron particle may be governed by the classical wave equation[4]

$$\frac{\partial^2 \phi}{\partial x^2} + \frac{\partial^2 \phi}{\partial y^2} + \frac{\partial^2 \phi}{\partial z^2} = \frac{1}{c^2}\frac{\partial^2 \phi}{\partial t^2} \tag{2-13}$$

where ϕ is a periodic function equal to

$$\phi = \psi(x,y,z)e^{j(2\pi c/\lambda)t} \tag{2-14}$$

If Eq. (2-14) is substituted into (2-13) to eliminate the time dependence and de Broglie's equation $\lambda = h/mv$ is also substituted, the wave equation becomes

$$\frac{\partial^2\psi}{\partial x^2} + \frac{\partial^2\psi}{\partial y^2} + \frac{\partial^2\psi}{\partial z^2} = \frac{-4\pi^2 m^2 v^2}{h^2}\psi \tag{2-15}$$

The total energy E of an electron is the sum of its kinetic and potential energies V.

$$E = \tfrac{1}{2}mv^2 + V$$
$$mv^2 = 2(E - V) \tag{2-16}$$

Therefore, in terms of energy rather than velocity, Eq. (2-15) becomes

$$\frac{h^2}{8\pi^2 m}\left(\frac{\partial^2\psi}{\partial x^2} + \frac{\partial^2\psi}{\partial y^2} + \frac{\partial^2\psi}{\partial z^2}\right) + (E - V)\psi = 0 \tag{2-17}$$

This is the famous Schrödinger wave equation for three dimensions, wherein E is the total energy of the electron, independent of position, and V is its potential energy, a function of position. The solution of this equation for various boundary conditions provides the mathematical foundation on which the new quantum theory is based and furnishes a satisfactory explanation of the phenomena occurring in solid-state physics and chemistry.

In a sense, the Schrödinger wave equation describes the electromechanical behavior of the electron as it moves about the atom in the presence of potential fields due not only to its nucleus but to other electrons and atoms as well. The wave nature is imparted by the function ψ, the so-called wave function that is the solution of the equation. Strictly speaking, it is almost meaningless to attribute any physical significance to the wave function, for it is regarded only as a mathematical abstraction. According to quantum mechanics,[5] the square of the wave function $|\psi^2|$ is a probability density which is a measure of the amount of time the electron spends in each elemental volume of space occupied by its wave function. Since the electron must be somewhere in the region of the atom bounded by the wave function, $\int|\psi^2|\,dV$ must equal unity. Consequently, $|\psi^2|$ gives the time-average charge-density distribution produced by the electron and might be depicted pictorially as a "spherical cloud of charge," as shown in Fig. 2-4.

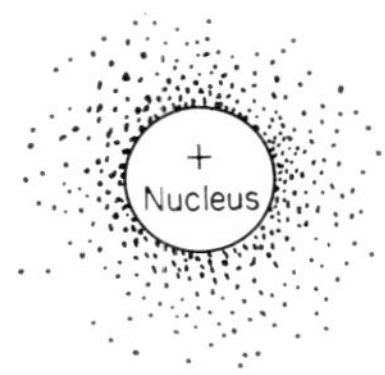

FIG. 2-4. Wave-function charge density. (*After W. Shockley.*)

The wave-function solutions of the Schrödinger equation, necessary for determination of the allowed energy states for atomic systems, are beyond the scope of this text. For the simplest case of the single electron of the hydrogen atom, it has been shown that results for the allowed electron energies similar to those derived from the Bohr theory are obtained.

2-7. Quantum States. Wave-mechanical analysis dictates that the ψ functions obtained for any atom are not characterized by a single quantizing number such as the n in the Bohr theory, but by a group of quantum numbers for different quantum states. By analogy, the solutions of the Schrödinger wave equation are similar to the solutions of the problem of a vibrating membrane, where only certain frequencies or normal modes will satisfy the particular boundary conditions. These modes may be described by a set of constants depicting the number of half wavelengths for each direction. In the same manner, the wave function is characterized by its number of wavelengths in the x, y, and z directions, yielding the three quantum numbers, n, l, and m. For each quantum number n, where $n = 1, 2, 3$, etc., there are values for $l = 0, 1, 2, 3, \ldots, (n - 1)$. Additionally, the third quantum number m will have values of $m = 0, \pm 1, \pm 2, \ldots, \pm l$. Thus there are $(2l + 1)$ different possible states for each quantum number.

In addition to its motion through space, the electron also rotates about its own axis. This rotation introduces a fourth degree of freedom and a fourth quantum number, the *spin*. For reasons associated with the theory of relativity, the fourth quantum number is found to have only two permitted values, denoted by $+\frac{1}{2}$ and $-\frac{1}{2}$. Thus, the concept of a "quantum state" is defined as follows:[5] A quantum state describes a possible mode of behavior of an electron. It is specified by stating its four quantum numbers, three describing the wave function and the fourth describing the spin. To simplify one's understanding of quantum numbers, they may be likened to positional coordinates for the spinning orbital electron, wherein n denotes the radial characteristic, l denotes the azimuth characteristic, m denotes the magnetic characteristic or the plane of orbit, and s denotes the direction of spin. The significant point to be noted here is that while the Bohr theory postulated discrete energy levels, quantum mechanics postulates the existence of several quantum states at each level.

As a matter of spectrographic nomenclature, quantum states when $l = 0$ are called s states, when $l = 1$ are called p states, when $l = 2$ they are d states, when $l = 3$ they are f states; additional states are in alphabetical order. The possible number of states that may exist in the atom, for each value of n, is shown in Table 2-1.

For $n = 5$ there are a total of 50 states and for $n = 6$ there are 72. Let us consider the hydrogen atom as an illustrative example of the table. In the lowest energy level, where $n = 1$, l and m are 0 and the quantum numbers are $(1,0,0,+\frac{1}{2})$ and $(1,0,0,-\frac{1}{2})$. Therefore, the single electron may occupy either of the two allowed 1s states.

If no restriction were imposed on the many-electron atom, it would be natural to expect that all the electrons would try to seek the lowest

TABLE 2-1. DISTRIBUTION OF QUANTUM STATES

Quantum state	*Total number per atom*	
1s	2	
2s	2	8
2p	6	
3s	2	18
3p	6	
3d	10	
4s	2	32
4p	6	
4d	10	
4f	14	

energy level. This, of course, would literally wipe out the periodic table of elements. The workings of nature are preserved by Pauli's exclusion principle, which simply states that no two electrons may occupy the same quantum state. Or, in other words, no two electrons in a given system may have the same four quantum numbers. Pauli's principle thus accounts for the manner in which electrons are arranged in "shells" about the nucleus, as shown in Fig. 2-5. Each concentric shell is filled by as many electrons as the number of allowed states permits. Strictly speaking, the energy-level schemes found in many of the elements are not as simple as implied by Fig. 2-5. In the periodic table, there are many elements whose next-to-outermost levels are only partially filled by electrons, owing to the complicated wave function generated by electron interactions. Tin, for example, has 4 electrons in the fifth level but only 18 electrons in the fourth level, instead of the 32 permitted by the number of allowed quantum states. Nevertheless, some of the other representative elements in the periodic table adhere to the aforementioned quantum state distribution, as shown in Table 2-2. Here, part of the periodic table has been extracted to illustrate these elements. It should be noted that these elements are the ones we will be most concerned with in relation to semiconductors and their impurities. The numbers in parenthesis are the atomic numbers or the numbers of electrons in the atom. Since the first two levels permit two and eight states respectively, then for silicon with fourteen electrons, the third level will be filled with four electrons. Similarly, for arsenic the first three levels are filled

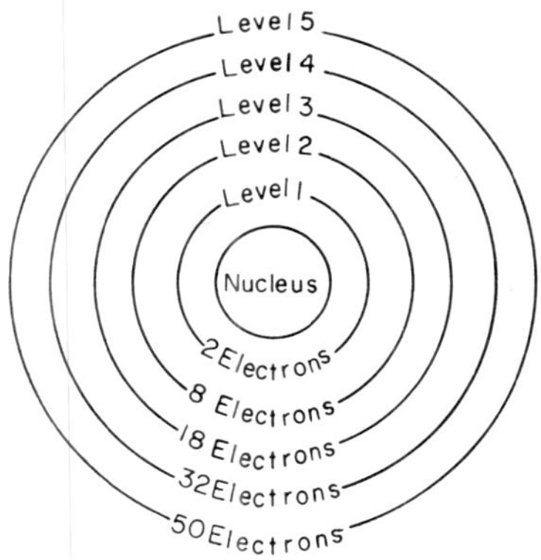

FIG. 2-5. Filled energy levels for many-electron atoms.

TABLE 2-2. PERIODIC TABLE FOR REPRESENTATIVE ELEMENTS IN SEMICONDUCTORS

No. of levels	Number of electrons in outermost level		
	3	4	5
2	Boron (5)	Carbon (6)	Nitrogen (7)
3	Aluminum (13)	Silicon (14)	Phosphorus (15)
4	Gallium (31)	Germanium (32)	Arsenic (33)
5*	Indium	Tin	Antimony

* These elements have only 18 electrons in the fourth level.

with two, eight, and eighteen electrons, leaving five electrons in the outermost orbit, making a total of thirty-three electrons.

Let us consider the allowed quantum-state picture, i.e., the number of l states for each n level for the outermost electrons (valence electrons) of *isolated* atoms of the group IV elements. For carbon, two of the electrons will completely fill the 2s states, while the other two will fill only two of the six 2p states. For silicon and germanium, the same applies to the 3s, 3p and 4s, 4p states, respectively. These states are shown pictorially in the energy-level diagrams of Fig. 2-6. Note that in each case, the s

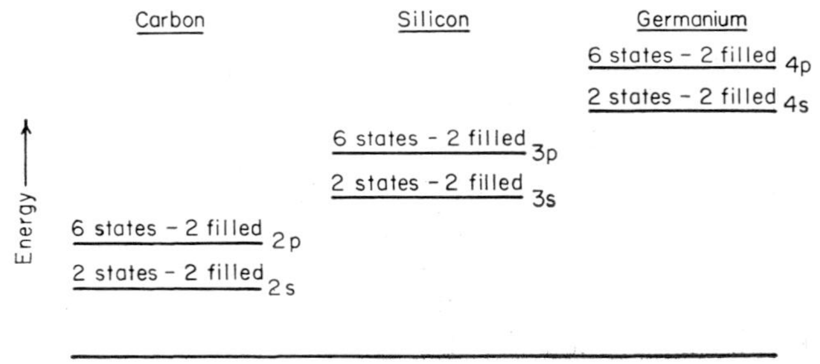

FIG. 2-6. Energy-level diagrams for valence electrons of isolated group IV atoms.

state is completely filled and the p state partially filled with but two electrons.

2-8. Theory of Energy Bands. To this point we have presented a qualitative description of how the wave-function solution of the Schrödinger wave equation establishes the quantum states and energy levels for electrons in isolated atoms. Now it remains to show what effects occur when we consider the electron wave functions moving about within the many close-lying atoms in solids. In crystalline solids, the individual atoms are arranged in a perfect array such that each atom is separated by the same lattice constant everywhere throughout the crystal. The simplest hypothetical array to consider would be the evenly spaced line of atoms in one dimension portrayed by Fig. 2-7*a*.[6] Here each atom is shown with a 1s and a 2s energy level. This arrangement produces an electric potential field which varies in a periodic manner as does that of Fig. 2-7*b*. For the sake of analysis, Kronig and Penney assumed a potential field having a rectangular shape as shown in Fig. 2-7*c*. By solving the Schrödinger wave equation for the boundary conditions imposed by this rectangular potential distribution, we obtain solutions (called Bloch functions) which are satisfied only for certain allowed electron energies as a function of the lattice constant. The actual crystal has a normal lattice constant, but mathematically we may permit the lattice constant to vary. For very large lattice constants it is

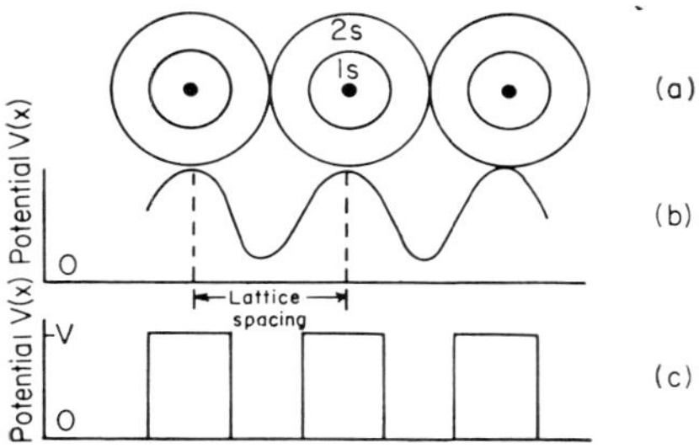

FIG. 2-7. Potential functions $V(x)$ for linear array of atoms.

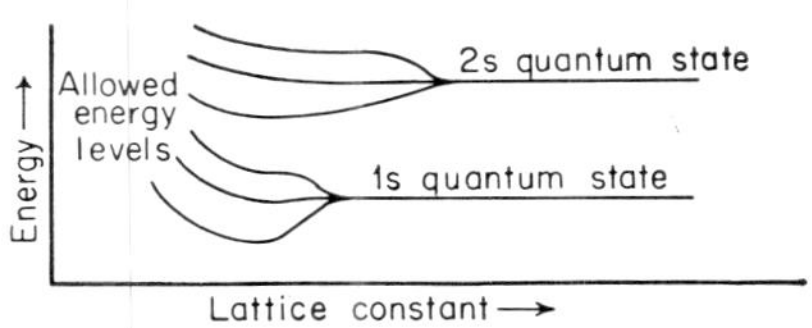

FIG. 2-8. Split energy levels for linear array of three atoms.

found that the energy levels obtained are identical to those for isolated atoms. As the lattice is decreased, approaching the normal value, these discrete energy levels will each *split* into several allowed values. This is shown in Fig. 2-8 for the three-atom array of Fig. 2-7*a*. Each atom, when isolated at large lattice constants, is shown to have the single 1s and 2s quantum states. When the three atoms are coupled closer together, each state splits into three allowed energy values. Mathematically, this is due to interactions or coupling of the electron wave functions for each atom as they are brought closer to the wave functions produced by the potential fields of adjacent atoms. The energy-level diagram shows the 2s states or the upper energy levels splitting at larger separations, since in these states the electrons are farther from the nuclei and thus interact sooner.

The fact that each quantum state for this three-atom array splits into only three levels may best be explained by the analogy used by Shockley.[5] Consider three identical weighted springs, each having the same single mode of vibration or the same single energy level. If the three springs are lightly coupled together with, for example, a weak spring as shown in Fig. 2-9, the coupled system will possess three modes of vibration, neither of which is the same as the mode of the isolated springs; it might be said that the system has split into three energy levels. Thus, on this basis we may conclude that for a group of n identical atoms, n levels may be expected to exist. However, for very large values of n, the split energy levels lie very close together. When n is of the order of 10^{22} (the approximate number of atoms per cubic centimeter of a solid) there are 10^{22} levels. In this case, the levels are so close in value as to be indistinguishable and may be considered as bands of allowed energies or

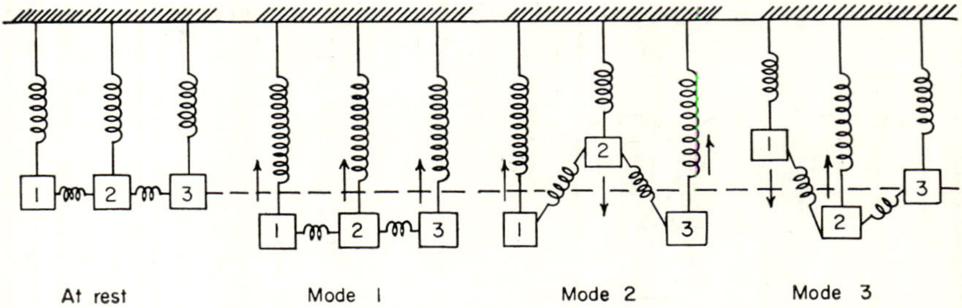

FIG. 2-9. Mechanical analogy of split energy levels. (*After W. Shockley.*)

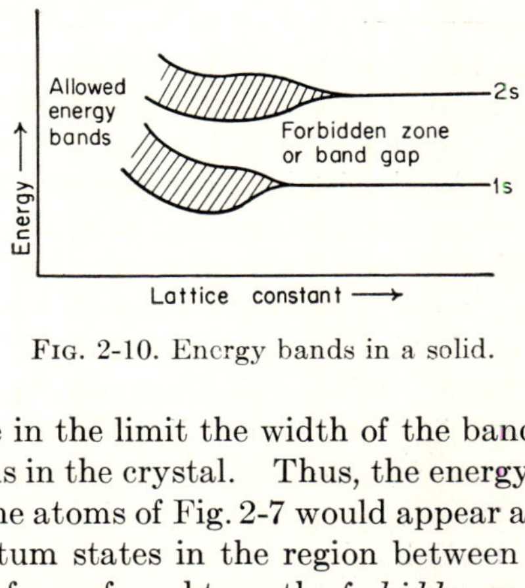

FIG. 2-10. Energy bands in a solid.

energy bands, where in the limit the width of the bands is independent of the number of atoms in the crystal. Thus, the energy-level diagram for a solid composed of the atoms of Fig. 2-7 would appear as shown in Fig. 2-10. There are no quantum states in the region between the allowed energy bands and it is therefore referred to as the *forbidden energy zone* or *band gap*.

It may be pointed out here that the Kronig-Penney analysis is a simplified approximation for an imaginary one-dimensional atomic array. An exact analysis of the nature of the electron wave function in a three-dimensional periodic potential field requires the theory of Brillouin zones, which is beyond the scope of this text.

2-9. Energy Bands in Crystals. As will be seen in the next chapter, much of the electrical conductivity properties of semiconductors is learned from the study of the energy-band structure and the distribution of quantum states. Since the properties of carbon are akin to those of silicon and germanium, it would be well at this point to study its band picture as a representative semiconductor material.

If the atoms of carbon are brought together to form a crystal, viz., diamond, the resulting energy bands appear as in Fig. 2-11.[7] As mentioned previously, the four valence electrons in the isolated atom are distributed evenly between the 2s and 2p quantum states. Since there are two allowed states at 2s and six allowed states at 2p, the 2s level is completely filled and the 2p level is partially filled. Of course, each atom

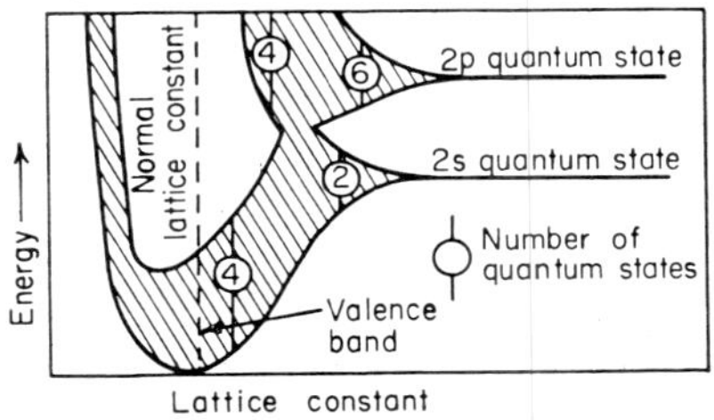

FIG. 2-11. Energy bands in diamond. (*From W. Shockley, "Electrons and Holes in Semiconductors," D. Van Nostrand Company, Inc., Princeton, N.J.*, 1950.)

also contains two electrons in the 1s state, but these are bound so tightly to the nucleus that no appreciable interaction occurs. They can be considered, therefore, as unaffected by their being in the crystal.

As shown in the diagram, the energy levels split into bands but the eight allowed states per atom split equally between the upper and lower bands. There is a quantum conservation theorem which states that the number of quantum states in split bands must be the same as the number of atomic quantum states from which the bands are produced. Furthermore, for each atom, the four allowed states in the lower band become completely filled with electrons, leaving the upper band completely empty. This occurrence is of utmost significance with regard to semiconductors, as will be seen later. Physically, it implies that as the lattice spacing approaches normal, the electrons of one atom interact so strongly with those of another that every two electrons finally pair up and become bonded. The symmetry is evident from the fact that each of the four electrons of one atom "sees" one electron from each of four neighboring atoms, forming a tetrahedral structure for the lattice.

These electron pairs are called valence bonds; therefore the lower energy band is referred to as the *valence-bond band*. Shockley points out that the average energy of the valence-bond band is much lower than the original 2s energy level; this drop in energy is associated with the binding energy of the crystal.

PROBLEMS

2-1. Using the Bohr theory, calculate the energy in electron-volts necessary to ionize a single hydrogen atom.

2-2. A single impurity atom having five valence electrons is substituted for an atom of germanium in a crystal. Assuming that the behavior of the excess electron is similar to one in a hydrogen atom, calculate the energy in electron-volts necessary to ionize the impurity atom. The dielectric constant for germanium is 16. Calculate the radius of orbit for the excess electron.

2-3. For isolated atoms of both germanium and silicon, write down the quantum numbers (n,l,m,s) for the outermost valence electrons only. Assume that all the electrons of the atom have filled the lowest energy states.

REFERENCES

1. Richtmyer, F. K., and E. H. Kennard: "Introduction to Modern Physics," pp. 175–177, McGraw-Hill Book Company, Inc., New York, 1947.
2. Hausman, E., and E. P. Slack: "Physics," pp. 630–636, D. Van Nostrand Company, Inc., Princeton, N.J., 1939.
3. *Ibid.*, p. 717.
4. Richtmyer and Kennard, *op. cit.*, pp. 259–271.
5. Shockley, W.: "Electrons and Holes in Semiconductors," chap. 5, pp. 122–134, D. Van Nostrand Company, Inc., Princeton, N.J., 1953.
6. Kittel, C.: "Introduction to Solid State Physics," pp. 250–255, John Wiley & Sons, Inc., New York, 1953.
7. Shockley, *op. cit.*, p. 133.

3

Semiconductors

3-1. Application of Energy-band Theory. We have seen that the electromechanical interactions within a crystalline solid lead to bands of allowed energies and that the electrons may occupy only those quantum states dictated by the energy bands. These results were obtained by examining the nature of the electron wave function, obtained as solutions to the Schrödinger wave equation, which characterizes the motion of the electrons through the periodic potential fields within the atomic lattice. In the group IV elements such as carbon, silicon, and germanium, the outermost electrons were shown in bound states, completely filling the lower energy or valence-bond band and leaving the upper band empty.

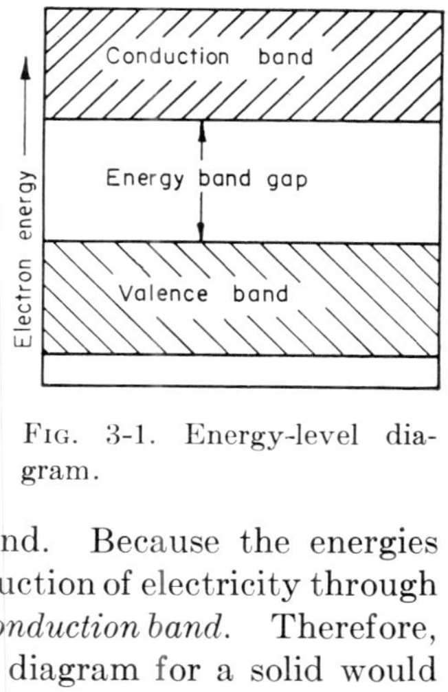

FIG. 3-1. Energy-level diagram.

Strictly speaking, the valence states are completely filled only at the absolute zero of temperature, 0° Kelvin. At room temperature, thermal agitation will impart sufficient energy to permit some of the electrons to break the bonds and excite them into the unfilled energy states in the upper energy band. Because the energies associated with these electrons permit the conduction of electricity through the solid, the upper energy band is called the *conduction band*. Therefore, at normal lattice constants, the energy-level diagram for a solid would appear as shown in Fig. 3-1. The band gap represents those energies for which there are no allowed quantum states. It must be emphasized that the energy-band picture shown is highly simplified. Actual wave-mechanical analysis would reveal a complex structure involving maximum and minimum values for each band edge.[1,*] Nevertheless, the concept implied by Fig. 3-1 is adequate for this discussion.

* References, indicated in the text by superscript figures, are listed at the end of the chapter.

Suppose that for a given solid the valence band is completely full, while the conduction band is completely empty. No electron in the valence band can take on an energy value within the same band, for this would violate Pauli's exclusion principle. Furthermore, no electron can take on the additional energy necessary to place it in the band gap since there are no allowed energies permitted in the forbidden zone. As long as the valence band remains filled, those electrons cannot contribute to conduction, since this implies an acceleration or increase in energy due to an electric field and would necessitate the occupancy of a close-lying state. However, an electron may jump to any of the unfilled conduction-band states if the energy increase accepted is sufficiently large to do so. This energy increase could be of the form of heat radiation or high electric fields. Thus, if thermal agitation excited some electrons into the conduction band, then the conduction band would be partially filled, and it is apparent that these electrons would contribute to conduction, since there are other unoccupied states in the same band. As will be seen later, electrical conduction is also possible in the valence-bond band because of the empty states left by the excited electrons.

3-2. Conduction Properties of Solids. The concept of occupancy of allowed quantum states in the energy bands gives us a qualitative

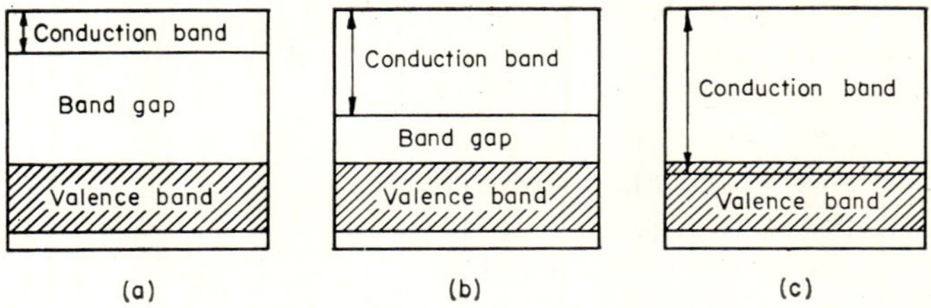

FIG. 3-2. Energy-level diagrams for various solids. (*a*) Insulator; (*b*) semiconductor; (*c*) metal.

explanation for the differences in electrical conductivity observed among insulators, semiconductors, and metals. Since electrical conductivity is a measure of the number of electrons available for electric-field acceleration, the nature of the band pictures for the three types should be indicative of conductivity. Figure 3-2 shows the energy-level diagrams for insulators, semiconductors, and metals. The fundamental difference is the size of the energy band gap.

In the case of the insulator, the band gap is so large that even at room temperature hardly any of the electrons can acquire sufficient energy to lift them to the energy levels within the empty conduction bands. In semiconductors the band gap is much narrower, and at room temperature an appreciable number of electrons may be thermally excited to the conduction band, attributing conductivities to these solids. In metals

(Fig. 3-2*c*), the energy bands overlap one another in such a manner that many states are left unfilled and available for occupancy by electrons. This gives very high conductivities for metals even at very low temperatures. Kittel arbitrarily classified the conductivities for these three classes of solids as follows:[2]

Insulators: 10^{-22} to 10^{-14} mho-cm^{-1}
Semiconductors: 10^{-9} to 10^{2} mho-cm^{-1}
Metals: 10^{5} mho-cm^{-1} and greater

In many solids, including semiconductors, the conduction bands are not as simple as those shown in Fig. 3-2, but consist of several overlapping conduction bands. However, this is not of direct consequence for semiconductors since we need be concerned only with those states near the bottom of the conduction bands. Similarly, the only states of interest in the valence-bond band are those lying near the top of the band. We are interested only in those energies near the edges of the bands because thermal energies can excite electrons by only about a fraction of an electron-volt of energy, whereas the bands may be several electron-volts wide.

3-3. Semiconductor Crystals. As previously indicated, the group IV elements, with four valence electrons in the outermost orbits, comprise

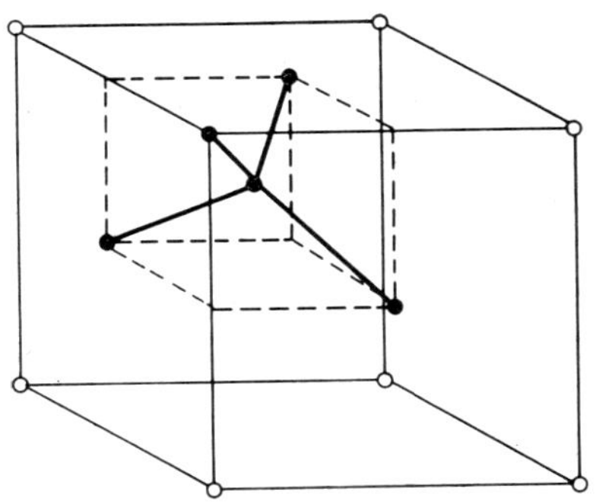

FIG. 3-3. Diamond-type structure of semiconductor crystals. (*After W. Shockley.*)

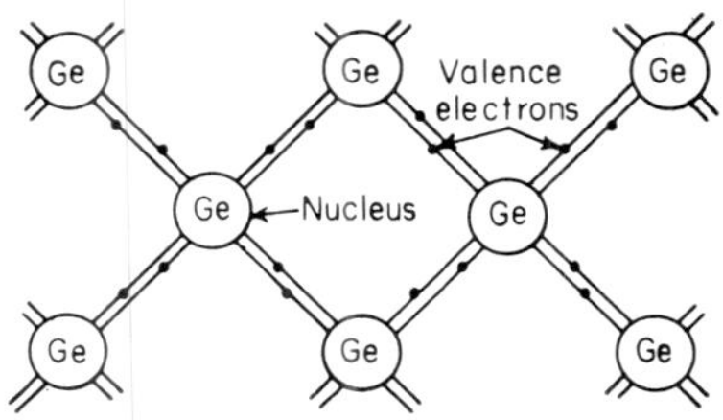

FIG. 3-4. Two-dimensional representation of germanium lattice.

the semiconductor solids. These elements are carbon, silicon, and germanium. Tin and lead, also group IV elements, have overlapping bands and therefore exhibit metallic conduction properties. In the crystalline form, each of the atoms forms covalent or electron-pair bonds with four other atoms, creating the tetrahedral structure of the diamond-type crystal. This is illustrated in Fig. 3-3, where the structure is a cubic arrangement with each of the eight corner atoms forming a simple cubic lattice like the one sketched in dotted lines. More simply, in two dimensions, the arrangement would appear as shown in Fig. 3-4 for a germanium crystal. Here each germanium atom forms an electron-pair bond with four neighbors.

In the energy-band picture, the energies associated with the bonded electrons lie in the valence-bond bands, and the energy necessary to break the bonds corresponds to the energy-band gap. Measurements have been made of the band gaps for semiconductors, and the present-day values are as follows (0°K):

Carbon	*Silicon*	*Germanium*
6–7 ev	1.21 ev	0.785 ev

It can be seen that the band gap for carbon is quite enormous, indicating that at room temperature a pure diamond crystal would behave very much like an insulator. Furthermore, since the band gap of silicon is larger than that of germanium, silicon is able to withstand higher temperatures than germanium, as far as conductivity is concerned. Thus, silicon and germanium comprise the semiconductor materials in major use today. Their physical properties are given in Table 3-1.[3]

TABLE 3-1. PROPERTIES OF GERMANIUM AND SILICON

Property	Germanium	Silicon
Melting point, °C	936	1420
Density at 25°C, g/cm³	5.323	2.330
Thermal expansion coefficient at 25°C, 1/°C	6.1×10^{-6}	4.2×10^{-6}
Thermal conductivity at 25°C, cal/(sec)(cm)(°C)	0.14	0.20
Specific heat at 0–100°C, cal/(g)(°C)	0.074	0.181
Atomic weight	72.60	28.08
Atomic number	32	14
Lattice constant at 25°C, cm	5.657×10^{-8}	5.429×10^{-8}
Atoms/cm³	4.42×10^{22}	4.96×10^{22}
Volume compressibility, cm²/dyne	1.3×10^{-12}	0.98
Dielectric constant	16	12
Energy band gap at 300°K, ev	0.72	1.1

Since the band gap of a semiconductor is a function of the lattice constant (see Fig. 2-11), it would be expected that the band gap depend somewhat on temperature. It has been found for germanium and silicon that the energy-band gap decreases slightly with temperature at the rate of 2.23×10^{-4} ev/°K for germanium[7] and 3.6×10^{-4} ev/°K for silicon.[8] For example, at 0°K, E_G for germanium is 0.78 ev, and at room temperature, 300°K, $E_G = 0.72$ ev. The basic relationship is given as

$$E_G(T) = E_G(0°\text{K}) - \beta T \tag{3-1}$$

where β has the aforementioned respective values, and T is the temperature in degrees Kelvin.

3-4. Hole Conduction. It was stated that when an electron from the valence band is excited to a state in the conduction band, it leaves behind a vacant state in the valence band. In the case of a semiconductor, this corresponds to a broken bond, as shown in Fig. 3-5 for the germanium crystal. This vacant or empty state is referred to as a *hole*, and it is evident that a hole will be created for each electron jumping into the conduction band. Since a hole in the valence band is essentially an unfilled state it is possible for an electron within the band to jump into the hole, leaving another hole behind. This process may be continuously repeated, with the valence-band electrons creating a hole for each one they fill. Thus it is possible for conduction to occur in the valence-bond band, since the electrons may take on the necessary energy supplied by an electric field to place them into the empty hole states.

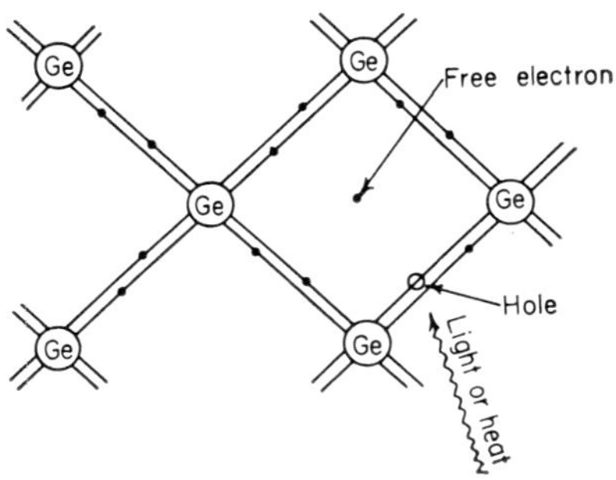

FIG. 3-5. Broken bond creating hole-electron pair.

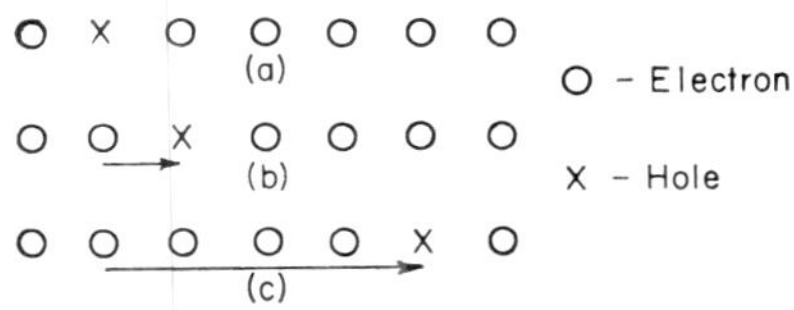

FIG. 3-6. Analogy of hole conduction.

It is convenient, however, to look upon the conduction process in terms of hole movement rather than electron movement; that is, the holes move in the direction opposite to the direction of electron movement. This phenomenon may be simplified by visualizing, for example, a number of valence electrons arranged in a row with one of them missing, as shown in Fig. 3-6*a*. The missing electron represents a hole; as each adjacent electron moves to the left to fill the hole, the hole moves to the right as in *b*. After successive electron jumps, the hole has moved far to the right as in Fig. 3-6*c*. The movements constitute a transfer of negative charge to the left, or in effect, a transfer of positive charge to the right. Thus, the movement of holes is analogous to the movement of positive charges, in contrast to the movement of electrons, which are negative charges.

Extensive experimentation and theoretical analysis has shown that this is the case for semiconductors.[4] The observed conduction process is a combined movement of electrons in the conduction band and holes in the valence band. Furthermore, according to quantum mechanics, the behavior of either type of "particle" (referred to as a carrier) within the crystal is very much the same as that of the classical free particle, providing corrections are made for both the sign of charge and the mass. It has been proven that although the effective masses of the electron (m_n) and the hole (m_p) are different, they are both of the same order of magnitude as the free electron mass m. These differences are related to the energies and velocities involved for applied electric and magnetic fields, and have not as yet been accurately measured.

In summary, it has been demonstrated that for a pure semiconductor crystal at room temperature, a number of electrons will be excited into the conduction band, leaving an equal number of holes in the valence band. Furthermore, both the holes and electrons contribute to conduction as positive and negative charges, respectively. It now remains to calculate these concentrations or densities as functions of temperature and band gap.

3-5. Fermi-Dirac Distribution Function. Under conditions of thermal equilibrium the distribution of the electrons among the quantum states cannot be completely governed by classical statistics such as the Maxwell-Boltzmann distribution law. This would apply to problems concerning, for example, the distribution of velocities of gas molecules, where any number of molecules may have the same energy. In semiconductors, electrons cannot have the same energy states because of the limitations imposed by the Pauli exclusion principle. To account for this restriction, Fermi-Dirac statistics must be employed. The Fermi-Dirac distribution, function $f(E)$, which gives the probability that a quantum state with energy E is occupied by an electron, is given by the formula

$$f(E) = \frac{1}{1 + \epsilon^{(E-E_F)/kT}} \tag{3-2}$$

where E_F is called the Fermi energy or level and k is Boltzmann's constant. The kT term arises from statistical mechanical theory and corresponds to the thermal energy of a one-dimensional harmonic oscillator. It is seen that when the energy E is equal to the Fermi level, the value of $f(E)$ is one-half.

The significance of this function may be understood by referring to the curves of Fig. 3-7, which are plots of $f(E)$ with E with temperature. At absolute zero ($T = 0°K$) no electrons will occupy any of the conduction states; therefore, the probability of occupancy $f(E)$ for $E \geq E_F$ will be zero. The probability of occupancy of the valence states (i.e., where $E \leq E_F$) will be unity. Under those conditions, the Fermi level will be at the midpoint of the forbidden zone.

At room temperature ($T = 300°K$) the function becomes more diffuse, appearing as shown in Fig. 3-7*b*. Since some electrons will be thermally excited in this case, it would be expected that for the energies associated with the bottom of the conduction band E_C, a small probability of occupancy would exist. At the top of the valence band E_V, the probability of occupancy is likewise decreased from unity, since the valence band is partially emptied. It should be noted that in both cases *a* and *b*, the Fermi level is just a mathematical definition associated with the distribution function and does not imply an allowed energy level. Lastly, if the temperature were increased further, the $f(E)$ function would appear as shown by the dotted curve in Fig. 3-7*b*.

The value of kT is 0.026 ev for $T = 300°K$; if the band gap is 0.72 ev, as for germanium, the energy difference between the conduction band and the Fermi level (viz., $E_C - E_F$) is 0.36, or approximately $14kT$.

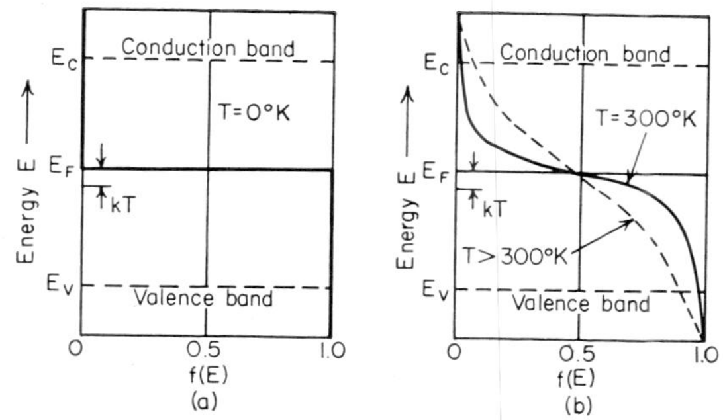

FIG. 3-7. Fermi-Dirac distribution functions superimposed on energy-band diagrams.

For this situation, where $E - E_F \gg kT$, Eq. (3-2) may be given by the approximation

$$f(E) = \epsilon^{-(E-E_F)/kT} \tag{3-3}$$

since the exponential term is greater than unity. Equation (3-3) gives the value of that fraction of the quantum states at energies E occupied by electrons. With this being so, $1 - f(E)$ corresponds to the fraction left vacant or occupied by holes. By algebraic manipulation,

$$1 - f(E) = \frac{1}{1 + \epsilon^{(E_F-E)/kT}} \tag{3-4}$$

By the same approximation (viz., $E_F - E \gg kT$) the distribution function for holes at the top of the valence band becomes

$$1 - f(E) = \epsilon^{-(E_F-E)/kT} \tag{3-5}$$

Equations (3-3) and (3-5) will be used in the following section to calculate the equilibrium concentration of carriers in the semiconductor.

3-6. Equilibrium Concentration of Carriers. It is apparent that if the distribution of the quantum states with energy and the probability of occupancy of those states were known, the free electron density n and the hole density p could be determined from the product. If $N(E)$ is defined as the number of quantum states per unit energy per unit volume of the crystal, then the density of states S in a particular energy range would be

$$S = \int_{E_1}^{E_2} N(E)\, dE \tag{3-6}$$

Therefore the density of free carriers would be given by

$$n = \int_{E_1}^{E_2} f(E)N(E)\, dE \tag{3-7}$$

Although the wave mechanics of the Brillouin-zone analysis is not treated here, the theory gives the expression for $N(E)$ as [5]

$$N(E) = \frac{4\pi}{h^3}(2m)^{3/2}(E - E_o)^{1/2} \tag{3-8}$$

This equation gives the density of quantum states per unit energy for energies slightly above the minimum energy E_o.

Applying Eqs. (3-3) and (3-8) to (3-7) to calculate the density of electrons in the conduction band, we obtain

$$n = \int_{E_C}^{\infty} \epsilon^{-(E-E_F)/kT} \frac{4\pi}{h^3}(2m)^{3/2}(E - E_C)^{1/2}\, dE \tag{3-9}$$

where E_C is the bottom energy of the conduction band. If the substitution $E_1 = E - E_C$ is made and the constant terms are factored out, the integral becomes[6]

$$n = \epsilon^{-(E_C-E_F)/kT} \frac{4\pi}{h^3}(2m)^{3/2} \int_0^{\infty} E_1^{1/2} \epsilon^{-E_1/kT}\, dE_1 \tag{3-10}$$

This may be integrated with the following result:

$$n = 2\left(\frac{2\pi mkT}{h^2}\right)^{3/2} \epsilon^{-(E_C-E_F)/kT} \tag{3-11}$$

This is of the form $n = N_C f(E_C)$, where N_C may be interpreted to be the effective density of quantum states in the conduction band, or

$$N_C = 2\left(\frac{2\pi mkT}{h^2}\right)^{3/2} = 4.82 \times 10^{15} T^{3/2} \qquad \text{cm}^{-3} \tag{3-12}$$

For the density of holes in the valence band, the same method of analysis gives similar results as follows:

$$p = 2\left(\frac{2\pi mkT}{h^2}\right)^{3/2} \epsilon^{-(E_F-E_V)/kT} \tag{3-13}$$

and

$$N_V = 2\left(\frac{2\pi mkT}{h^2}\right)^{3/2} = 4.82 \times 10^{15} T^{3/2} \qquad \text{cm}^{-3} \tag{3-14}$$

where N_V is the effective density of quantum states in the valence band.

Thus it is seen from Eqs. (3-11) and (3-13) that, for a perfect crystal, the density of electrons and holes available for conduction at a given temperature obeys complete statistical symmetry. However, the derivations assumed the effective masses to be equal to the free electron mass m. Equations (3-11) and (3-13) are corrected for this by multiplying them by $(m_n/m)^{3/2}$ and $(m_p/m)^{3/2}$, respectively. A very important result is obtained by multiplying Eqs. (3-11) and (3-13), viz.,

$$np = N_C N_V \epsilon^{-(E_C - E_V)/kT} = N_C N_V \epsilon^{-E_G/kT} \tag{3-15}$$

where $-E_G$ is the band gap equal to $-(E_C - E_V)$. If the correction is made for effective masses and the temperature variation of the band gap given by Eq. (3-1), then (3-15) becomes

$$np = 2.33 \times 10^{31} \left(\frac{m_n m_p}{m^2}\right)^{3/2} T^3 \epsilon^{-E_{G0}/kT} \epsilon^{\beta/k} \tag{3-16}$$

It is evident that for a given semiconductor material, the equilibrium

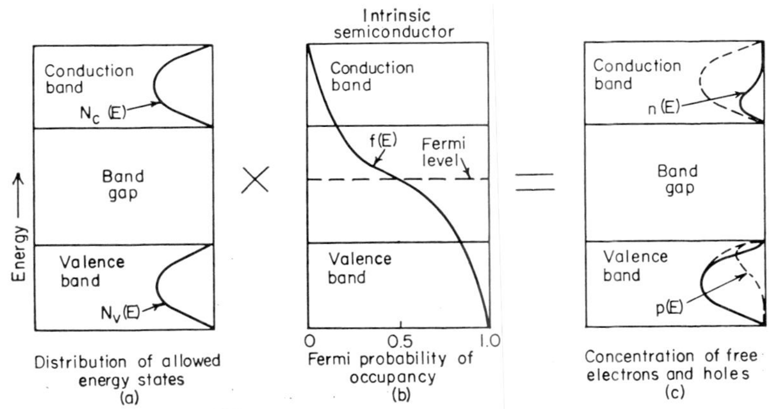

FIG. 3-8. Graphic computation of free carrier concentrations.

product np is independent of the Fermi level E_F and is dependent only on temperature. This relationship, as will be seen later, is of utmost significance with regard to semiconductors. It implies that regardless of what densities n or p may be, the product is always constant for a given temperature.

It would be appropriate here to further clarify the application of the Fermi function in determining the free carrier concentrations. In Fig. 3-8 is a graphical portrayal of the foregoing analysis. For the purpose of illustration, let us assume that the distribution of allowed states as a function of energy within the bands, as given by (3-8), is arbitrarily of the form shown in Fig. 3-8*a*. Note that for this case, which applies to the *perfect* crystal, there are no allowed states in the band gap. Now, to determine which of these states are filled with carriers, we need to know what their probability of occupancy is. This is given by the Fermi function, plotted in Fig. 3-8*b*, for a specific temperature. Multiplication of function (*a*) by (*b*) therefore yields the concentration of free carriers in each of the bands, shown in Fig. 3-8*c*. In this example, the hole concentration in the valence band is equal to the allowed density of states less the electron density there. Since there is a free electron for each hole, the carrier distributions in the bands must be equal. Note that although the Fermi function indicates a probability of occupancy in the band gap, there are no allowed states and therefore zero concentration of carriers.

3-7. Intrinsic Conduction. The previous sections have considered only the pure, perfect semiconductor crystal which is free of any impurities and structural defects such as lattice vacancies, dislocations, and disorders. Since such a defect would disturb the periodic potential field in the vicinity of the defect, it would be expected that the energy-band structure in the region would similarly be distorted. Thus the quantum-state distribution is altered, introducing additional carriers into the bands. This results in conductivities higher than that which is theoretically attainable for a perfect crystal.

Nevertheless, for the perfect crystal, the total number of electrons is equal to the number of holes, or

$$n = p$$

and the conduction due to n and p is referred to as *intrinsic*. Equating (3-11) and (3-13), with the respective effective mass correction, we obtain

$$N_C\left(\frac{m_n}{m}\right)^{3/2}\epsilon^{-(E_C-E_F)/kT} = N_V\left(\frac{m_p}{m}\right)^{3/2}\epsilon^{-(E_F-E_V)/kT}$$

Solving for E_F, we obtain

$$E_F = \frac{E_V + E_C}{2} - \tfrac{3}{4}kT\ln\frac{m_n}{m_p} \qquad (3\text{-}17)$$

Subtracting E_V from both sides of Eq. (3-17), we get

$$E_F - E_V = \frac{E_C - E_V}{2} - \tfrac{3}{4}kT\ln\frac{m_n}{m_p}$$

$$E_F - E_V = \frac{E_G}{2} - \tfrac{3}{4}kT\ln\frac{m_n}{m_p} \qquad (3\text{-}18)$$

This equation gives the position of the Fermi level, in electron-volts, with respect to the valence-band edge. If $m_n = m_p = m$, the free electron mass, then $E_F - E_V = E_G/2$, which means that for an intrinsic semiconductor the Fermi level lies midway in the band gap, as was described in Sec. 3-5.

Although conductivity (or resistivity), per se, will be discussed in the next chapter, it is apparent that it is related to the free carrier densities

n and p. If the conduction is intrinsic, i.e., $n = p = n_i$, then

$$n_i^2 = np = 2.33 \times 10^{31} \left(\frac{m_n m_p}{m^2}\right)^{3/2} T^3 \epsilon^{-E_{G0}/kT} \epsilon^{\beta/k} \tag{3-19}$$

Present experimental measurements of high-quality crystals by Morin and Maita have yielded the following empirical expression for Eq. (3-19):[7,8]

$$np = 3.10 \times 10^{32} T^3 \epsilon^{-0.785/kT} \qquad \text{germanium} \tag{3-20}$$

$$np = 15 \times 10^{32} T^3 \epsilon^{-1.21/kT} \qquad \text{silicon} \tag{3-21}$$

These relationships are valid up to temperatures of about 300°C and are

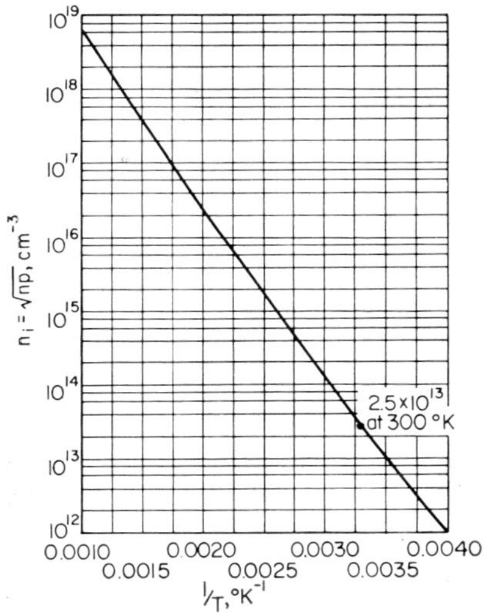

$n_i = 1.76 \times 10^{16} T^{3/2} \epsilon^{-0.785/2kT}$

FIG. 3-9. Temperature variation of n_i for intrinsic germanium. (*F. J. Morin and J. P. Maita, Phys. Rev., vol.* 94, *p.* 1525, 1954.)

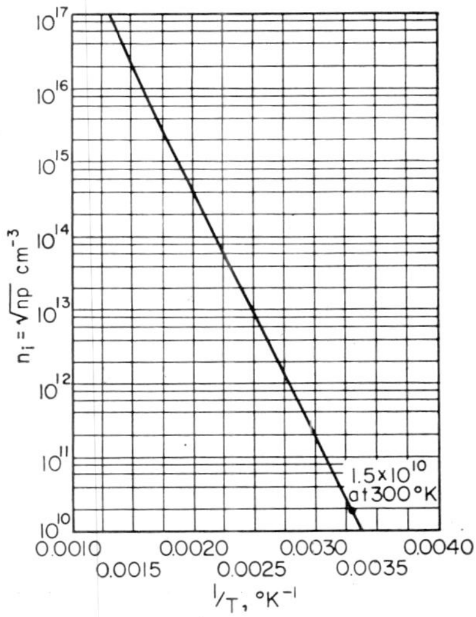

$n_i = 3.87 \times 10^{16} T^{3/2} \epsilon^{-1.21/2kT}$

FIG. 3-10. Temperature variation of n_i for intrinsic silicon. (*F. J. Morin and J. P. Maita, Phys. Rev., vol.* 96, *p.* 28, 1954.)

shown in Figs. 3-9 and 3-10 where $n_i = (np)^{1/2}$ is plotted as a function of temperature. Additionally, the coefficients of (3-20) and (3-21) include the possible values for the hole and electron masses as well as the band-gap temperature dependence.

3-8. Impurity Conduction in Semiconductors. It should be evident that the perfect, flawless, impurity-free crystal would be the exception rather than the rule in semiconductor technology. Even when prepared under the most carefully controlled laboratory conditions, the ultimate crystal exhibiting absolute intrinsic conduction is hard to come by. Furthermore, as was indicated in Chap. 1, practical junction-transistor structures are dependent upon impurity-doped semiconductors in order to obtain the appropriate conductivities in the various regions. The practice of intentionally introducing impurities into purified semiconductor crystals is of basic importance to semiconductors. In this and the following sections, we will see how the number of electrons excited into the conduction band, or holes into the valence band, may be increased considerably by the presence of certain impurity atoms. The electrical conductivity that results from impurity doping is referred to as *impurity conduction* or sometimes as *extrinsic conduction.*

Because of the covalent structure of the four valence electrons of semiconductor atoms, the impurities of interest are those of the group III and group V elements (periodic table, Table 2-2). It turns out that in the

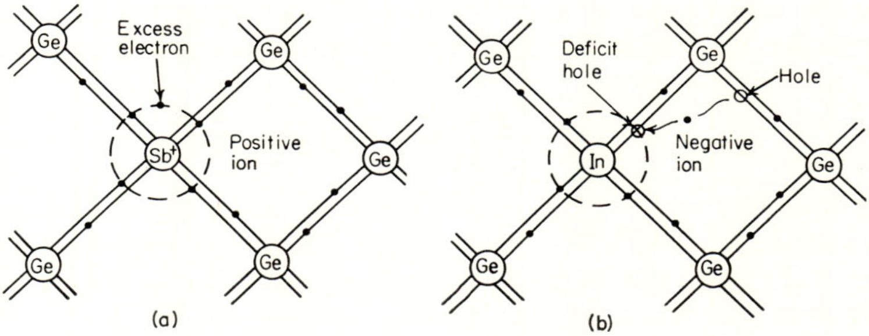

FIG. 3-11. Impurity conduction in semiconductors. (*a*) Excess conduction due to donor impurities (group V); (*b*) deficit conduction due to acceptor impurities (group III).

process of doping crystals, the atoms of these impurities will be substituted in place of the semiconductor atoms in the crystal lattice. This is in contrast to some cases where atoms might be positioned interstitially, that is, in the space between lattice atoms.

Let us consider, for example, a group V element such as antimony, which has five electrons in the outer orbit. If an atom of antimony is substituted for a germanium atom in a crystal, four of the five electrons will satisfy the covalent bond requirements of the neighboring germanium atoms. The fifth electron, however, remains unbound and is left to "wander" throughout the crystal. This is shown in Fig. 3-11*a*. Since each antimony impurity atom, or in general each group V atom, donates an excess electron to the crystal, they are called *donor impurities.* The loss of the electron leaves the impurity atom with a net positive charge, since only four electrons are available to neutralize the donor nucleus.

On the other hand, if a group III element such as gallium, which has three electrons in its outer orbit, is substituted, only three of the four neighboring covalent bonds will be satisfied. This creates a deficit bond or hole, as shown in Fig. 3-11*b*. In order to satisfy the fourth bond, the group III impurity atom will readily accept an electron from the crystal. For this reason, these elements are called *acceptor impurities.* The acceptance of an electron must, of course, create a hole elsewhere in the crystal. Also, the presence of the accepted electron establishes a net negative charge for the acceptor atom.

It should be emphasized that the donor and acceptor impurity atoms appear as fixed positive and negative ions, respectively. In the crystal, they are immobile and do not conduct. Only the extra electrons and holes are mobile and contribute to impurity conductivity.

3-9. Ionization Energies of Impurity Atoms. Wave-mechanical analysis of the motion of the excess electron about the donor impurity atom has shown that the situation is similar to that of the single electron revolving about the nucleus of a hydrogen atom.[9] It was demonstrated, using the Bohr theory, that the energy binding the electron to the hydrogen nucleus is

$$E = \frac{2\pi^2 mq^4}{h^2} = 13.6 \text{ ev}$$

This, however, was the case for free space. In a semiconductor, the charge of the nucleus of the impurity atom is shielded by the dielectric constant of the surrounding medium, which lessens the force of attraction on the excess electron. If the equation is corrected for this effect, the impurity ionization energy becomes

$$E_i = \frac{2\pi^2 mq^4}{\kappa^2 h^2} = \frac{13.6}{\kappa^2} \text{ ev}$$

where κ is the dielectric constant. For germanium, $\kappa = 16$; for silicon, $\kappa = 12$. As a result, the impurity ionization energy for germanium would be 0.053 ev, and for silicon 0.095 ev. At normal temperatures, the excess electron will readily acquire these energies and effectively become unbound from the impurity atom. Similar reasoning will also apply to the holes created by acceptor impurities. However, it would be expected that the ionization energies of acceptors be slightly different from that of donors because of the differences in effective mass between holes and electrons. Experimental measurements of impurity ionization energies give remarkably close but slightly different results. The most recently obtained values for group III and V elements are presented in Table 3-2. The ionization energies decrease as the impurity concentration reaches high densities of the order of 10^{18} atoms/cm^3. This is explained by the fact that the electrostatic attraction of neighboring ionized impurity atoms enhances the escape of the excess electron or hole.

The presence of impurities will modify slightly the distribution of quantum states within the crystal. At very low temperatures, where

TABLE 3-2. IMPURITY IONIZATION ENERGIES[10,11]

Type	Element	E_i, ev	
		Germanium	Silicon
Acceptor	Boron	0.0104	0.045
	Aluminum	0.0102	0.057
	Gallium	0.0108	0.065
	Indium	0.0112	0.160
Donor	Phosphorus	0.0120	0.039
	Arsenic	0.0127	0.049
	Antimony	0.0097	0.039

the excess electrons are bound to the donor atoms, each donor atom will remove a state from the conduction band and establish it as an allowed state of lower energy. This lower energy level is referred to as the *donor impurity energy level* and is below the bottom of the conduction band by an amount equal to the donor ionization energy. Similarly, for an acceptor impurity, each atom raises a state from the valence band to an acceptor-impurity energy level, just equal to the acceptor ionization energy. These impurity energy levels are depicted in Fig. 3-12, where E_D and E_A are the respective levels. The donor ionization energy $E_{iD} = E_C - E_D$ and $E_{iA} = E_A - E_V$; these energies are called *impurity band gaps*. Thus, at normal temperatures, the bound electrons readily jump the impurity band gap and fill empty states in the conduction band, leaving the donor atoms positively ionized. By the same reasoning, electrons from the valence band will become bound to the acceptor atoms, leaving holes in the valence band available for conduction. Here each acceptor atom is negatively ionized.

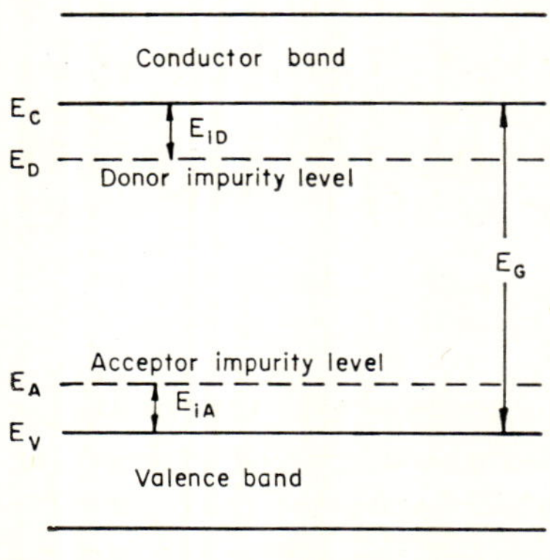

FIG. 3-12. Energy-level diagram for impurity semiconductor.

3-10. Electrical Charge Neutrality. Whether donor, acceptor, or both are present in a crystal, the total net charge within the crystal must be zero; i.e., the crystal must be electrically neutral. Each positive donor ion is neutralized by its excess electron and each negative acceptor ion is neutralized by a hole. At very low temperatures, some of the excess carriers will be bound to the impurity atoms, but charge neutrality is still not disturbed. For high temperatures, in addition to the free excess carriers and the ionized impurity atoms, electron-hole pairs will be generated by thermal agitation, as previously described for intrinsic conduction. Since there is an electron from each positive hole, it becomes evident that no charge unbalance occurs. If both donors and acceptors are present, free electrons from the donor atoms will fill the vacancies of the acceptor atoms, leaving an excess of carriers determined by the greater of the impurity densities.

This picture of charge neutrality may best be exemplified by Fig. 3-13, which shows a simplified distribution of the free carriers and the impurity atoms. The symbols are defined as follows:

N_D = density of donor impurity atoms
N_A = density of acceptor impurity atoms
n = density of electrons in conduction band
p = density of holes in valence band
n_D = density of electrons bound to donors
p_A = density of holes bound to acceptors

In this case, $N_D = 7$ atoms and $N_A = 4$ atoms. For the temperature involved, $n_D = 2$ and $p_A = 2$ impurity atoms that are unionized. Also, three thermally generated hole-electron pairs are shown. Therefore,

$$(N_D - n_D) - (N_A - p_A) = n - p \tag{3-22}$$

where n and p include the hole-electron pairs. If it is assumed that all the impurity atoms are fully ionized, (3-22) becomes

$$N_D - N_A = n - p \tag{3-23}$$

which states that the net excess of donors over acceptors is just equal to the excess of electrons. For the example of Fig. 3-13, the net excess is three electrons and three donor ions. The same reasoning applies to an excess of holes and acceptors. Thus, Eqs. (3-22) and (3-23) are basic in charge neutrality.

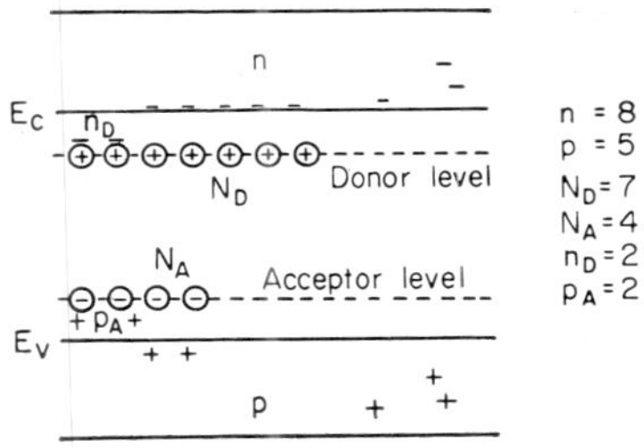

FIG. 3-13. Simplified picture of carrier distribution in impurity semiconductor.

3-11. Semiconductor Nomenclature. At this point it would be appropriate to establish the terminology that is often used in reference to semiconductors. As a matter of definition, if there exists an excess of donor impurities over acceptor impurities within a crystal (i.e., $N_D > N_A$) the semiconductor is called *n type*. This refers to the fact that conduction is due to concentration of negative charge carriers or electrons in the conduction band. On the other hand, if $N_A > N_D$ such that there is an excess of acceptors over donors so that there is an excess of hole carriers in the valence band, the crystal is called *p type*. It follows that if an impurity balance exists, that is, if $N_D = N_A$, the excess holes equal the excess electrons. For this condition, the conduction is intrinsic in nature, and the crystal is said to be *compensated.* In the next chapter, we will see that a compensated crystal will not have the same conductivity as a pure intrinsic crystal.

This nomenclature is summarized in Table 3-3. It should be noted that in an n-type semiconductor the electrons are called *majority* carriers. In a p-type semiconductor, the holes would be the majority carriers. The opposite carriers in n- and p-type crystals are called *minority carriers.* As will be seen later in the text, minority-carrier conduction is of prime importance in junction transistors.

TABLE 3-3

Type	n	p
Impurity	Donor	Acceptor
Majority carriers	Electrons	Holes
Minority carriers	Holes	Electrons

3-12. Fermi Level in Impurity Semiconductors. In an intrinsic semiconductor, that is, one free of impurities, the Fermi level will lie at the middle of the band gap, indicating that there are equal concentrations of holes and electrons due to thermal generation at a particular temperature. If a given density N_D of donor impurity atoms, for example, is added to the crystal, assuming that for the temperature involved all the impurity atoms become ionized, there will exist an excess of $N = N_D$ electrons in the conduction band. Since the allowed density of states in the band is N_C, the fraction N_D/N_C of these states will be filled, thereby reducing the number of states available for occupancy by thermally agitated electrons from the valence band. In other words, it is probable that the donor electron will fill up the lower energy states in the conduction band, making it more difficult for electrons to jump the gap, since they must now acquire sufficient energy to fill the higher states. In effect, this reduces the concentration of generated electron-hole pairs for that temperature. Therefore, for constant temperature, as the donor density increases, the hole concentration (minority carrier) in the valence band will decrease, thereby making the crystal more n type. If the temperature is increased, more electron-hole pairs will be generated, but because all the donors are ionized (saturation), a point will be reached where the thermally generated electron concentration in the conduction band will be very much greater than the donor-electron density. For this condition, the electron concentration approximately equals the hole concentration in the valence band and the crystal becomes intrinsic. Identical reasoning may be applied to acceptor impurities, where the semiconductor becomes more p type with impurity density and the electron minority-carrier concentration decreases with same. There also, the conduction approaches intrinsic as temperature is increased.

Since the Fermi level is a measure of the probability of occupancy of the allowed states, it is apparent that it must be adjusted to account for the impurity effects. What we have said above was that the presence of donor impurities increases the probability of occupancy of the conduction states by electrons and decreases the probability of holes occupying the valence state. This means the Fermi level must rise above the midpoint of the band gap. For acceptors the opposite effects are obtained and the Fermi level falls below the midpoint. These results are shown in

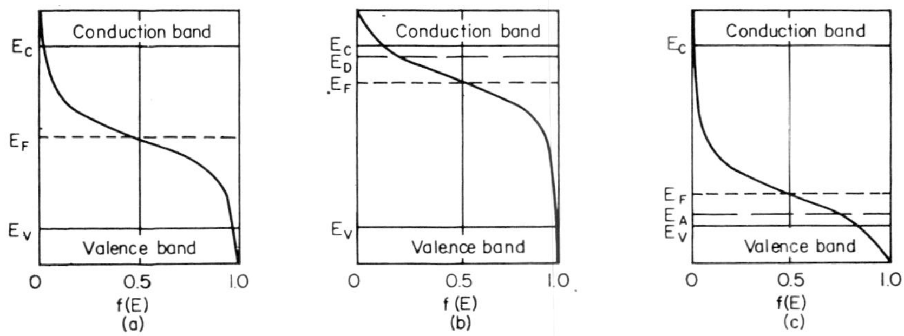

FIG. 3-14. Positions of Fermi level for impurity semiconductors. (*a*) Intrinsic; (*b*) n type; (*c*) p type.

Fig. 3-14 for each of three possible cases. We may conclude that as the crystal becomes more n type, E_F rises closer to the conduction band; as the crystal becomes more p type, E_F falls closer to the valence band. For a given impurity density of either type, as the temperature increases, the Fermi level will return to the intrinsic position.

As for the intrinsic case, in order to calculate the carrier concentration in an impurity semiconductor for a particular temperature, it is necessary to know both the distribution of states and the position of the Fermi level. It was stated that the number of impurity atoms will remove an equal number of states from the energy bands and establish them at the impurity energy levels E_D and E_A. For most impurity concentrations encountered, $N_D \ll N_C$ and $N_A \ll N_V$, so that for all practical purposes the effective densities of states in an impurity semiconductor are the same as N_C and N_V respectively. Thus, the total electron concentration in the conduction band is still

$$n = N_C \epsilon^{-(E_C - E_F)/kT} \tag{3-24}$$

and the total hole concentration is

$$p = N_V \epsilon^{-(E_F - E_V)/kT} \tag{3-25}$$

where E_F may now have any value, depending on temperature and impurity density. To determine E_F we must know the number of bound donor electrons n_D, or more specifically, the number of filled states at the donor-impurity energy level E_D. We apply Eq. (3-7) and obtain

$$n_D = \frac{N_D}{1 + \epsilon^{(E_D - E_F)/kT}} \tag{3-26}$$

If $E_F = E_D$ in (3-26), then $n_D = N_D/2$, which means that half the states are occupied or half the donor atoms are ionized. When $E_F > E_D$, then $n_D = N_D$ and none of the excess electrons is free. Likewise, the number of holes bound to acceptors is

$$p_A = \frac{N_A}{1 + \epsilon^{(E_F - E_A)/kT}} \tag{3-27}$$

It is a good assumption that for room temperature (25°C) or above, all the donor and acceptor impurities in germanium and silicon are fully ionized.

Equations (3-24) to (3-27) are functions of E_F, the Fermi level. Therefore, we cannot calculate n and p until E_F is known. However, by the charge neutrality Eq. (3-22), viz.,

$$(N_D - n_D) - (N_A - p_A) = n - p$$

or

$$n + n_D - N_D = p + p_A - N_A \tag{3-28}$$

we have a single equation for all the quantities involved. Since E_C, E_D, E_A, E_V, N_D, and N_A are known for a semiconductor, we can substitute the correct expressions in (3-28) and solve for E_F at some temperature T. Substitution yields

$$N_C\epsilon^{-(E_C - E_F)/kT} + \frac{N_D}{1 + \epsilon^{(E_D - E_F)/kT}} - N_D = N_V\epsilon^{-(E_F - E_V)/kT} + \frac{N_A}{1 + \epsilon^{(E_F - E_A)/kT}} - N_A \tag{3-29}$$

Once Eq. (3-29) is solved for E_F*, (3-24) and (3-25) are used to determine n and p. The manner in which E_F varies with temperature and impurity density for germanium is plotted in Fig. 3-15. The impurity band gap is given as 0.01 ev, although it is not shown to scale. Silicon would behave in a similar manner except that the band-gap values would be larger.

For $N_D = 10^{15}$ cm^{-3}, at the very low temperatures near absolute zero, the donor states are mostly bound and the number of carriers due to thermal activation is very small. This is the bound range and E_F lies very close to E_D. As the temperature is increased above 100°K, all the

* This is a transcendental function which is difficult to solve. Shockley presents an excellent graphical method of solving Eq. (3-29) for E_F. See page 465 of Ref. 4 at the end of this chapter.

donors become ionized and the Fermi level falls to account for the increasing number of electrons being excited into the conduction band. Approximately 100 to 400°K would be the saturation range. At 500°K or above, the hole concentration approaches the total electron concentration and the conduction range becomes intrinsic. An identical, but oppositely symmetrical, Fermi-level dependence is shown by the dashed curve for an equal density of acceptor atoms, $N_A = 10^{15}$ cm^{-3}. If the impurity density were increased to 10^{17}, the curve would appear as shown in Fig. 3-15. For room temperature 300°K, the Fermi level is higher

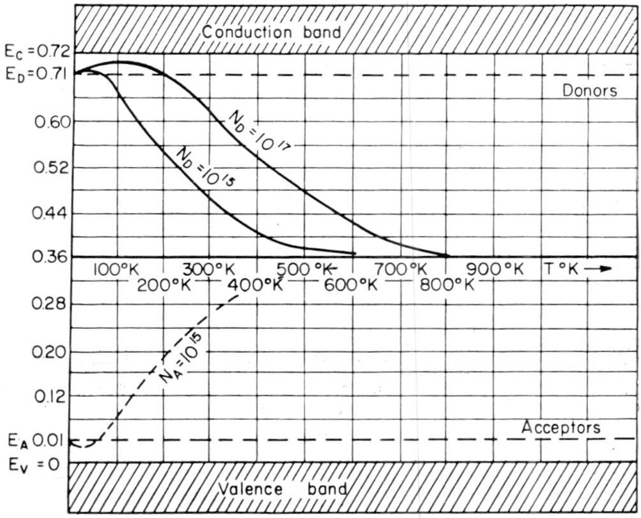

FIG. 3-15. Fermi-level dependence with temperature in germanium.

(more n type) for the increased-density case, since a greater proportion of the available conduction-band states are filled. Further, the intrinsic range is reached at higher temperatures, since greater hole densities are required.

3-13. Minority-carrier Concentration. A graphical method similar to that of Fig. 3-8 may be used to illustrate the effect of impurities on the free carrier concentrations in a semiconductor. In Fig. 3-16 is shown the graphical computation for an n-type impurity semiconductor. In Fig. 3-16a are shown the densities of the allowed energy states $N_C(E)$ and $N_V(E)$ in each of the energy bands. Also shown are the energy states for the donor impurities in the crystal. Although the figure is not drawn to scale, it must be remembered that the density of donor states is very much smaller than the density of states in the conduction band. To account for the number of electrons contributed to the conduction band by the donor impurities, the Fermi level rises above the middle of the band gap, as shown in Fig. 3-16*b*. As before, the product of (*a*) and (*b*) yields the carrier concentrations shown in (*c*). In this case, the increased probability of occupancy in the conduction band increases the electron concentration there. The hole concentration (minority carriers), however, is reduced considerably. Actually, for the function of (*b*), there exists a probability that some electrons will be found at the donor energy levels. These are electrons which are not ionized and are bound to the impurity atoms. At room temperatures and above, this concentration would be practically negligible.

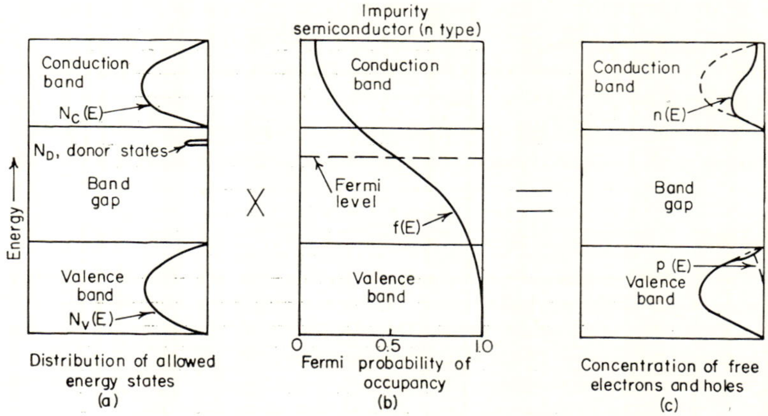

FIG. 3-16. Graphic computation of free carrier concentrations.

As was explained in Sec. 3-12, the same equations used for determining the carrier concentrations in the intrinsic case may be used for the impurity or extrinsic case. Equations (3-24) and (3-25) are repeated below.

$$n = N_C \epsilon^{-(E_C - E_F)/kT} \tag{3-30}$$

$$p = N_V \epsilon^{-(E_F - E_V)/kT} \tag{3-31}$$

Note that the position or value of the Fermi level E_F is what determines the magnitude of n or p. The product np for an impurity semiconductor will be, as in the intrinsic case, independent of Fermi level. From Eq. (3-15), we have again

$$np = N_C N_V \epsilon^{-E_G/kT} = n_i^2 \tag{3-32}$$

This is an important result, for it enables one to determine the minority-carrier concentrations in an impurity semiconductor. For example, consider an n-type semiconductor having a concentration of N_D donor impurities. If complete ionization of the impurities is assumed, there would be N_D electrons in the conduction band. There would also be a smaller number of electrons due to thermal generation. According to Eq. (3-32), therefore,

$$(n + N_D)p = n_i^2 \tag{3-33}$$

where $n = p$, since these are generated pairs. For typical doping levels, however, $N_D > n$, and (3-33) may be approximated as

$$N_D p \approx n_i^2 \qquad \text{n type} \tag{3-34}$$

In the same manner,

$$N_A n \approx n_i^2 \qquad \text{p type} \tag{3-35}$$

where N_A corresponds to the concentration of holes in the valence band, equal approximately to the acceptor impurity density. In Eqs. (3-34) and (3-35), p and n are the minority-carrier concentrations. It is apparent that the minority-carrier concentration will vary inversely with the impurity level. As we shall see in later chapters, the minority carrier plays an important role in transistor conduction. In Chap. 4, its variations with resistivity and temperature will be considered in detail.

PROBLEMS

3-1. For perfect impurity-free crystals of germanium and silicon, calculate the equilibrium value of the intrinsic carrier concentration n_i for a temperature of 127°C. Refer to Eq. (3-19) and assume that for the effective masses, $m = m_p = m_n$, where m is the free electron mass. Boltzmann's constant k equals 8.63×10^{-5} ev/°K.

3-2. Show that the thermal equilibrium product np for an impurity semiconductor is independent of the Fermi level E_F. Assume that $m = m_p = m_n$.

3-3. n-type samples of germanium and silicon crystals each have a uniform density N_D of 10^{15} atoms/cm^3 of antimony. (This is an impurity of about 1 part in 10 million.) For each semiconductor material, determine:

a. The temperature at which half the donor-impurity atoms are ionized. Assume that all the electrons come from the impurity levels, that is, thermal generation from the valence band is negligible. What is the Fermi-level energy as measured with respect to the valence-band edge $E_V = 0$?

b. The temperature in degrees C at which the conduction becomes intrinsic. Use the empirical curves of n_i versus $1/T$ and define the intrinsic point where $n_i = 10N_D$.

c. The equilibrium minority-carrier concentration at room temperature ($T = 300$°K). Assume that the total electron concentration in the conduction band is equal to the donor-impurity density and that the latter are fully ionized.

3-4. What should the donor-impurity concentration (atoms per cubic centimeter) be to completely fill the allowed number of states in the conduction band at room temperature? Assume complete ionization and no thermal generation.

REFERENCES

1. Herman, F.: The Electronic Energy Band Structure of Silicon and Germanium, *Proc. IRE*, vol. 43, pp. 1703–1732, December, 1955.
2. Kittel, C.: "Introduction to Solid State Physics," p. 273, John Wiley & Sons, Inc., New York, 1953.

3. Pearson, G. L., and W. H. Brattain: History of Semiconductor Research, *Proc. IRE*, vol. 43, p. 1804, December, 1955.
4. Shockley, W.: "Electrons and Holes in Semiconductors," chap. 7, D. Van Nostrand Company, Inc., Princeton, N.J., 1953.
5. See, for example, Seitz, F.: "The Modern Theory of Solids," McGraw-Hill Book Company, Inc., New York, 1940.
6. Shockley, *op. cit.*, p. 464.
7. Morin, F. J., and J. P. Maita: Conductivity and Hall Effect in the Intrinsic Range of Germanium, *Phys. Rev.*, vol. 94, pp. 1525–1529, June, 1954.
8. Morin, F. J., and J. P. Maita: Electrical Properties of Silicon Containing Arsenic and Boron, *Phys. Rev.*, vol. 96, pp. 28–35, October, 1954.
9. Kittel, *op. cit.*, pp. 277–280.
10. Geballe, T. H., and F. J. Morin: Ionization Energies of Groups III and V Elements in Germanium, *Phys. Rev.*, vol. 95, pp. 1085–1086, 1954.
11. Morin, F. J., J. P. Maita, R. G. Shulman, and N. B. Hannay: Impurity Levels in Silicon, *Phys. Rev.*, vol. 96, p. 833, 1954.

4

Properties of Semiconductor Crystals

4-1. Introduction. In previous chapters we have shown how energy bands give rise to holes and electrons in a semiconductor and how the number of these free carriers are related to the impurity content and the temperature. In this chapter, we shall qualitatively describe the way these carriers move within the crystal under conditions of applied electric fields (or voltage), thereby establishing the electrical conduction properties of the material. Since current is actually the movement of a total amount of charge or carriers per unit time, the study will introduce those basic bulk parameters such as mobility, resistivity, diffusion constant, and lifetime which affect the current flow in semiconductors. As will be seen later, these bulk parameters are essential in determining the characteristics and performance of any semiconductor device. Therefore, in addition to the theory, this chapter will present much of the most recently accumulated experimental values and curves for these parameters for both germanium and silicon. These curves and data are basic to transistor design; they will be referred to often throughout the text and are to be used in all problem calculations.

4-2. Drift Motion of Carriers. We have seen that electrons and holes are negative and positive charges, respectively, each having a mass on the order of the free electron mass m. The concentration of these charge carriers, dictated mathematically by the Fermi-level position, is actually a statistical or thermal equilibrium result of many random processes within the crystal. Because of crystal imperfections such as impurities and lattice defects, the electrons, for example, are continually making transitions from one quantum state to another within the conduction band. Simultaneously, electrons are continually making transitional jumps back to the valence band, where they recombine with holes. However, for thermal equilibrium the rate of generation will be balanced by the rate of recombination, yielding the observed net concentrations. The composite effect of these random processes is that a random motion is imparted to the carriers, even in the absence of applied fields. The ease of motion or mobility of the holes and electrons, however, is limited by the occurrence of collisions within the crystal lattice. The effect of these collisions on the motion of the carrier is very much like that of a viscous medium on a body falling in it. The body will accelerate to a finite limiting velocity such that the terminal kinetic energy just equals the energy loss of friction in the viscous medium.

From electrostatics, if a charge q, such as that of the electron, is placed in an electric field of $\mathcal{E}$ volts/cm, the force acting on the electron is

$$F = q\mathcal{E} \tag{4-1}$$

and the acceleration, from $F = ma$, is

$$a = \frac{q\mathcal{E}}{m} \tag{4-2}$$

where m is the mass of the electron. For a given electric-field intensity, the acceleration would be constant, and the velocity of the electron would increase without limit. However, this is not the case, since the electron (or hole) will lose energy by collisions within the crystal. Strictly speaking, the collisions are actually abrupt deflections of the carrier's path of travel owing to localized disruptions of the periodic potential field within the crystal. One type of collision process is that of *lattice scattering*, due to the thermal vibrations of the atoms of the crystal lattice. Another is *impurity scattering*, which is due to electrostatic forces acting on the carrier from the ionized impurity charges. Other collision mechanisms such as scattering between carriers and scattering by lattice defects exist but are relatively insignificant in most cases. With the application of an electric field, the total effect of the collisions is to limit the velocity of the electron to some finite average value such that the free path between collisions is traversed in a specific mean free time. This relationship is given by

$$v_D = at \tag{4-3}$$

where v_D is the average velocity, and t is the mean free time. If Eq. (4-2) is substituted in Eq. 4-3, we obtain

$$v_D = \left(\frac{qt}{m}\right)\mathcal{E} = \mu\mathcal{E} \qquad \text{cm/sec} \tag{4-4}$$

where the quantity in parenthesis is μ, the mobility. The mobility is defined as the average carrier velocity per unit electric field, and is expressed in units of centimeter^2 per volt-second.

The foregoing description of the free carrier dynamics may best be understood by referring to Fig. 4-1, where the motion of a single electron is shown.[1,*] In (*a*), the random motion of the electron in the crystal is shown, where the electron takes off in any arbitrary direction after each collision. The lengths of the arrows correspond to the mean free paths and are drawn equal for simplicity. If an electric field $\mathcal{E}$ is applied, the component of motion added by the field would be as shown in (*b*). The direction is opposite to that of the field since the negative charge makes the force negative. When the motions of (*b*) are superimposed on (*a*), the field accelerates the electron between collisions, thereby curving the mean free paths in the indicated direction, as in (*c*). The net effect is to impose a drift velocity to the electron motion.

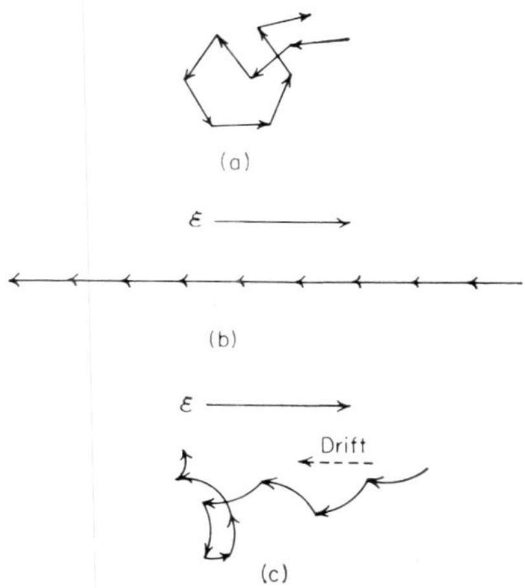

FIG. 4-1. Effect of electric field in superimposing drift on random motion. (*a*) Random motion of electron between collisions; (*b*) motion introduced by electric field; (*c*) combination of (*a*) and (*b*). (*After W. Shockley.*)

The mobility in Eq. (4-4) is referred to as the *drift mobility*, and as the equation indicates, the drift velocity is linearly related to the mobility and the electric field. Actually, this relationship holds only for low electric-field intensities. As the field intensity increases, the electrons gain energy and become "hotter." This increases their frequency of collisions with the lattice. It is found[2] that there is a range of electric field over which the drift velocity varies as $\mathcal{E}^{1/2}$. Under these conditions, the electrons are transferring energy to the acoustical modes of lattice vibration. As the field reaches very large values, the drift velocity becomes independent of field and a limiting velocity on the order of 7×10^6 cm/sec is reached. In Fig. 4-2 are curves summarizing Ryder's data[2] on drift velocity as a function of electric field for n- and p-type crystals of germanium and silicon.

4-3. Drift Mobility. It should be apparent that as the number of carrier collisions per unit time increases, the drift mobility will decrease. Furthermore, because μ is a function of the carrier mass, the mobility of electrons would be different from that of holes, since the effective masses are different from energy considerations. The drift mobilities of the carriers are also influenced by the two principal collision mechanisms, namely, impurity-atom scattering and lattice-vibration scattering. If μ_I is designated as the mobility due to impurities alone, and μ_L is lattice

* References, indicated in the text by superscript figures, are listed at the end of the chapter.

mobility, it is a fair approximation to write[3]

$$\frac{1}{\mu} = \frac{1}{\mu_I} + \frac{1}{\mu_L} \tag{4-5}$$

for the total drift mobility. We shall now examine each of these scatter mechanisms independently.

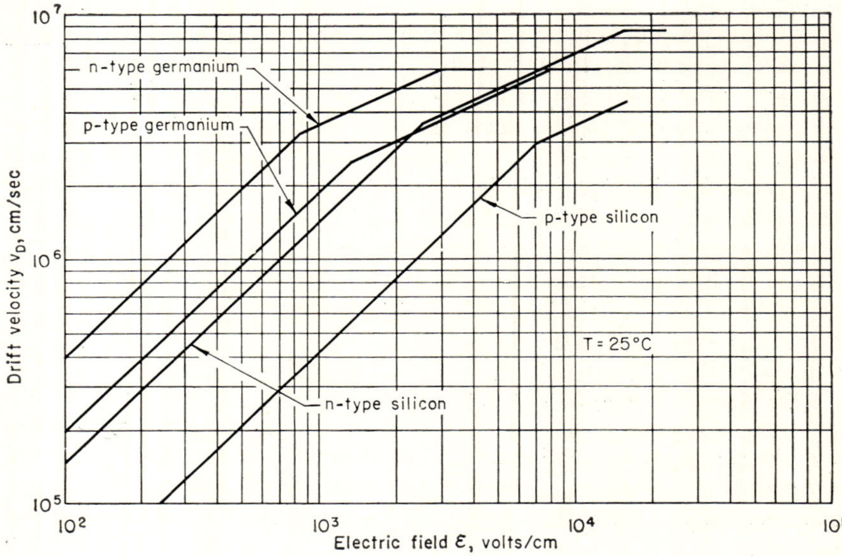

FIG. 4-2. Drift velocity as a function of electric field. (*After E. J. Ryder.*)

The theory of impurity-scattering mobility was investigated by Conwell and Weisskopf and their analysis gives the following formula:[4,5]

$$\mu_I = \frac{8\sqrt{2}\,\kappa^2(kT)^{3/2}}{\pi^{3/2}N_I q^3 m_{\text{eff}}^{1/2} \ln\,(1 + 3\kappa k T^2/q^2 N_I^{1/3})} \tag{4-6}$$

where T = temperature, °K

κ = dielectric constant

N_I = total density of *all* ionized impurities

m_{eff} = effective mass of electron or hole

Thus, the impurity mobility will decrease with increasing impurity concentration, but increase with temperature as $T^{3/2}$. Physically, as the number of charge centers increases, the probability of collision becomes greater, making the mobility smaller. Further, as the temperature increases, the average velocities (or energies) of the carriers increase, making the charge centers less effective, thereby increasing the mobility.

Lattice-scattering mobility is associated with the collisions of the carriers with the vibrating atoms of the crystal lattice. For μ_L the theory shows a temperature dependence as follows:

$$\mu_L \propto T^{-1.5} \tag{4-7}$$

The reason for the decrease of lattice mobility with temperature is twofold:

1. Since the mean free path for a given temperature is not a function of the carrier's speed but of its direction, the faster the carriers move due to heating the more collisions per unit time will be made, thereby decreasing the mobility.

2. As the temperature increases, the lattice vibrations become more vigorous, thereby decreasing the distance between collisions and the mobility.

Actual measurements of drift mobility in high-quality crystals of germanium and silicon, where the impurity concentrations were so small that only lattice mobility would be predominant, show definite departures from the theoretically predicted $T^{-1.5}$ law. Recent measurements give the following relationships for lattice mobility:

$$\left.\begin{aligned} \mu_{Ln} &= 4.90 \times 10^7 T^{-1.66} \quad (4\text{-}8)\\ \mu_{Lp} &= 1.05 \times 10^9 T^{-2.33} \quad (4\text{-}9) \end{aligned}\right\} \text{germanium}^{6,7}$$

$$\left.\begin{aligned} \mu_{Ln} &= 2.1 \times 10^9 T^{-2.5} \quad (4\text{-}10)\\ \mu_{Lp} &= 2.3 \times 10^9 T^{-2.7} \quad (4\text{-}11) \end{aligned}\right\} \text{silicon}^{8}$$

where the subscripts n and p refer to mobility for electrons and holes, respectively. It suffices to say that the deviations from the theory are associated with the effective mass values and certain assumptions made concerning the nature of the energy surfaces related to the Brillouin-zone theory.[9]

The composite effects of both impurity and lattice mobilities are shown in the curves of Figs. 4-3 and 4-4. These data were reported by Prince,[10,11] Backenstoss,[12] and others,[14] as obtained from drift-mobility measurements of minority carriers in samples of various impurity concentrations. The measurement is based on the technique devised by Haynes and Shockley. With voltage applied across the ends of the semiconductor sample, minority carriers are injected at one point and collected at another point a distance l away. By knowing the values of l, V, and the drift time for the carriers between the points, the value of drift mobility may be calculated. The curves give the values of minority-carrier drift mobility as a function of impurity concentration at room temperature for germanium and silicon. It is seen that for both germanium and silicon, the effects of impurity scattering become predominant at concentrations of 10^{15} cm^{-3} or more. Below this value, lattice scattering controls the mobility. For neither germanium nor silicon does the drift mobility increase appreciably above the values at concentrations of $N = 10^{14}$ cm^{-3}.

In the foregoing discussion we have been concerned with the magnitudes of the drift mobilities of *minority* carriers. Of interest also would

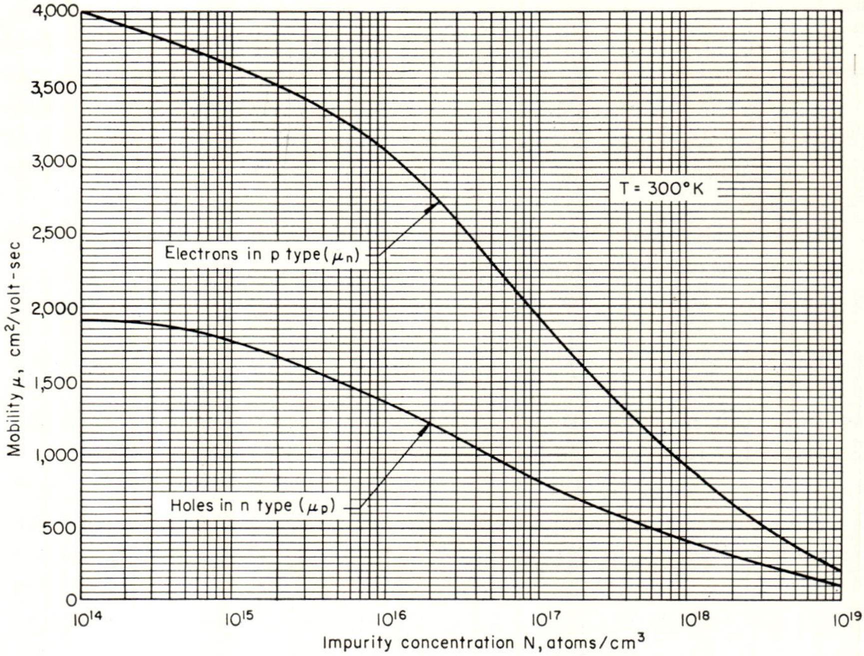

FIG. 4-3. Minority carrier drift mobilities in germanium.

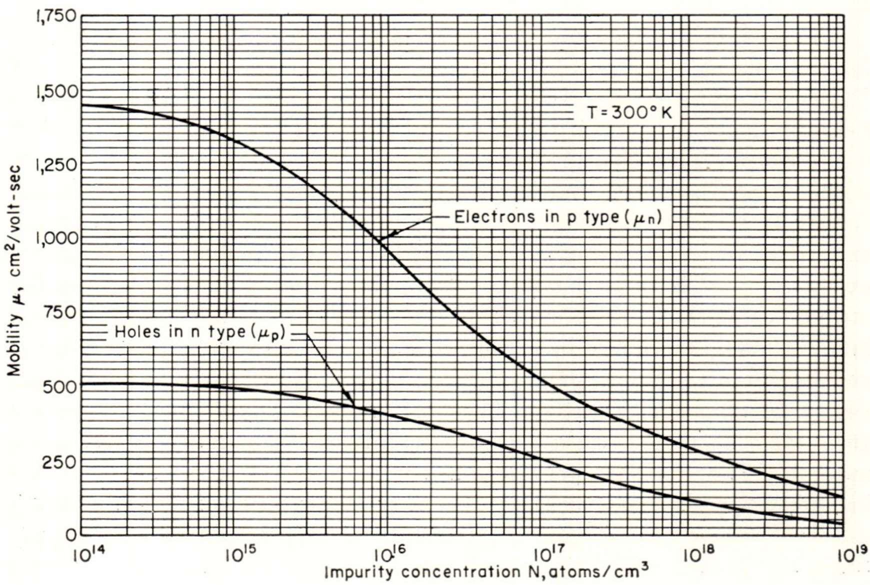

FIG. 4-4. Minority carrier drift mobilities in silicon.

be the mobility values for the *majority* carriers in the semiconductor. We recall that drift mobility is limited by the total number of charged scattering centers that a carrier would experience within the crystal. In the lattice-scattering range, where the impurity concentrations are small, the majority-carrier mobilities are approximately the same as the minority carrier values. This is so because both carriers see almost the same number of charge centers. For example, in an n-type crystal, the minority holes see N_D donors, N_A acceptors, and $N_D - N_A$ majority

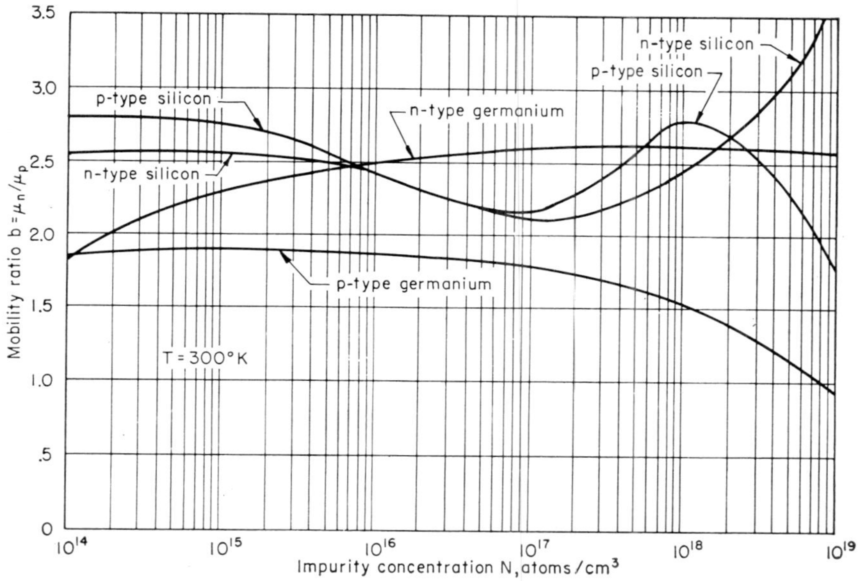

FIG. 4-5. Calculated variation of mobility ratio with impurity concentration. b values calculated from $\rho(N)$ and $\mu(N)$ curves.

carriers, or a total of $2N_D$ scattering centers. The majority electrons, however, see only N_D donors and N_A acceptors. Thus, for small net impurity concentrations in the lattice range, $2N_D$ would approximately equal $N_D + N_A$. On the other hand, for very-low-resistivity samples where $N_D \gg N_A$, the majority electrons see only half as many charge centers as the minority carrier, and would therefore have a slightly higher mobility in n type. Applying the same reasoning to the p-type case, we can conclude that for high impurity concentrations (approximately greater than 10^{15}), the drift mobility of a carrier is slightly higher when it is a majority carrier than when it is a minority carrier.

To estimate the majority-carrier values we make use of the calculated ratios $b = \mu_n/\mu_p$ for n- and p-type crystals of germanium and silicon. These are given in Fig. 4-5 as functions of impurity concentration. The coefficient b is the ratio of the electron mobility to the hole mobility

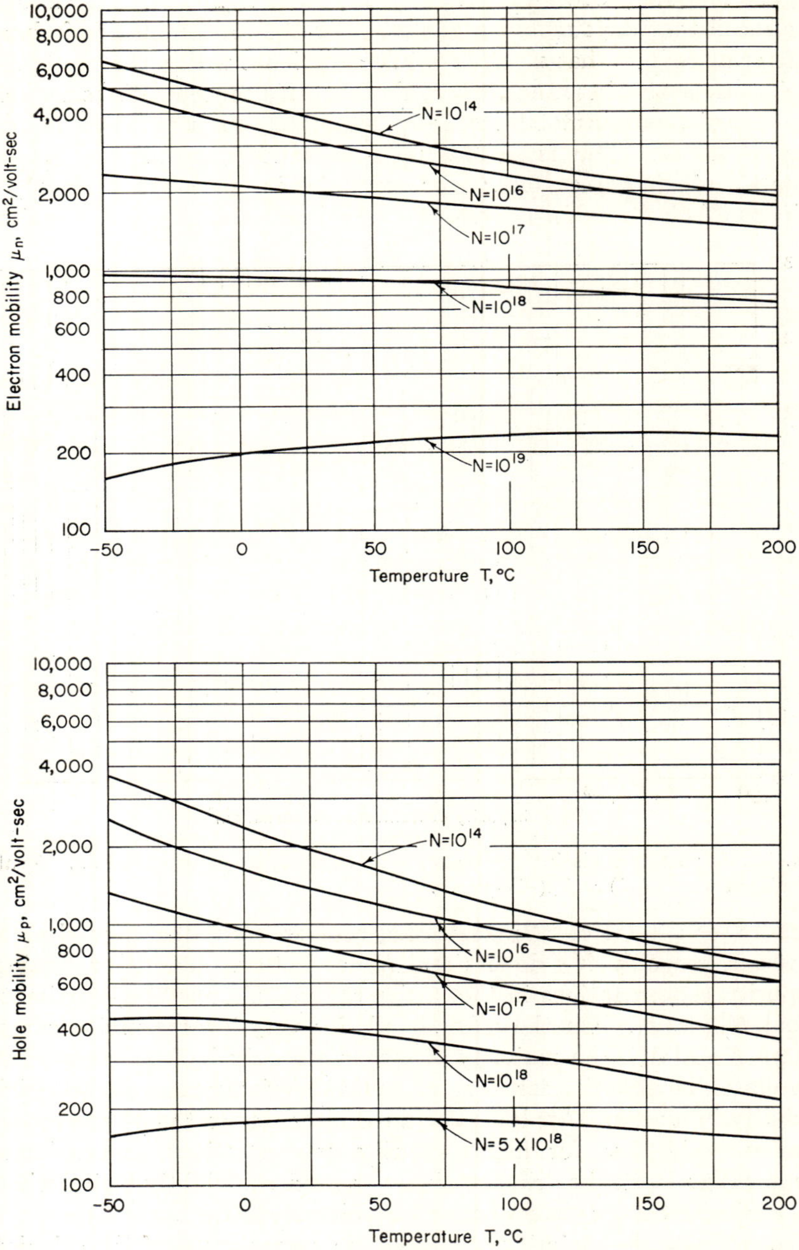

FIG. 4-6. Temperature dependence of mobility in germanium. (*After W. W. Gärtner.*)

within a particular type semiconductor. To illustrate the use of these curves, we consider a simple example. Suppose we have an n-type sample of germanium with $N = 2 \times 10^{16}$ cm^{-3}, and the majority-carrier (electron) mobility is required. From the μ_p curve of Fig. 4-3 we obtain $\mu_p = 1{,}200$ cm^2/volt-sec, which is the minority-carrier mobility. Also, from the n-type germanium curve of Fig. 4-5 we get $b = \mu_n/\mu_p = 2.53$. Therefore the majority-carrier mobility μ_n is $b\mu_p$, or

$$2.53 \times 1{,}200 = 3{,}036 \text{ cm}^2/\text{volt-sec.}$$

Note that from Fig. 4-3 the minority-carrier value is 2,750. Thus, the b curves are used in conjunction with the minority-carrier values to obtain the majority-carrier values.

The temperature dependence of drift mobility, also of extreme importance in semiconductor design, may be obtained by combining the temperature dependence of Eq. (4-6) with those of Eqs. (4-8) through (4-11), using the simple relationship given by Eq. (4-5). Equation (4-5), however, is not a very good approximation, and was improved upon.[14] With the more complex relationship as a basis, the temperature dependence of μ was calculated for both germanium and silicon[15] and the results are shown in Figs. 4-6 and 4-7. It should be noted that for low impurity concentrations, the mobility decreases considerably as a result of the effects of lattice scattering. As the impurity doping increases, however, the variation with temperature becomes less and less. These curves, which reflect a combination of theory and experiment, are essential to the calculation of the temperature behavior of transistor parameters, since nearly all parameters can be related to mobility, carrier concentration, and lifetime.

4-4. Current Flow in Semiconductors. Electric current is defined as the flow of a total amount of charge per unit time within an electric field. From Eq. (4-4), for a constant electric field $\mathcal{E}$, the drift velocity for electrons is $v_{Dn} = -\mu_n\mathcal{E}$ and for holes is $v_{Dp} = \mu_p\mathcal{E}$. The minus sign for v_{Dn} arises from the negative charge of the electron and indicates that electrons should flow in directions opposite to those of the holes for the same field. This is shown in Fig. 4-8, where a voltage V is impressed on a semiconductor of length l and cross-sectional area A. By convention the direction of electric field is shown to be the same as the current I, viz., from plus to minus. It is equal in magnitude to $\mathcal{E} = V/l$ volts/cm. Since the crystal would contain a density of n electrons/cm^3, a total charge density of $-qn$ coulombs/cm^3 would drift opposite to the field and a charge of $+qp$ coulombs/cm^3 would drift in the same direction as the field. The current density J for electrons would be

$$J_n = \frac{I_n}{A} = -qnv_{Dn} \qquad \text{amp/cm}^2 \tag{4-12}$$

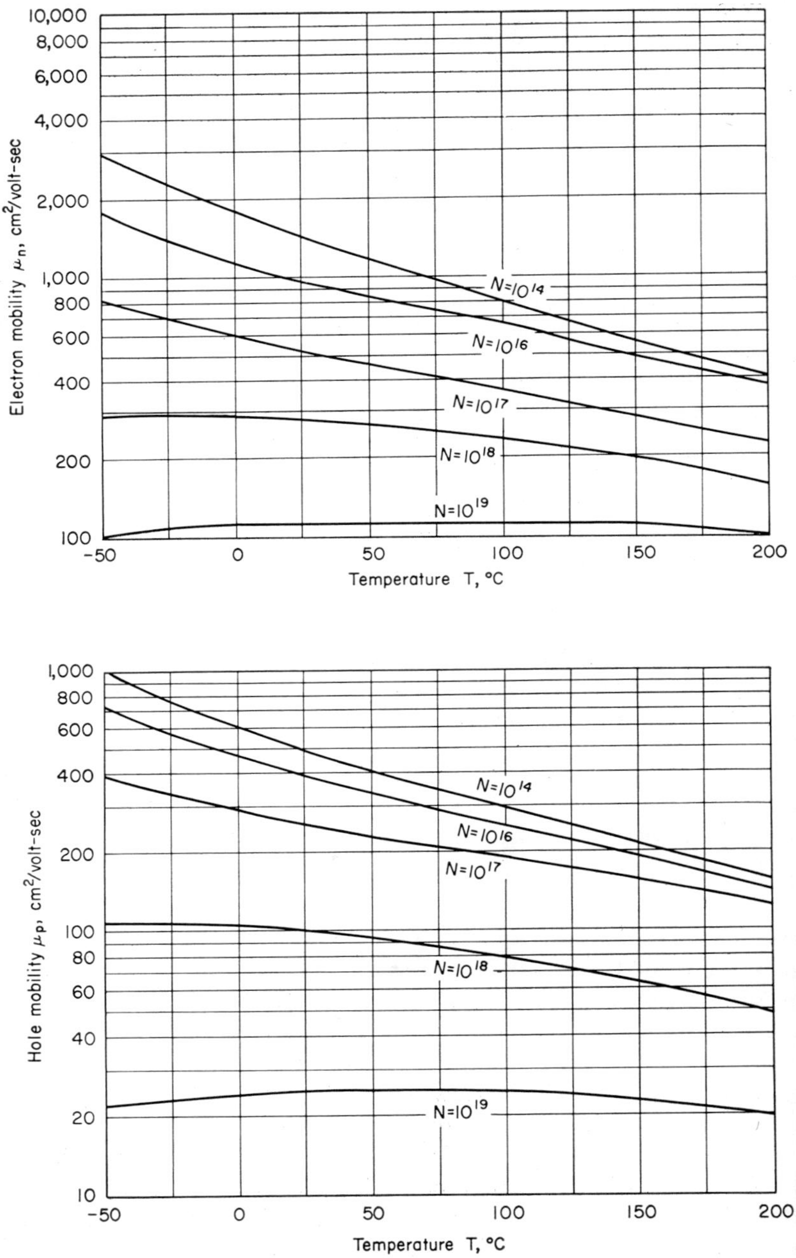

FIG. 4-7. Temperature dependence of mobility in silicon. (*After W. W. Gärtner.*)

and for holes

$$J_p = \frac{I_p}{A} = +qpv_{Dp} \qquad \text{amp/cm}^2 \tag{4-13}$$

Since a transfer of negative charge to the right is the same as a transfer of positive charge to the left, the total current density would be the sum of the components, or

$$J = J_n + J_p = -qnv_{Dn} + qpv_{Dp} \tag{4-14}$$

Substituting the correct expressions for v_{Dn} and v_{Dp} in (4-14), we obtain

$$J = -qn(-\mu_n\mathcal{E}) + qp\mu_p\mathcal{E}$$
$$J = \mathcal{E}q(\mu_n n + \mu_p p) \tag{4-15}$$

Multiplying Eq. (4-15) by A to obtain I, and substituting V/l for $\mathcal{E}$, we obtain

$$I = \frac{VA}{l} q(\mu_n n + \mu_p p) \tag{4-16}$$

This equation gives the total current that would flow for the impressed voltage V. The total current is made up of an electron current and a hole current. Application of Ohm's law to obtain the resistance in ohms changes Eq. (4-16) to

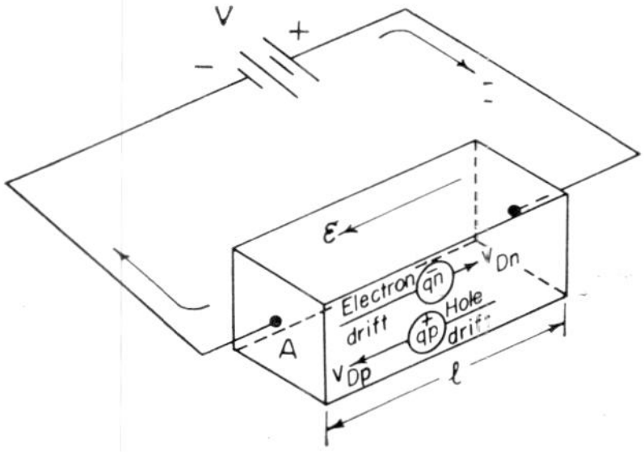

FIG. 4-8. Contributions of carrier drift to total current flow.

$$R = \frac{V}{I} = \frac{l}{A}\frac{1}{q(\mu_n n + \mu_p p)} \qquad \text{ohms}$$

This is of the form $R = \rho l/A$ where ρ is the resistivity, expressed in units of ohm-centimeter.

$$\rho = \frac{1}{q(\mu_n n + \mu_p p)} \qquad \text{ohm-cm} \tag{4-17}$$

Equation (4-17) is important to semiconductors. The greater the concentration of holes and/or electrons due to either impurities or thermal generation, and the more mobile they are, the lower the resistivity of the semiconductor. The dependence on impurities, mobility, and temperature will be discussed later.

It should be noted that the reciprocal of resistivity is the conductivity $\sigma = 1/\rho$, expressed in units of mhos per centimeter. Thus,

$$\sigma = q(\mu_n n + \mu_p p) \qquad \text{mho cm}^{-1} \tag{4-18}$$

4-5. Resistivity. The formula for the resistivity of a semiconductor is given by

$$\rho = \frac{1}{q(\mu_n n + \mu_p p)} \qquad \text{ohm-cm} \tag{4-19}$$

Since ρ is related to the mobility and the carrier concentration, it would be of interest to note the manner in which resistivity behaves as a function of temperature. Such a curve is drawn in Fig. 4-9 for a semiconductor having room-temperature resistivities in the range of 1 to 10 ohm-cm. At very low temperatures near absolute zero, few of the impurity atoms are ionized and the resistivity is quite high. As T increases toward 0°C more impurities ionize, thereby decreasing ρ. In this range, the mobility is predominantly due to impurity scattering, varying inversely as N_I. Over the temperature range of about 0 to 200°C, where all the impurities are fully ionized and the carrier concentration is somewhat constant, lattice scattering is usually dominant. Thus, the resistivity will increase as the lattice mobility decreases with temperature. At the higher temperatures, the impurity concentration becomes swamped by thermal generation of carriers and the resistivity decreases considerably. This is the intrinsic range for the impurity semiconductor.

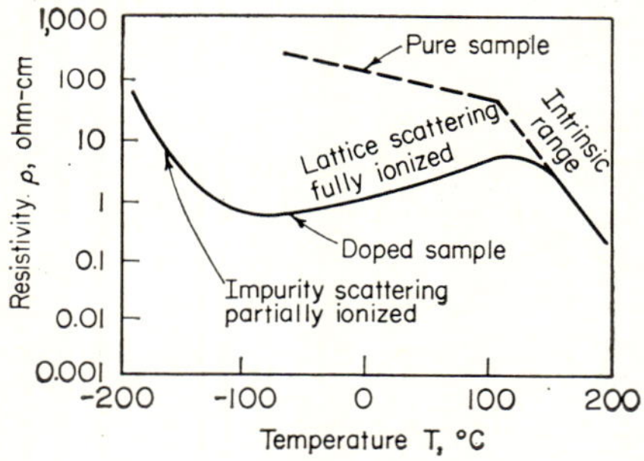

FIG. 4-9. Example of typical temperature dependence of resistivity in semiconductors.

The temperature dependence of resistivity for a pure semiconductor is also shown in Fig. 4-9. In this case, since the impurity level is negligible, the resistivity is principally controlled by the lattice mobility. Although the lattice mobility decreases with temperature, the effect is overcome by the thermal generation of hole-electron pairs. Consequently, the resistivity decreases with temperature, or in other words, intrinsic semiconductors are characterized by a *negative* temperature coefficient. This offers another explanation as to why semiconductors differ from metals. A metal has a positive temperature coefficient of resistance. This is because the electron concentration does not change with temperature since the bands are overlapping. However, since mobility decreases with temperature due to scattering by the thermal vibrations of the metal atoms, we observe an increase in resistance with temperature for metals.

It may be recalled that in the intrinsic range the number of electrons equals the number of holes and

$$n = p = n_i \tag{4-20}$$

Therefore, Eq. (4-19) can be written, for the intrinsic resistivity, as

$$\rho_i = \frac{1}{qn_i(\mu_n + \mu_p)} = \frac{1}{qn_i\mu_p(1 + b)} \tag{4-21}$$

The empirical relationships for n_i as a function of temperature, given in Chap. 3, are repeated here for reference:

$$n_i = \sqrt{np} = 1.76 \times 10^{16} T^{3/2} \epsilon^{-0.785/2kT} \qquad \text{germanium} \qquad (4\text{-}22)$$

$$n_i = \sqrt{np} = 3.87 \times 10^{16} T^{3/2} \epsilon^{-1.21/2kT} \qquad \text{silicon} \qquad (4\text{-}23)$$

At room temperature (300°K) the intrinsic resistivities determined from the above equations are approximately:

$$\rho_i \approx 47 \text{ ohm-cm} \qquad \text{germanium}$$

$$\rho_i \approx 214{,}000 \text{ ohm-cm} \qquad \text{silicon}$$

This is striking evidence of the contribution of the larger energy-band gap for silicon. In an intrinsic semiconductor, although $n = p$, the electron mobility μ_n is greater than the hole mobility μ_p, and therefore the conduction is n type.

For impurity semiconductors at normal temperatures, it is a good assumption to consider $n \approx N_D - N_A$ for n-type material and $p \approx N_A - N_D$ for p type. Since the product np is a constant for a particular temperature, the equilibrium minority-carrier concentration for either type is very small. For n type, $n = N_D - N_A$ is greater than p; for p type, $p = N_A - N_D$ is greater than n. The resistivities may therefore be approximated as follows:

$$\rho_n = \frac{1}{q\mu_n(N_D - N_A)} = \frac{1}{qb\mu_p(N_D - N_A)} \qquad \text{n type} \qquad (4\text{-}24)$$

$$\rho_p = \frac{1}{q\mu_p(N_A - N_D)} = \frac{1}{q(\mu_n/b)(N_A - N_D)} \qquad \text{p type} \qquad (4\text{-}25)$$

These are good assumptions for impurity semiconductors and may be used for purposes of calculation at or close to room temperature.

It should be evident that the total impurity concentration has a pronounced effect on the resistivity of a semiconductor because of the impurity mobility dependence. It is possible for two semiconductor crystals having the same net impurity difference (e.g., $N_D - N_A$) to have different resistivities. This would occur if the total impurity level $N_D + N_A$ is greater in one case than in the other. From the drift-mobility relations we see that μ decreases with impurity level, which increases the resistivity. Similarly, for a compensated crystal where $N_D = N_A$, the resistivity is higher than for the pure intrinsic case, providing the impurity density is sufficiently high to lower the mobility.

On the basis of uncompensated impurity concentrations, one can plot resistivity as a function of either donors or acceptors for both germanium and silicon using the previous mobility data. This is given in Fig. 4-10 and is probably the most often used curve in transistor design.[10–13]

4-6. Minority-carrier Concentrations. Although the magnitude of the equilibrium minority-carrier concentration may be neglected for most

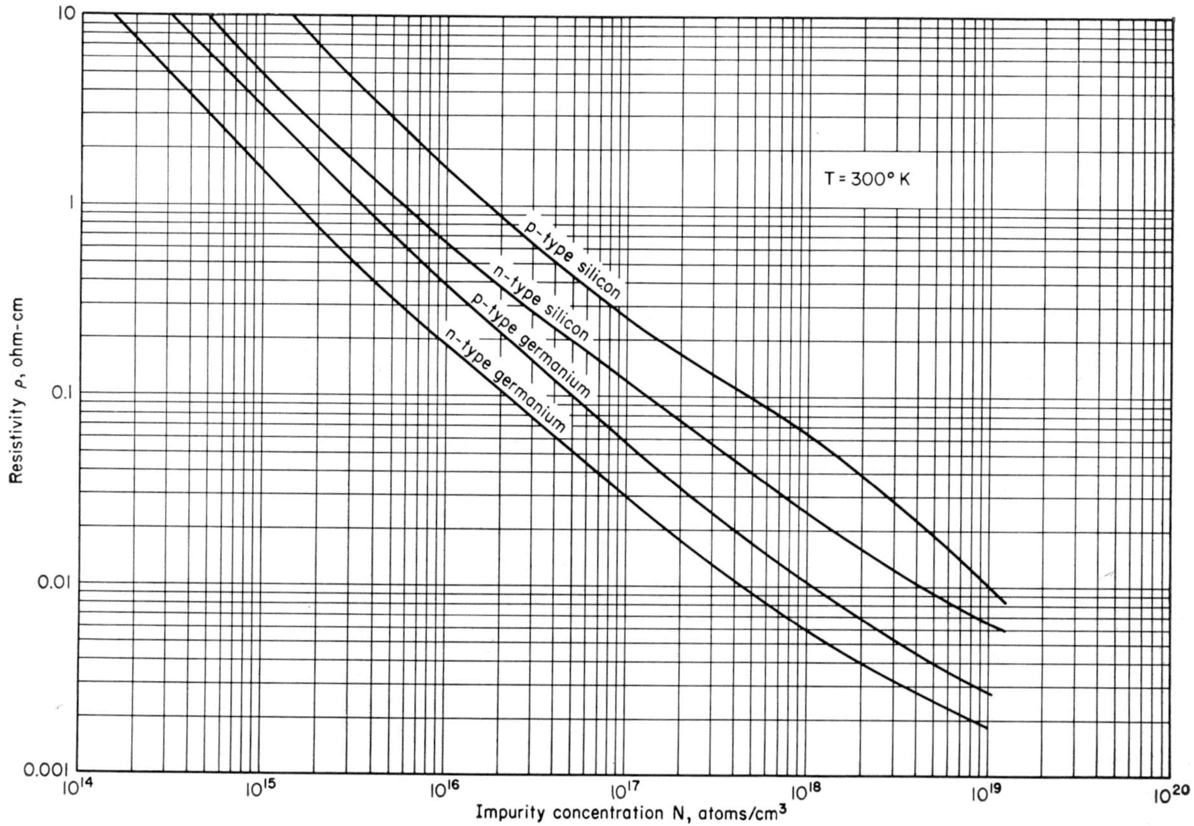

FIG. 4-10. Resistivity as a function of impurity concentration.

resistivity calculations, it is important to know how it varies with temperature and resistivity. As will be seen in subsequent chapters, the minority-carrier concentration plays a dominant role in p-n junctions and transistors. The terms p_n (minority hole concentration in n-type material) and n_p (minority electron concentration in p-type material) appear often in design equations.

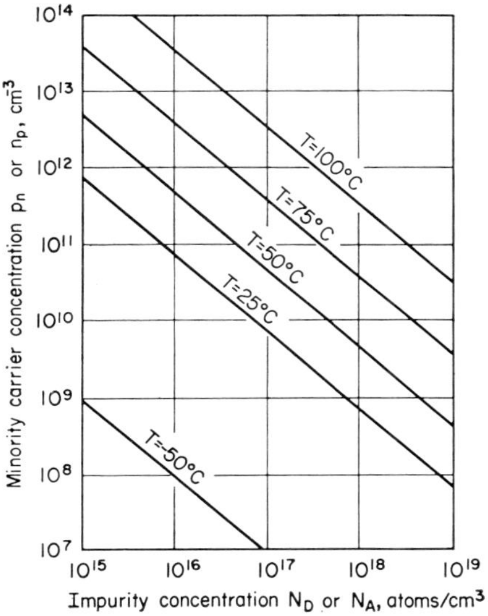

FIG. 4-11. Minority-carrier temperature dependence in germanium.

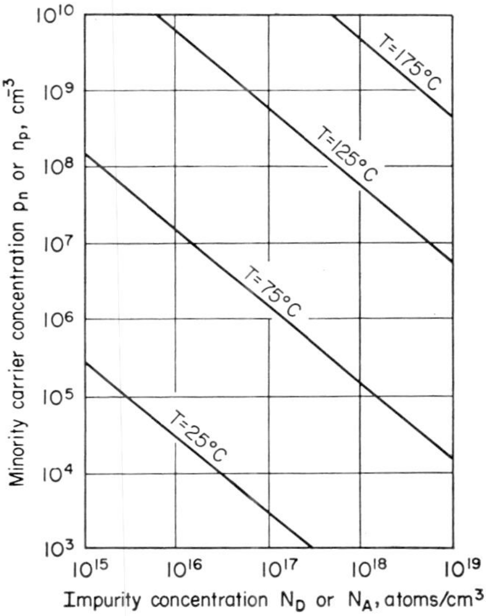

FIG. 4-12. Minority-carrier temperature dependence in silicon.

We have shown in Chap. 3 that, because of the relationship $n_i^2 = np$, we can write, for n-type semiconductors,

$$p_n \approx \frac{n_i^2}{N_D} \tag{4-26}$$

where N_D is the donor-impurity concentration. Similarly, for p-type semiconductors, we have

$$n_p \approx \frac{n_i^2}{N_A} \tag{4-27}$$

where N_A is the acceptor-impurity concentration. Although these relationships are approximations, they are nevertheless reasonably accurate for impurity concentrations greater than 10^{15} atoms/cm^3 and for temperatures less than 100°C for germanium and 200°C for silicon. Therefore, with Eqs. (4-26) and (4-27), the dependence of minority-carrier concentrations on temperature is determined simply by considering the temperature dependence of n_i^2 (see Figs. 3-9 and 3-10). This is shown by the curves of Figs. 4-11 and 4-12, which yield p_n and n_p as functions of impurity concentration for different ambient temperatures. It should be noted that at the higher temperatures, the minority-carrier concentrations become practically constant with resistivity. This is the condition of intrinsic conduction, wherein thermal generation causes the minority-carrier concentration to become comparable in magnitude to the majority-carrier concentration.

4-7. Diffusion Motion of Carriers. In addition to motion by drift in an electric field, electrons and holes may move within the semiconductor crystal by *diffusion* in the absence of a field. Consider, for example, a low-resistivity n-type crystal where the equilibrium concentration of majority carriers (electrons in this case) is large. If, by some means, an excess of holes or minority carriers was created within a localized region of the crystal, this excess hole concentration would gradually spread out or diffuse in all directions. The rate of diffusion would be dependent on the concentration gradient, the difference between the excess minority-carrier concentration and the equilibrium concentration within the crystal. This diffusion process is analogous to the flow of heat in a long rod, where the rate of flow is proportional to the temperature gradient between the ends of the rod.

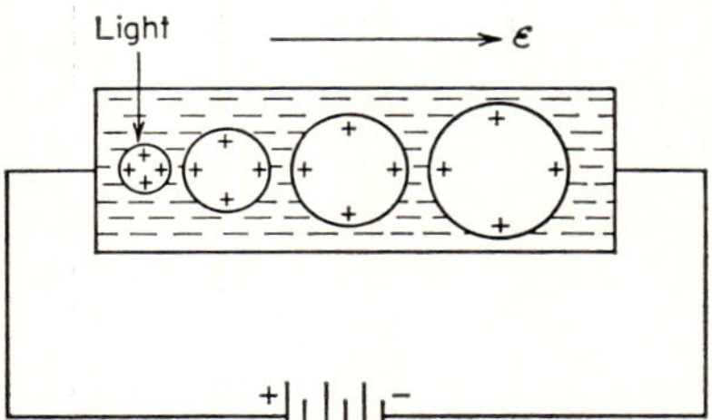

FIG. 4-13. Diffusion of holes in an n-type semiconductor.

For semiconductors, excess or nonequilibrium concentrations may be created by light, local heatings, etc. For example, light energy will create additional hole-electron pairs within the region of incidence. If the generated density is small compared to the initial concentration of electrons, then the effect of the light would be to alter the hole concentration appreciably, since the equilibrium hole concentration in an n-type semiconductor would be quite small initially. Nevertheless, electrical neutrality is maintained, as both the small hole and electron concentrations diffuse throughout the crystal. If an electric field is applied to the crystal by external means, then a drift motion is superimposed on the diffusion.[16] This effect is shown one-dimensionally in Fig. 4-13, where the holes only are shown having both diffusion and drift motion. Actually, the diffusion electrons will drift in the opposite direction, but this effect will be negligible compared to the majority-carrier drift currents.

Although diffusion mechanisms are three-dimensional problems, let us consider a one-dimensional case as an illustration. Suppose, for the above example, the distribution of the injected hole concentration is given as $p = p(x)$. Then it follows that the concentration gradient is given as dp/dx. Since the diffusion current density is directly proportional to the gradient, we may immediately write

$$J_p = -C \frac{dp}{dx} \qquad \text{amps/cm}^2$$

where C is a constant. The sign is negative because the holes diffuse in a direction away from where p is increasing. It should be apparent that C includes the magnitude of the hole charge q. Thus,

$$J_p = -qD_p \frac{dp}{dx} \tag{4-28}$$

where D_p is the diffusion constant in units of centimeters² per second. This applies to holes diffusing in n-type material. Similarly, for electrons in one dimension,

$$J_n = +qD_n \frac{dn}{dx} \tag{4-29}$$

where D_n is the diffusion constant for electrons in p-type material. The sign becomes plus because of the minus charge of the electron. Therefore, if electrons are diffusing in a direction opposite to holes, the total diffusion current is given as

$$J = J_n + J_p = q\left(D_n \frac{dn}{dx} - D_p \frac{dp}{dx}\right) \tag{4-30}$$

More generally, Eq. (4-30) may be written for three-dimensional flows as

$$J = J_n + J_p = q(D_n \nabla_n - D_p \nabla_p) \tag{4-31}$$

where ∇ (called *del*) is a vector operator which makes ∇_p equal to $i \frac{\partial p}{\partial x} + j \frac{\partial p}{\partial y} + k \frac{\partial p}{\partial z}$. These three components make up a concentration gradient vector located in any x, y, z space.

The values for the diffusion constants are directly related to mobility by the Einstein relationship, which is

$$D = \mu \frac{kT}{q} \qquad \text{cm}^2\text{/sec} \tag{4-32}$$

where μ = mobility
k = Boltzmann's constant
T = absolute temperature

This relationship holds for equilibrium conditions and for small diffusion densities. At room temperature (300°K) kT/q equals 0.026 volts. As an example, for n-type germanium having a hole mobility of $\mu_p = 1{,}850$, the corresponding hole diffusion constant would be $0.026 \times 1{,}850$ or

$$D_p = 48 \text{ cm}^2\text{/sec}$$

The temperature dependence of the diffusion constant is readily obtained from the temperature variation of μ and Eq. (4-32).

4-8. Lifetime. If no means existed in a semiconductor to return the system to thermal equilibrium the injected carriers would diffuse indefinitely. This unfortunately is not the case, since it is observed that the minority carriers eventually recombine with the majority carriers. Experimental observations have shown that when a small density of carriers is injected, the density decreases with time proportionally as $\epsilon^{-t/\tau}$. When $t = \tau$ the carriers have decayed to $1/\epsilon$ of the original value. Thus, τ is referred to as the *lifetime* of the carrier and is an inverse measure of the recombination rate.* For holes, lifetime is denoted as τ_p, and for electrons as τ_n. It is defined that the average distance a carrier will diffuse before recombining is:

$$L_p = \sqrt{D_p \tau_p} \quad \text{cm} \quad \text{holes} \tag{4-33}$$

$$L_n = \sqrt{D_n \tau_n} \quad \text{cm} \quad \text{electrons} \tag{4-34}$$

where L is called the *diffusion length*. These terms will appear often in device-design equations.

If no imperfections other than the group III or group V impurities existed in a semiconductor crystal, it would be expected that the lifetime would be limited only by the simplest recombination process, wherein the electron and hole collide and release a quantum of radiation. It may be recalled that this is the inverse of pair generation due to heating. From physical reasoning, the rate of recombination should be proportional to the product np of the respective concentrations, since the probability of an electron recombining is proportional to both the numbers of electrons and holes. Let this rate be Rnp. If e is the rate of thermal emission of holes, then for *thermal equilibrium*, the net rate of recombination must be zero, or

$$Rnp - e = 0 \tag{4-35}$$

and

$$e = Rnp = Rn_i^2 \tag{4-36}$$

If the crystal is intrinsic, i.e., $n = p = n_i$, and a small carrier density $\Delta c = \Delta p = \Delta n$ is injected, then the net rate of recombination for the departure from equilibrium would be

$$\frac{d(\Delta c)}{dt} = R(n + \Delta n)(p + \Delta p) - Rn_i^2 \tag{4-37}$$

Simplifying this result, we obtain

$$\frac{d(\Delta c)}{dt} = R(p + n)\,\Delta c \tag{4-38}$$

* τ should not be confused with the mean free time associated with mobilities, since a carrier may experience many collisions before it recombines.

Since dt is the incremental time during which the excess carriers recombine, the lifetime τ from Eq. (4-38) is

$$\tau = \frac{1}{R(p + n)} \tag{4-39}$$

and from Eq. (4-36) we obtain, finally,

$$\tau = \frac{n_i^2}{e(p + n)} \tag{4-40}$$

For intrinsic semiconductors (4-40) becomes $\tau = n_i/2e$; for n type $\tau_p = n_i^2/en$; for p type $\tau_n = n_i^2/ep$. Values of the emission rates e at 300°K as obtained from measurements of the absorption spectra are given as 3.5×10^{13} cm^{-3}-sec for germanium[17] and 2×10^9 cm^{-3}-sec for silicon.[18] With these values substituted in Eq. (4-40), the theoretical room-temperature lifetimes listed in Table 4-1 are calculated.

TABLE 4-1. CALCULATED ROOM-TEMPERATURE LIFETIMES

Carrier concentration	Lifetime	
	Ge	Si
Intrinsic	0.39 sec	3.7 sec
10^{15} cm^{-3}	20 msec	0.11 msec

4-9. Recombination Centers. The theoretical lifetime values calculated in the previous section on the basis of the simple, direct recombination process are actually very much larger than those that are measured in germanium and silicon crystals. Typical values for germanium may range from 400 to 2,000 μsec, and for silicon from 30 to 500 μsec. Since these lifetimes are considerably lower than the calculated values, it is apparent that some other processes must be responsible for the observed recombination of electrons and holes.

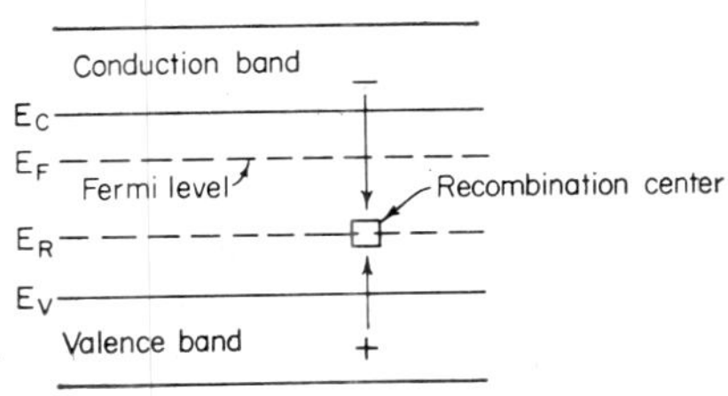

FIG. 4-14. Recombination center in a semiconductor.

According to the theory, it is suggested that certain allowed energy states exist in the region of the forbidden-energy band. These states are called recombination centers and are attributed to foreign impurities or perhaps structural defects in the crystal. Such a state might appear as shown in Fig. 4-14. The action of this center is to capture either a free electron from the conduction band or an electron from the valence band, leaving a hole behind. Likewise, the filled center may be emptied by capturing a hole or releasing its electron back to the conduction band. Recombination occurs when the center captures a free electron and holds it until emptied by capturing a hole. Thus, the recombination center may be considered as a "stepping stone" in the gap between the bands. If the imperfection that creates the centers happens to be a foreign metal

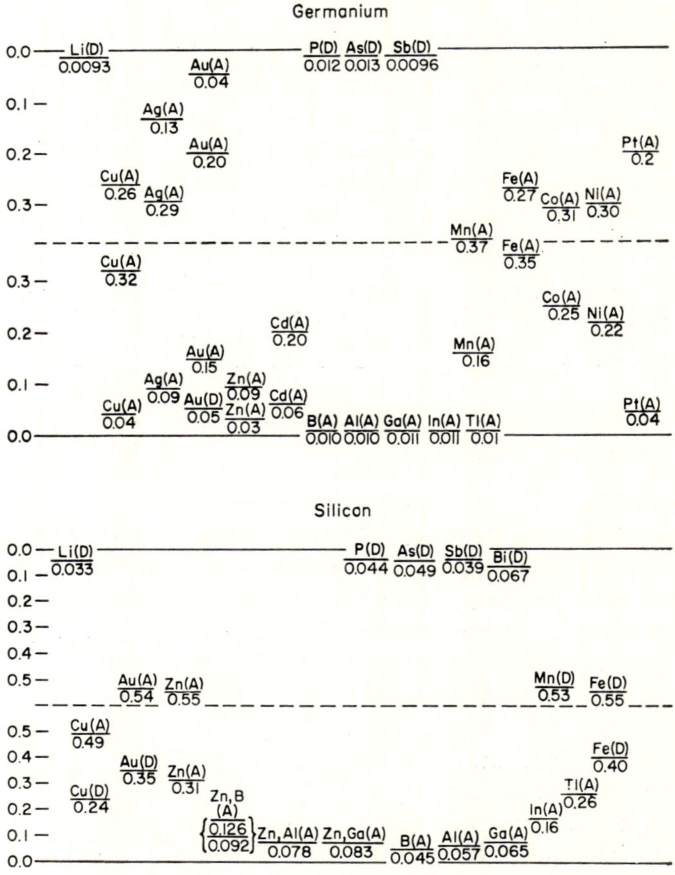

FIG. 4-15. Energy levels of impurities in semiconductors. (*After E. M. Conwell.*)

such as nickel or copper, the position of the center in the band gap would depend on the allowed energy level for the impurity in question. This is similar to the situation of the donor and acceptor impurities, except that the energy states for most metals are deeper in the band gap. Some energy levels for typical impurity centers in germanium and silicon, as summarized by Conwell,[9] are shown in Fig. 4-15. Some of the common metals of interest here are iron, gold, nickel, and copper. It should be noted from the figure that these impurities may have other levels, as indicated by the dashed recombination centers.

The effect of the recombination centers on the carrier lifetimes is given by the Shockley-Read-Hall theory.[18–21] The analysis is based on probability considerations employing Fermi-Dirac statistics. It should be apparent that the chance of a carrier recombining at a recombination center is dependent on several factors, viz., (1) the concentration of recombination centers in the crystal, (2) the concentration of the free carrier in question, which, of course, is related to the resistivity, (3) the capture probability of the centers, and (4) the concentration of centers that are normally filled under equilibrium conditions. The last factor is dependent on the position of the centers in the band gap with respect to the Fermi level, since this would determine the degree of occupancy of the centers. These factors may be expressed analytically. Since the net rate of recombination is equal to the capture rate less the emission rate, it is shown that the lifetime expression is

$$\tau = \frac{\tau_{pr}(n + n_r) + \tau_{nr}(p + p_r)}{n + p} \tag{4-41}$$

The complete derivation of this equation is given in the appendix. The analysis has two restrictions in that the center concentration is assumed to be small and that the injected carrier density is low. In Eq. (4-41), τ is the lifetime, n and p are the respective electron and hole concentrations for the particular semiconductor, and n_r and p_r are the respective electron and hole concentrations that would exist if the Fermi level coincided with the energy level E_R of the recombination centers. In other words,

$$n_r = N_C\epsilon^{-(E_C-E_R)/kT} \tag{4-42}$$

$$p_r = N_V\epsilon^{-(E_R-E_V)/kT} \tag{4-43}$$

Finally, τ_{pr} and τ_{nr} are the limiting lifetimes that would be observed in heavily doped crystals, i.e., τ_{pr} is the lifetime of holes in highly n-type material, and τ_{nr} is the lifetime of electrons in highly p-type material. Both these lifetimes are limited by the number of centers and are given by

$$\tau_{pr} = \frac{1}{N_R C_p} \tag{4-44}$$

$$\tau_{nr} = \frac{1}{N_R C_n} \tag{4-45}$$

where N_R is the concentration of centers in centimeters^{-3}, and C_p and C_n are the capture probabilities of the centers for holes and electrons, respectively, in centimeters^3 per second. The capture probabilities differ because of differences in the effective mass values for holes and electrons.

An illustration of the Shockley-Read-Hall lifetime formula is given in Fig. 4-16 for germanium, where it is assumed that the limiting electron lifetime is 100 μsec and the hole lifetime is 1 μsec. The curve is for the case where the recombination centers lie 0.2 ev above the middle of the band gap ($E_G = 0.72$ ev). Experimental lifetime measurements of good quality germanium have shown good correlation with this curve. The lifetime is about 2,500 μsec for intrinsic material and decreases with increasing concentration (decreasing resistivity) in either direction to a level of 100 μsec for p type, and 1 μsec for n type. A physical explanation of the shape of the curve may be given on a qualitative basis. When the semiconductor is strongly p type, the Fermi level is close to the valence band

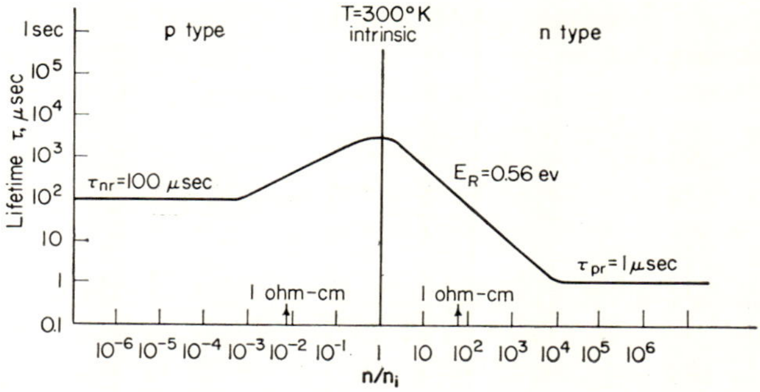

FIG. 4-16. Variation of lifetime with impurity content in germanium. (*After R. N. Hall.*)

and the recombination centers are almost completely empty. Therefore, the lifetime is limited wholly by the capture of electrons by the unfilled centers. This corresponds to the term τ_{nr}. As the crystal becomes less p type the Fermi level rises toward the middle of the band gap. For this condition, more centers become filled with electrons and the hole concentration decreases, thereby increasing the lifetime for electrons. The curve attains a maximum at the intrinsic point and begins to decrease again in the n-type range. Here, the Fermi level is rising closer to the conduction band, and the centers are becoming filled more and more. Thus the hole lifetime becomes affected by the hole capture of filled centers. For strongly n-type material, the lifetime is limited wholly by hole capture by the many filled recombination centers.

The discussion thus far was for room temperature. However, because the electron-hole concentrations will vary with temperature, there would be a change in lifetime with temperature. At low temperatures the lifetimes are limited by the plateau values. As the temperature increases, the Fermi levels approach the band-gap center, and the lifetimes increase accordingly to where the conduction becomes intrinsic. At this point, the hole-electron product (n_i^2) starts increasing and proportionately enhances the recombination rate, thereby decreasing the lifetime. The variation of lifetime with temperature may be calculated by factoring in the temperature dependence of the appropriate terms in Eq. (4-41).

Another point of interest on the subject of lifetime concerns the variation of lifetime with relative carrier injection density. It will be seen that this is an important consideration in interpreting the performance of junction transistors operating at high currents. In the discussion thus far, the treatment was based on the assumption that the injected carrier density was small compared to the free carrier concentration in the crystal. According to the Shockley-Read-Hall theory, it can be seen from Eq. (4-41) that as n and p become large τ approaches the sum of τ_{pr} and τ_{nr}. Therefore, under conditions of large injection, the high-level lifetime is given as

$$\tau_\infty = \tau_{pr} + \tau_{nr} \tag{4-46}$$

For the germanium example used previously, τ_∞ would be 101 μsec. In the typical range between the equilibrium and infinite injection levels, Eq. (4-41) must be modified as follows:

$$\tau = \frac{\tau_{pr}(n + n_r + \Delta c) + \tau_{nr}(p + p_r + \Delta c)}{n + p + \Delta c} \tag{4-47}$$

where Δc is the injected carrier level. If we define $\upsilon = \Delta c/(n + p)$ as the injection ratio and τ_o as the low-level equilibrium lifetime, then (4-47) will become

$$\tau = \frac{\tau_o + \upsilon\tau_\infty}{1 + \upsilon}$$

or

$$\frac{\tau}{\tau_\infty} = \frac{\tau_o/\tau_\infty + \upsilon}{1 + \upsilon} \tag{4-48}$$

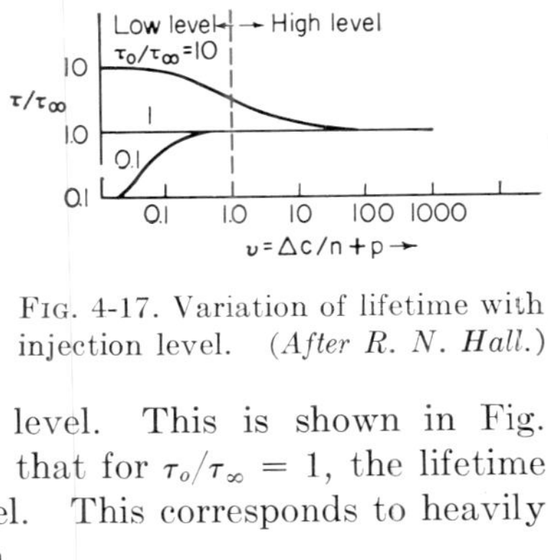

FIG. 4-17. Variation of lifetime with injection level. (*After R. N. Hall.*)

This expression enables us to show the variation of the low-to-high-level lifetime ratio as a function of injection level. This is shown in Fig. 4-17 for different values of τ_o/τ_∞. Note that for $\tau_o/\tau_\infty = 1$, the lifetime variation is independent of injection level. This corresponds to heavily doped material in the order of 1 ohm-cm.

The explanation of the effects of recombination centers on lifetime given in this section was restricted primarily to germanium. The situation for silicon is somewhat different in that less is known about the impurity recombination centers in silicon. This is attributed mainly to the problem of obtaining silicon crystals that are relatively free of structural imperfections and unwanted impurities. Furthermore, the lifetime processes in silicon are complicated by the presence of another type of recombination center, called a *trap*. Traps differ in the respect that they can capture carriers but cannot be emptied by capturing carriers of opposite type. The trap is emptied only by releasing the initial carrier. Trapping effects appear in germanium at temperatures of about −80°C, whereas in silicon they occur at room temperature.[22,23] The action of a trap in p-type silicon, for example, is to capture an electron from the conduction band and hold it for a relatively long period of time before releasing it to recombine with a hole at a normal recombination center. This results in very long apparent lifetimes which may vary from milliseconds to as much as several minutes. At high injection levels, the trapping effects are virtually eliminated, since a large number of carriers are available to fill the traps. With all the traps filled, only the shorter-lifetime recombination mechanisms prevail.

4-10. Semiconductor Surfaces. It has been fairly well established, although not completely understood, that the electrical properties of a free semiconductor surface differ from those of the bulk interior. This is somewhat to be expected, since the surface represents an abrupt boundary for the outermost layer of semiconductor atoms and should therefore disrupt the orderly energy distributions in the region. Even if the surface were absolutely perfect, that is, completely free of any atomic dissymmetry, it would be hard to imagine the absence of any energy disorder. Certainly the orbital electrons of the surface atoms would behave differently from those in the interior because of the finite termination. From a practical point of view, of course, the surface is not perfect. We know, from the chemistry of semiconductors such as germanium and silicon, that these elements are very active in air and form oxide layers quite readily. Furthermore, it is possible for these layers to absorb chemical ions or perhaps water or gas molecules, depending on the kind of environment the surface is exposed to. Thus, the chemistry of the semiconductor surface itself would clearly justify any differences in electrical properties from the bulk. Considerable experimental work on semiconductor surface studies has shown this to be the case. It has been found that because of these chemical and structural disorders, the carrier concentrations and conductivity at the surface may not be the same as in the interior. Also, the surface exhibits a recombination rate for carriers which may be different from the lifetime of the bulk material.

Many ideas have been propounded to explain these surface phenomena, and it is only lately that any progress has been made toward establishing a widely accepted theory. The most recent conclusions, as surveyed by Kingston, will be presented here.[24] It is generally accepted, from the early work of J. Bardeen, that at the surface there exist a number of *surface energy states* having energies that fall within the forbidden band gap of the semiconductor.[25] These surface states have been categorized into two types, viz., *layer* states and *interface* states, each caused by different effects. It is believed that the layer states are due to the characteristics of the oxide layer and arise from absorbed ions or molecules or perhaps chemical imperfections. Experimental evidence has demonstrated that the nature of the layer states is sensitive to the ambient (moisture, gases, ionic compounds, etc.) which the surface is exposed to. The interface states, as originally suggested by Brattain and Bardeen,[26] are similar in behavior to the recombination centers of the Shockley-Read-Hall theory. These states occur at the interface or boundary between the semiconductor and the oxide layer. The interface states are found to be independent of ambient, but apparently reflect the quality of the initial surface treatment *before* the oxide layer is formed. Thus, they would be dependent on the kind of chemical or electrolytic etching treatment which the semiconductor was subjected to. (Etching is an important process, basic to transistor technology, for preparing semiconductor surfaces.) Any etch imperfections beneath the oxide layer could cause departures from surface uniformity and could drastically change the densities and energies of the interface surface states.

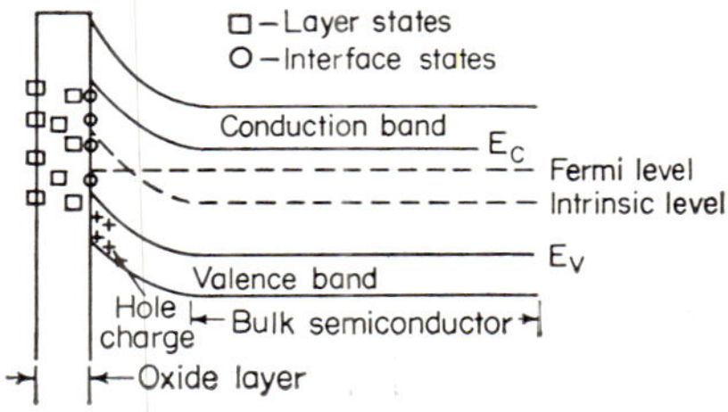

FIG. 4-18. Theoretical energy-band picture of an n-type semiconductor at the surface.

Using the existence of the layer and interface states as a basis, we can establish a theoretical model of the surface and predict its electrical properties. This is shown in Fig. 4-18, using an n-type semiconductor for illustration. The interface states are shown as circles and the oxide-layer states are designated by squares. A number of layer states greater than the number of interface states is shown, as is believed to be the case for properly treated surfaces. Consequently, we can assume that any electric charge observed at the oxide layer is due primarily to the ionic charges trapped in the layer states. If the net charge is negative, some redistribution of charges is necessary, in order to keep the surface electrically neutral. The charge neutrality is achieved by the attraction of holes from the semiconductor bulk, as shown in the diagram. However, because of the added concentration of holes in the bulk region adjacent to the interface, the Fermi level must be adjusted to account for the new carrier distribution in the region. It is more appropriate, however, to keep the Fermi energy as a constant reference, and therefore the energy bands are drawn bent up in Fig. 4-18. Just at the interface, note that the Fermi level is closer to the valence band, indicating that the conductivity of the semiconductor reverses from n type to p type as the surface is approached. The amount the energy bands bend is a function of the density and energy distribution of the layer states, which is directly affected by the ambient. It is also possible for the bands to bend down, becoming more n type, due to the attraction of electrons by positive ionic charges. The regions of conductivity reversal are called *inversion layers,* and are illustrated in Fig. 4-19. In later chapters we shall consider how these inversion layers contribute to surface leakage currents.

As was indicated earlier, the density of the interface states is usually much smaller than that of the layer states, such that it does not contribute appreciably to the formation of inversion layers. However, it is postulated that the interface states have capture probabilities many orders of magnitude greater than the layer states. Therefore, any

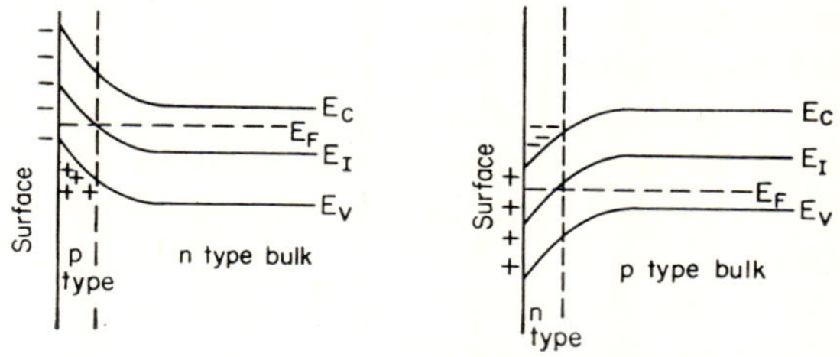

FIG. 4-19. Inversion layers in semiconductors.

observed recombination of carriers at the surface is attributed mainly to recombination at the interface centers. Surface recombination is expressed in terms of a *surface recombination velocity* s in centimeters per second. This is defined as the number of carriers recombining per second per unit surface area divided by the excess concentration over the equilibrium value at the surface. The magnitude of s for any surface is shown to be dependent upon the same considerations that were applicable in determining the bulk lifetime. Thus, the density of the interface states, the energy of the states, the capture probabilities, the surface concentrations of the holes and electrons, and the degree of occupancy of the interface states all contribute to the surface recombination velocity. It should be apparent that for a given interface state density and energy, s would be affected by the nature of the oxide layer, since the latter determines the free carrier distribution at the surface due to the band bending.

We can now state the conclusions for this theory of semiconductor surfaces. The density and energy levels of the interface states are determined by the method of initial surface treatment and are independent of any ambient effects. The ambient affects the nature of the ionic charge in the oxide-layer states, which in turn alters the conductivity and type of the bulk layer just beneath the surface. The resulting carrier distribution, in conjunction with the nature of the interface states, determines the surface recombination velocity.

It will become evident that a high recombination rate s for a semiconductor surface can be harmful to certain aspects of device performance, since useful carriers are lost by surface recombination. This effect manifests itself in the form of an effective lifetime which may be considerably lower than the bulk lifetime. For a rectangular crystal having side dimensions B and C, Shockley has shown that the surface lifetime τ_s is related to the surface recombination velocity as[27]

$$\frac{1}{\tau_s} = s\left(\frac{1}{B} + \frac{1}{C}\right) \qquad s \to 0 \tag{4-49}$$

Note that this expression holds only for small values of s approaching zero. Thus we see that for samples of small dimensions and high surface recombination velocities, the surface lifetime may be quite small. In considering the actual lifetime of a semiconductor, therefore, we cannot neglect surface effects, since a number of injected carriers may be recombining at the surface. The effect of the surface lifetime on the effective lifetime is given by

$$\frac{1}{\tau_{\text{eff}}} = \frac{1}{\tau_V} + \frac{1}{\tau_s} \tag{4-50}$$

where τ_V is the lifetime of the semiconductor interior. This equation is used quite often as a basis for determining the surface recombination velocities of samples when the other parameters are known by direct measurement.

Typical values of the surface recombination velocity for properly etched germanium surfaces have been found to be as low as 100 cm/sec. A sandblasted surface, on the other hand, may have values of s as high as 10^5 cm/sec. In this case, the high recombination rate is associated with the high density of interface states generated by the mechanical imperfections of the surface lattice structure. It may be said that the present understanding of semiconductor surfaces is based mostly on experimental work with germanium. There is good reason to believe that the surface behavior of silicon would follow a similar pattern.

PROBLEMS

4-1. Calculate the position of the Fermi level in electron-volts for an n-type sample of germanium having a resistivity of 2 ohm-cm at room temperature. Assume that the donors are fully ionized and that $n = N_D - N_A$. (HINT: Consider the ratio n/n_i.)

4-2. Bars are cut from single crystals of intrinsic germanium and silicon to a length of 0.125 in. and an area of 0.002 in.^{2} What are the respective resistances at temperatures of 27 and 100°C?

4-3. A small concentration of minority carriers is injected at one point in a homogeneous semiconductor crystal. They drift to another point 1 cm away in a time of

125 μsec. If the crystal is 10 cm long and the voltage across it is 100 volts, determine the drift mobility for the carriers.

4-4. A germanium pellet contains a density of 2×10^{14} atoms/cm^3 of phosphorus. What is the resistivity and type at room temperature? What density of gallium atoms must be added to obtain a resistivity of 0.6 ohm-cm of opposite type? What percentage does the total impurity density constitute?

4-5. The base region of a p-n-p silicon transistor is 3 ohm-cm, n type, and has a thickness of 0.0005 in. What should the lifetime (for holes) in this region be to give a diffusion length (for holes) ten times the base thickness at room temperature?

4-6. Calculate the hole lifetime in microseconds of the 2 ohm-cm n-type germanium crystal of Prob. 4-1 at room temperature, for the case where the crystal contains an additional small concentration of copper recombination centers ($E_R = 0.28$ ev). Assume that the limiting lifetimes are $\tau_{pr} = 1$ μsec and $\tau_{nr} = 100$ μsec.

REFERENCES

1. Shockley, W.: Transistor Electronics: Imperfections, Unipolar and Analog Transistors, *Proc. IRE*, vol. 40, p. 1294, November, 1952.
2. Ryder, E. J.: Mobility of Holes and Electrons in High Electric Fields, *Phys. Rev.*, vol. 90, pp. 766–769, June, 1953.
3. Shockley, W.: "Electrons and Holes in Semiconductors," p. 287, D. Van Nostrand Company, Inc., Princeton, N.J., 1953.
4. Conwell, E. M.: Properties of Silicon and Germanium I, *Proc. IRE*, vol. 40, p. 1331, November, 1952.
5. Conwell, E. M., and V. F. Weisskopf: Theory of Impurity Scattering in Semiconductors, *Phys. Rev.*, vol. 77, pp. 388–390, 1950.
6. Morin, F. J., and J. P. Maita: Conductivity and Hall Effect in the Intrinsic Range of Germanium, *Phys. Rev.*, vol. 94, pp. 1525–1529, June, 1954.
7. Morin, F. J., and J. P. Maita: Electrical Properties of Silicon Containing Arsenic and Boron, *Phys. Rev.*, vol. 96, pp. 28–35, October, 1954.
8. Ludwig, G. W., and R. L. Watters: Drift and Conductivity Mobility in Silicon, *Phys. Rev.*, vol. 101, pp. 1699–1701, March, 1956.
9. Conwell, E. M.: Properties of Silicon and Germanium II, *Proc. IRE*, vol. 46, pp. 1281–1300, June, 1958.
10. Prince, M. B.: Drift Mobilities in Semiconductors. I. Germanium, *Phys. Rev.*, vol. 92, pp. 681–687, November, 1953.
11. Prince, M. B.: Drift Mobilities in Semiconductors. II. Silicon, *Phys. Rev.*, vol. 93, pp. 1204–1206, March, 1954.
12. Backenstoss, G.: Conductivity Mobilities of Electrons and Holes in Heavily Doped Silicon, *Phys. Rev.*, vol. 108, pp. 1416–1419, December, 1957.
13. Tyler, W. W., and T. J. Soltys: General Electric Research Laboratory Memo Report P-193, Schenectady, New York.
14. Debye, P. P., and E. M. Conwell: Electrical Properties of n-type Germanium, *Phys. Rev.*, vol. 93, pp. 695–706, February, 1954.
15. Gärtner, W. W.: Temperature Dependence of Junction Transistor Parameters, *Proc. IRE*, vol. 45, pp. 662–680, May, 1957.
16. Haynes, J. R., and W. Shockley: Investigation of Hole Injection in Transistor Action, *Phys. Rev.*, vol. 75, p. 691, 1949.
17. Dash, W. C., R. Newman, and E. A. Taft: *APS Bull.*, vol. 301, p. 53, 1955.
18. Bemski, G.: Recombination in Semiconductors, *Proc. IRE*, vol. 46, pp. 990–1004, June, 1958.

19. Hall, R. N.: Germanium Rectifier Characteristics, *Phys. Rev.*, vol. 83, p. 228, 1951.
20. Shockley, W., and W. T. Read, Jr.: Statistics of the Recombinations of Holes and Electrons, *Phys. Rev.*, vol. 87, pp. 835–842, 1952.
21. Hall, R. N.: Electron-hole Recombination in Germanium, *Phys. Rev.*, vol. 87, p. 387, 1952.
22. Haynes, J. R., and J. A. Hornbeck: Trapping of Minority Carriers in Silicon. I. p-type Silicon, *Phys. Rev.*, vol. 97, pp. 311–321, 1955.
23. Haynes, J. R., and J. A. Hornbeck: Trapping of Minority Carriers in Silicon. II. n-type Silicon, *Phys. Rev.*, vol. 100, pp. 606–615, 1955.
24. Kingston, R. H.: Review of Germanium Surface Phenomena, *J. Appl. Phys.*, vol. 27, pp. 101–114, February, 1956.
25. Bardeen, J.: Surface States and Rectification at a Metal Semiconductor Contact, *Phys. Rev.*, vol. 71, pp. 649–744, 1947.
26. Brattain, W. H., and J. Bardeen: Surface Properties of Germanium, *Bell System Tech. J.*, vol. 32, pp. 1–41, January, 1953.
27. Shockley, W.: "Electrons and Holes in Semiconductors," pp. 318–324, D. Van Nostrand Company, Inc., Princeton, N.J., 1953.

5

p-n Junction Theory

5-1. Introduction. The development thus far is intended to give a sufficient understanding of the inherent carrier mechanisms involved in semiconductor metals such as germanium and silicon, so that the electrical behavior of devices made from these materials may be explained in terms of these basic properties. In any semiconductor device, whether it be a rectifier or transistor, the simple p-n junction is a fundamental unit the understanding of which is prerequisite to the understanding of transistors. This chapter will therefore present a complete theoretical analysis of the p-n junction in terms of the flow of holes and electrons under applied potentials, thus giving rise to the expressions for the voltage-current characteristics. Adequate qualitative descriptions will be given throughout to render further physical meaning to the mathematics.

5-2. Potential Representation of Energy Levels. It was shown in Chap. 3 that the equilibrium concentrations of carriers in a semiconductor are given by the following relationships:

$$n = N_C \epsilon^{-(E_C - E_F)/kT} \qquad \text{n type} \tag{5-1}$$

$$p = N_V \epsilon^{-(E_F - E_V)/kT} \qquad \text{p type} \tag{5-2}$$

where N_C, N_V = effective densities of states in energy bands
E_C, E_V = energy values for band edges
E_F = Fermi level

Also, for the intrinsic case,

$$n_i = \sqrt{N_C N_V}\, \epsilon^{-E_G/2kT} \tag{5-3}$$

where E_G corresponds to the band gap energy. Note that in this result N_C would equal N_V, since we are disregarding differences in effective masses. If Eqs. (5-1) and (5-2) are each divided by (5-3), we obtain

$$\frac{n}{n_i} = \epsilon^{-(E_C - E_F - E_G/2)/kT} \tag{5-4}$$

$$\frac{p}{n_i} = \epsilon^{-(E_F - E_V - E_G/2)/kT} \tag{5-5}$$

Using the valence-band energy as a reference, we set $E_V = 0$ and $E_C = E_G$, and (5-4) and (5-5) become

$$n = n_i\epsilon^{(E_F - E_G/2)/kT} \qquad \text{n type} \tag{5-6}$$
$$p = n_i\epsilon^{(E_G/2 - E_F)/kT} \qquad \text{p type} \tag{5-7}$$

These results show that the carrier concentrations may now be expressed in terms of the deviation of the Fermi level from the middle of the band gap, which is the intrinsic position. It is apparent that when $E_F = E_G/2$, $n = p = n_i$.

Since energy is often expressed in electron-volts or qV, where q is the charge of the electron and V is the potential in volts, it becomes desirable to express both (5-6) and (5-7) in the same manner. From physical

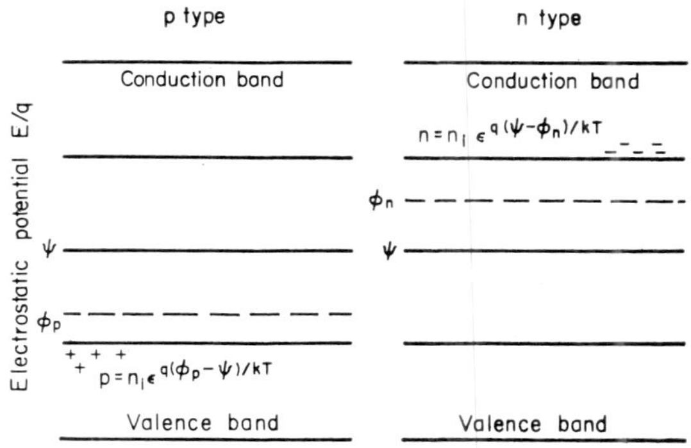

FIG. 5-1. Potential diagram for equilibrium semiconductor.

reasoning, we may say that the energy necessary to bring an electron from the valence band to the conduction band is the same as the work necessary to bring a charge q across a potential V equal to the difference in "potential" between the energy bands. Thus we may arbitrarily assign electrostatic potentials to any of the levels of energy in the energy-band picture. For purposes of analysis, we shall make the following definitions:

$$E_F = -q\phi \qquad \text{ev} \tag{5-8}$$
$$E_G/2 = E_{Fi} = -q\psi \qquad \text{ev} \tag{5-9}$$

where ϕ and ψ are electrostatic potentials corresponding to the Fermi level and the intrinsic level, respectively. The sign is negative because of the negative charge of the electron. Substituting the potential representations into (5-6) and (5-7), we get

$$n = n_i\epsilon^{q(\psi - \phi_n)/kT} \tag{5-10}$$
$$p = n_i\epsilon^{q(\phi_p - \psi)/kT} \tag{5-11}$$

The discussion to this point may best be summed up by the composite illustration shown in Fig. 5-1. The important result to be gleaned here is that ψ may take on any absolute value such as $(\psi + V)$, but ϕ will also change accordingly.

5-3. p-n Junction in Equilibrium. Let us now consider the case where a junction is formed by a p-type crystal and an n type such that the two regions comprise a homogeneous physical structure, but at the boundary there is an abrupt transition from p- to n-type impurity conduction. In the p region there exists a large concentration of holes, approximately equal to the excess acceptor-impurity density, and the Fermi level lies closer to the valence band. In the n region, on the other hand, there is a large concentration of electrons, approximately equal to the excess donor-impurity density, and the Fermi level lies closer to the conduction

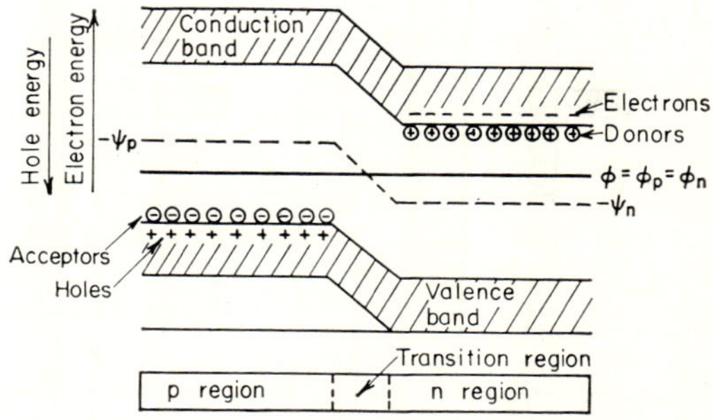

FIG. 5-2. Energy diagram for a p-n junction in equilibrium illustrating constant Fermi level.

band. When the two regions are in intimate contact, with no voltages externally applied, it is necessary that the Fermi level be constant all the way across the junction. If it were not so, the electrons on one side of the junction would on the average have higher energy than those on the other side, and there would be a transfer of energy until the levels became equal. From thermodynamic considerations, Shockley has likened the Fermi level to a chemical potential, and has shown that it must be constant for a system in equilibrium.[1,*] Therefore the energy-band diagram for a p-n junction would appear as shown in Fig. 5-2. It is seen that in order to maintain the constant Fermi potential ϕ across the junction (transition region), there must be the indicated shift in the energy bands. In terms of the respective energy scales, both the holes and electrons occur in regions of low potential energy.

For the electrostatic potential ψ, a definite difference in potential, equal to $|\psi_n - \psi_p|$, exists across the transition region. This means that an electric field must also exist in the region. The fact that an electric

* References, indicated in the text by superscript figures, are listed at the end of the chapter.

field exists within a p-n junction at equilibrium may be explained by considering what happens initially when the two regions are brought into contact, assuming that no potential exists. Before contact, both regions are electrically neutral, with the respective ionized impurity densities just balanced by the carrier concentrations. In contact, the holes in the p region will diffuse across the junction into the n region, where the hole concentration is small. Likewise, electrons from the n region will diffuse into the p region, where the electron concentration is small. Since the impurity ions are fixed in the lattice and are not free to move, as a result of the diffusion there will be regions of unneutralized charge density on

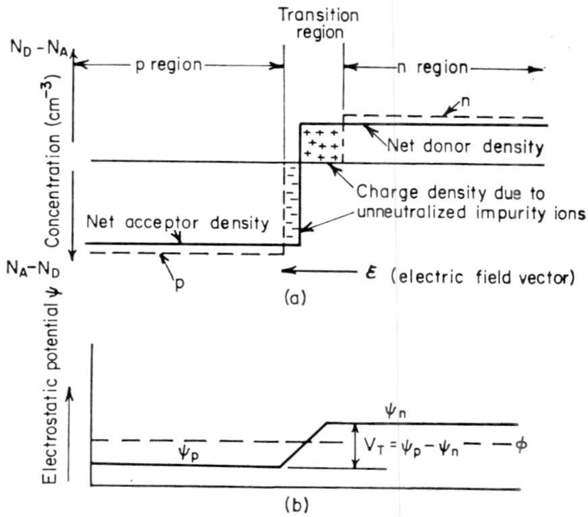

FIG. 5-3. Charge dipole and electrostatic potential in a p-n junction at equilibrium.

both sides of the junction. This is shown in Fig. 5-3a in the form of a charge dipole (darkened lines) which is negative in the p region and positive in the n region. This charge distribution will establish an opposing electric field $\mathcal{E}$, which at equilibrium would just prevent the further diffusion of carriers. The equilibrium condition may be regarded as the flow of two equal and opposite currents across the junction, such that the net current in the transition region is zero. In this case, one component would be due to the diffusion of carriers and the other would be a drift component due to the built-in electric field. In Fig. 5-3a, the hole and electron concentrations are shown reduced to zero at both sides of the charge layer. An analysis of the charge distribution, using Poisson's equation, will be performed later to show that the electrostatic potential curve of Fig. 5-3b is obtained. Also, the expression for the capacitance of the transition region will be derived.

To determine the expression for the equilibrium electrostatic or contact potential V_T, we make use of Eqs. (5-10) and (5-11) by taking the natural logarithm of each equation and solving.

$$\psi_n = \phi + \frac{kT}{q} \ln \frac{n_n}{n_i} \tag{5-12}$$

$$\psi_p = \phi - \frac{kT}{q} \ln \frac{p_p}{n_i} \tag{5-13}$$

Then

$$V_T = \psi_n - \psi_p = \frac{kT}{q} \ln \frac{n_n p_p}{n_i^2} \tag{5-14}$$

In terms of impurity concentrations, the equilibrium contact potential becomes

$$V_T = \frac{kT}{q} \ln \frac{N_D N_A}{n_i^2} \tag{5-15}$$

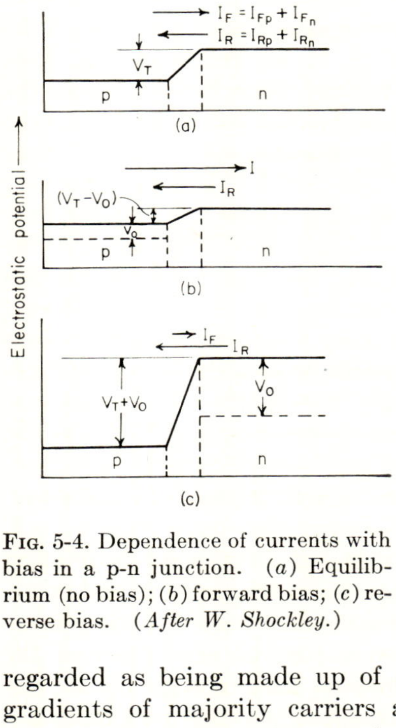

FIG. 5-4. Dependence of currents with bias in a p-n junction. (*a*) Equilibrium (no bias); (*b*) forward bias; (*c*) reverse bias. (*After W. Shockley.*)

From Eq. (5-15) we see that the contact potential is maximized for low values of resistivity on both sides of the junction. In other words, the greater the difference between the absolute Fermi-level potentials in each region, the larger the equilibrium contact potential. Theoretically, the maximum V_T attainable would approximate the band-gap potential for the semiconductor. It should be evident that for the same p-n junction resistivities, the contact potential for silicon would be larger than that for germanium.

5-4. p-n Junction with Applied Voltage.[2] As noted in the equilibrium case, the net current across the junction is zero. This current may be regarded as being made up of a component due to the concentration gradients of majority carriers across the junction and an equal and opposite component due to the minority carriers that diffuse to the transition region and drift across because of the built-in field. This is shown in Fig. 5-4*a*, where the vectors indicate the magnitudes of the currents ($I_F = I_R$). It should be emphasized that these currents are made up of both hole and electron components. Consider, for example, the hole components only. The hole current I_{Fp} arises from the concentration gradient of holes from the p to n region and corresponds to those holes which have sufficient energy to climb the potential hill (V_T)

and enter the n region, where they eventually recombine with electrons. Likewise, the equal and opposite hole current I_{Rp} consists of those holes in the n region which diffuse to the junction and drift across or slide down the potential hill into the p region. If the length of the n region is greater than the diffusion length L_p for holes in that region, it is apparent that only those diffusing holes within a distance L_p from the transition region reach the junction, since any holes further away than L_p recombine with electrons before reaching the junction. A similar treatment would apply to electrons, for which the potential hill is reversed.

When a voltage is applied to the p-n junction, the balance of currents just described is destroyed. Suppose a positive d-c voltage V_o is applied to the p region with respect to the n region. This would have the effect of raising the electrostatic potential of the p region, thereby lessening the height of the potential hill by the amount V_o, as shown in Fig. 5-4*b*. This lowering of potential would permit many more holes to cross the junction from the p region (also electrons from the n region), and the concentration-gradient current I_F would become much larger than the equilibrium value. However, the current I_R remains relatively unchanged, since the same number of diffusing carriers still find it easy to drift across the transition region. It should be obvious that the magnitude of I_F will be dependent on V_o; later we shall derive this relationship. We call this the case of a p-n junction under *forward bias*.

If the applied voltage is reversed, making the n region much more positive than the p region, we have the condition of a *reverse-biased* junction as shown in Fig. 5-4*c*. In this case, the height of the potential barrier is raised by an amount V_o, making it exceedingly difficult for the holes in the p region to climb the hill, and the forward current I_F reduces to a very small value as shown. The reverse current I_R, however, remains at the same value, since this is dependent only upon those holes in the n region which diffuse to the barrier within the distance of a diffusion length. Actually, the current I_F increases slightly for very small reverse potentials of the order of V_T, but quickly saturates to a constant value as V_o becomes large. This is attributed to the initial steepening of the potential hill and will be demonstrated later in the analysis.

On the basis of the qualitative description, it becomes evident that a p-n junction can be an excellent rectifying device, i.e., a device which passes current in one direction but not in the other. Suppose the p region was of very low resistivity (0.01 ohm-cm) such that the electron concentration (minority carriers) was negligible and the n region resistivity was of the order of 10 ohm-cm, such that p_n could not be neglected. Under forward-bias conditions, the forward current would consist almost wholly of hole flow, since the hole concentration in the p region would be very much greater than the electron concentration in the n region. Thus, we would expect very large currents to flow in the forward direction, depending on the applied voltage. On the other hand, under reverse or back-bias conditions, the current would be proportional to the minority-carrier concentrations in the regions. If n_p was negligible, as stated, then the reverse saturation current I_R would consist only of hole flow from the n region and would be proportional to p_n. Since $p_n \ll p_p$ for the chosen resistivities, the reverse current becomes very much smaller than the forward current and practically independent of voltage, thus making an excellent rectifying device. A typical voltage-current characteristic would appear as shown in Fig. 5-5.

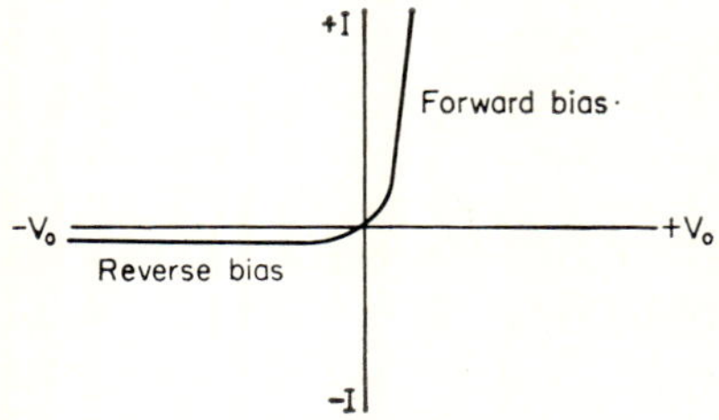

FIG. 5-5. Voltage-current characteristic for a p-n junction rectifier.

5-5. Carrier Currents and Injection Level. When forward or reverse voltages are applied to a p-n junction, the carrier concentrations in the respective regions must depart from the equilibrium values in order to account for the flow of currents. Since the position of the Fermi level is indicative of the carrier concentrations, one would expect the Fermi levels to change for the nonequilibrium case. Therefore it becomes necessary to determine the nature of the Fermi levels when voltage is applied to the p-n junction, since this enables one to determine the level of carriers injected into the respective regions. In particular, the magnitude of the minority-carrier injection level at the boundaries of the junction transition region is essential to the theoretical analysis of the p-n junction.

Let us now make some simplifying assumptions regarding our model of the p-n junction. Firstly, we assume that the resistances of the p and n regions are so low that any voltage applied is completely dropped across the transition region. This means that there cannot be any electric field outside the transition region. Secondly, we assume that no recombination of carriers takes place in the transition region. Thirdly, we assume that the minority-carrier diffusion lengths in the p and n regions are smaller than the physical thickness of same, so that the carriers recombine and establish equilibrium levels before the ends of the regions are reached. Finally, we state that the injection level of minority carriers is small compared to the majority-carrier concentration.

With these assumptions in mind, we can establish the flow of carrier currents and the nature of the Fermi levels. Figure 5-6 treats the case of a forward-biased junction. In Fig. 5-6a, a potential V_o is applied to the junction, raising both the electrostatic potential ψ_p and the Fermi level ϕ_p by an amount equal to V_o. Here the initial Fermi levels, viz., ϕ_p and ϕ_n, differ in potential by V_o, and are therefore joined together by the arbitrary lines as shown in Fig. 5-6*a*. These become the artificial potential functions ϕ_p and ϕ_n and are referred to by Shockley as quasi-Fermi levels.[3] They will serve to justify the values of carrier concentrations observed as a function of distance x under nonequilibrium conditions.

That the quasi-Fermi levels would have the shapes shown in Fig. 5-6*a* may be explained by examining the carrier currents in the junction[4] as

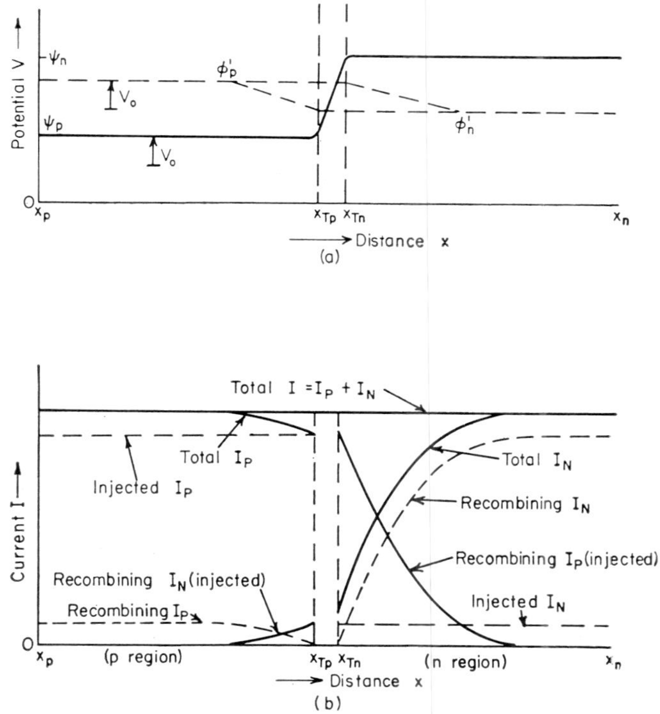

FIG. 5-6. Forward-biased junction. (*a*) Quasi-Fermi levels; (*b*) current distributions. (*After R. D. Middlebrook.*)

sketched in Fig. 5-6*b*. In the p region, at $x = x_p$, the total current I equals the hole current I_p, since any electrons injected into the p region recombine before they reach $x = 0$ As x_{Tp} is approached, I_p decreases slightly to account for the recombination of holes with electrons from the n region. However, the hole current at x_{Tn} will equal that at x_{Tp}, after which it will recombine with electrons in the n region and reach zero before x_n. Since the hole density injected into the n region is small compared to the electron density, we may conclude that the majority-carrier concentration in the n region remains practically unchanged from

x_{Tp} to x_n. Likewise, from x_p to x_{Tn} the majority-carrier concentration (holes) remains constant. We have seen from Eqs. (5-10) and (5-11) that the equilibrium densities n and p are related to $(\psi - \phi_n)$ and $(\phi_p - \psi)$, respectively. Thus, using the quasi-Fermi levels for nonequilibrium, we may write

$$n = n_i \epsilon^{q(\psi_n - \phi_n')/kT} \qquad \text{n region} \tag{5-16}$$

$$p = n_i \epsilon^{q(\phi_p' - \psi_p)/kT} \qquad \text{p region} \tag{5-17}$$

In order for n and p to be practically constant in their respective regions, as described above, ϕ_p' and ϕ_n' must have the shapes shown in Fig. 5-6*a*. Furthermore, since the injected carriers shown in Fig. 5-6*b* completely recombine or decay to the equilibrium concentrations, both ϕ_p' and ϕ_n' must change by V_o volts back to the equilibrium levels, as shown. Thus the quasi-Fermi levels as drawn are explained qualitatively from the behavior of the junction currents. It should be noted that the densities in the majority-carrier regions for the nonequilibrium case are practically the same as the equilibrium values, since in the n region neither ψ_n nor $\phi_n(\phi_n = \phi_n')$ has shifted in potential, and in the p region $\psi_p' = \psi_p + V_o$ and $\phi_p' = \phi_p + V_o$.

The same argument will apply to the reverse-biased junction, for which the quasi-Fermi levels and current distributions are shown in Fig. 5-7. Here again, the majority-carrier densities remain relatively undisturbed from the equilibrium values and therefore the quasi-Fermi levels are constant in those regions. Within a diffusion length of either side of the transition region, however, the quasi-Fermi levels will change abruptly to account for the decreased concentration of minority carriers in the region.

We can now make use of the quasi-Fermi-level argument to determine the carrier concentrations as a function of voltage at the transition region boundaries x_{Tn} and x_{Tp}. From Eq. (5-16),

$$n(x_{Tn}) = n_n = n_i \epsilon^{q(\psi - \phi_n')/kT} \tag{5-18}$$

and

$$p(x_{Tn}) = n_i \epsilon^{q(\phi_p' - \psi)/kT} \tag{5-19}$$

Solving Eq. (5-18) for $\epsilon^{-q\psi/kT}$, we obtain

$$\epsilon^{-q\psi/kT} = \frac{n_i}{n_n} \epsilon^{-q\phi_n'/kT} \tag{5-20}$$

Substituting Eq. (5-20) into (5-19),

$$p(x_{Tn}) = \frac{n_i^2}{n_n} \epsilon^{q(\phi_p' - \phi_n')/kT} \tag{5-21}$$

Since $p_n = n_i^2/n_n$ and $(\phi_p' - \phi_n') = V_o$, the applied potential, (5-21) becomes, finally,

$$p(x_{Tn}) = p_n \epsilon^{qV_o/kT} \tag{5-22}$$

In the same manner,

$$n(x_{Tp}) = n_p \epsilon^{qV_o/kT} \tag{5-23}$$

These last two equations are of basic importance to p-n junction theory and transistors, and simply state that *the concentration of injected minority carriers at a junction is directly proportional to the equilibrium minority-carrier concentration there and the exponential dependence of applied*

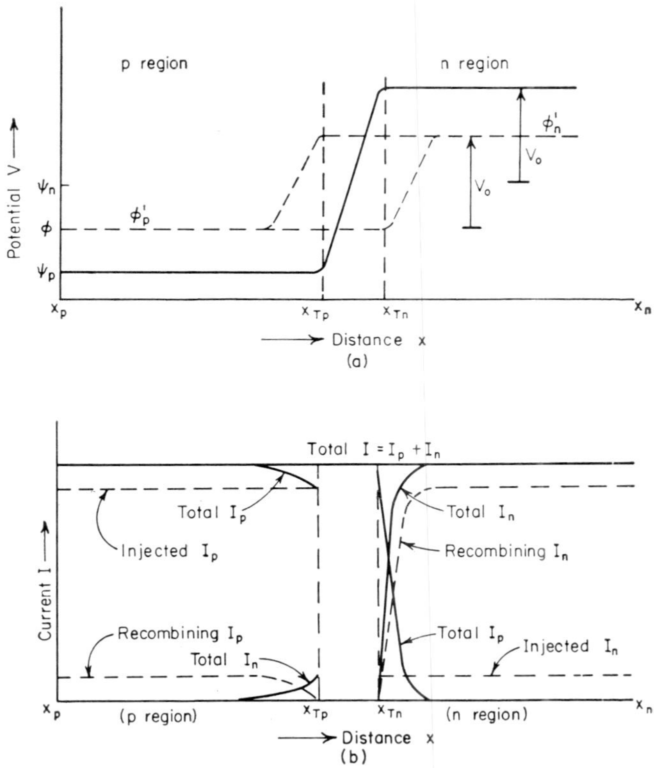

FIG. 5-7. Reverse-biased junction. (*a*) Quasi-Fermi levels; (*b*) current distributions. (*After R. D. Middlebrook.*)

voltage. Note that if V_o is negative and is of large magnitude compared to kT/q, then Eqs. (5-22) and (5-23) approximately equal zero.

5-6. The Continuity and Diffusion Equations. In order to properly analyze the p-n junction, it is first necessary to formulate the fundamental equation which completely describes the behavior of electrons and holes in any elemental volume of semiconductor material. This is the so-called continuity equation, which describes the way carrier concentrations vary with both time and space. Consider the elemental volume

$dx\,dy\,dz$, in which a carrier current is flowing and generation and recombination are also occurring. For continuity, the time rate of change of carriers in the volume must equal the excess of generation (or emission) over recombination and also the net flow across the surface. Mathematically, this is given as

$$\frac{\partial c}{\partial t} = e - R - \frac{1}{q}\left(\frac{\partial J}{\partial x} + \frac{\partial J}{\partial y} + \frac{\partial J}{\partial z}\right) \tag{5-24}$$

where J is the current density in amperes per centimeter². More generally, Eq. (5-24) may be written

$$\frac{\partial c}{\partial t} = e - R - \frac{1}{q}\nabla \cdot J \tag{5-25}$$

where the $\nabla \cdot J$ term is called the divergence of the current density and denotes partial differentiation in each of the three dimensions.

To evaluate the $(e - R)$ term, we make use of the lifetime considerations introduced in Chap. 4. From Eqs. (4-38) and (4-39) we obtain the relationship for the net rate of change of the excess carriers, which is

$$\frac{d(\Delta c)}{dt} = \frac{-\Delta c}{\tau} \tag{5-26}$$

The minus sign is introduced here to indicate that as the excess carriers decay to the equilibrium value, the rate of change decreases. Specifically, for electrons Eq. (5-26) becomes

$$\frac{d(\Delta n)}{dt} = \frac{-\Delta n}{\tau_n} = \frac{-n + n_p}{\tau_n} \tag{5-27}$$

where n_p is the equilibrium value. Also, for holes,

$$\frac{d(\Delta p)}{dt} = \frac{-\Delta p}{\tau_p} = \frac{-p + p_n}{\tau_p} \tag{5-28}$$

where p_n is the equilibrium value. Equations (5-27) and (5-28) describe the net rate of change of the respective carriers. Substituting these results into the general continuity equation, we have

$$\frac{\partial n}{\partial t} = \frac{-n + n_p}{\tau_n} + \frac{1}{q}\nabla \cdot J_n \tag{5-29}$$

$$\frac{\partial p}{\partial t} = \frac{-p + p_n}{\tau_p} + \frac{1}{q}\nabla \cdot J_p \tag{5-30}$$

To evaluate the current density term J we can recall from Chap. 4 that currents in semiconductors may consist of both drift (see Sec. 4-4) and diffusion (see Sec. 4-7) components. The drift term is dependent on the electric field; the diffusion term is dependent on the carrier concentration gradient. Thus we have

$$J_n = q\mu_n n\mathcal{E} + qD_n\nabla n \tag{5-31}$$
$$J_p = q\mu_p p\mathcal{E} - qD_p\nabla p \tag{5-32}$$

On the assumption that $\mathcal{E} = 0$ in our subsequent analysis of the p-n junction, we may neglect the drift current and consider only the diffusion part of Eqs. (5-31) and (5-32). Therefore Eqs. (5-29) and (5-30) become

$$\frac{\partial n}{\partial t} = \frac{-n + n_p}{\tau_n} + D_n\nabla^2 n \tag{5-33}$$
$$\frac{\partial p}{\partial t} = \frac{-p + p_n}{\tau_p} + D_p\nabla^2 p \tag{5-34}$$

where, by definition, $\nabla^2 n = \nabla \cdot \nabla n$, the second derivative in three dimensions. Thus we have the general diffusion equation in three dimensions, which, along with the proper boundary conditions, will be employed in analyzing the behavior of p-n junctions and transistors.

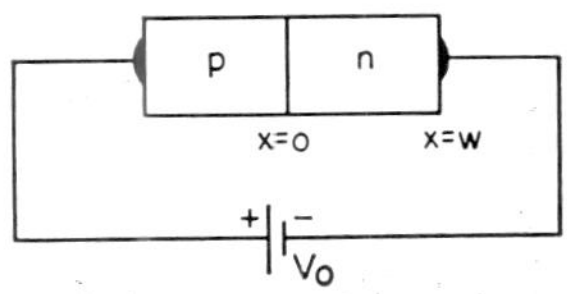

FIG. 5-8. Simple p-n junction.

5-7. D-C Voltage-current Analysis. With all of the foregoing discussions as foundation, we are now in an excellent position to analyze the p-n junction mathematically. Only the d-c case, giving rise to the voltage-current characteristic, will be treated here. Furthermore, only the carrier flow in one dimension will be considered and the assumptions indicated in Sec. 5-5 will apply. Consider, then, a simple model of a p-n junction of cross-sectional area A, where $x = 0$ at the junction and $x = W$ at the n region end, as shown in Fig. 5-8. The voltage V_o is applied across the end connections, which we will consider to be ohmic nonrectifying contacts.

It would be appropriate at this point to digress briefly to explain the significance of an ohmic contact to a semiconductor. Since we are concerned with the rectification properties of the actual p-n junction, we would certainly not want to have any rectification occurring at the connections to the respective regions. The connections should serve merely as electrical contacts. This is achieved by intentionally making the contacts regions of very high recombination velocities (i.e., low lifetime), approaching infinity. Thus, at the ohmic interface we would expect both the majority and minority carriers to remain at their equilibrium values. Any deviations from equilibrium would be restored by recombination. A typical contact of this sort is made by soldering to an abraded semiconductor surface. Another method of making ohmic contacts is to introduce a much more heavily doped layer of the same type as the semiconductor. These contacts are nonrectifying junctions and are referred to as pp+ junctions for p type and nn+ junctions for n type, where the plus sign denotes the heavy doping. The large majority-carrier concentration there will act to keep the carriers at their equilibrium levels. Also, in the region of the doped contact, the minority-carrier concentrations are quite small; therefore the possibility of injecting minority carriers is minimized. It will be assumed that for all ensuing discussions on p-n junctions and transistors, the end contacts are ohmic and noninjecting.

Now we may return to the problem of the simple p-n junction of Fig. 5-8. We have seen from the previous sections that for conditions of forward bias, a current, consisting of holes into the n region and electrons into the p region, will flow in the p-n junction. In order to establish analytically the variation of current with voltage, it becomes convenient to study the problem in terms of the individual hole and electron flows and then combine the two for the final result. We shall consider the hole flow first.

Firstly, we can assume that the holes move principally by diffusion through regions free of electric fields, i.e., $\mathcal{E} = 0$. In other words, the resistivities of the p and n regions are low enough so that very little voltage drop occurs in these regions, as compared to the drop across the junction. Secondly, we assume that most of the current flow is due to diffusion in the x direction only. It has been found that a one-dimensional analysis correlates very well with experimental observations of actual rectifiers. Applying these conditions to the general diffusion equation for holes, Eq. (5-32), the diffusion equation for one dimension, becomes

$$\frac{\partial p}{\partial t} = \frac{p_n - p}{\tau_p} + D_p \frac{d^2 p}{dx^2} \tag{5-35}$$

where τ_p = lifetime of holes in n region
D_p = diffusion constant for holes
p_n = equilibrium minority-carrier concentration

For the steady-state or d-c solution, $\dfrac{\partial p}{\partial t} = 0$ and Eq. (5-35) becomes

$$\frac{d^2 p}{dx^2} - \frac{p - p_n}{D_p \tau_p} = 0 \tag{5-36}$$

Substituting the relationship that $D_p \tau_p = L_p^2$, the diffusion length, we have

$$\frac{d^2 p}{dx^2} - \frac{p - p_n}{L_p^2} = 0 \tag{5-37}$$

This is a second-order differential equation whose solution is

$$p - p_n = A\epsilon^{x/L_p} + B\epsilon^{-x/L_p} \tag{5-38}$$

To determine the coefficients A and B, we must make use of the boundary conditions. Referring to Fig. 5-8, we can immediately state that

$$x = 0 \qquad p = p_n \epsilon^{qV_o/kT} \tag{5-39}$$

$$x = W \qquad p = p_n \tag{5-40}$$

It should be recalled that Eq. (5-39) was derived from quasi-Fermi-level considerations. It defines the concentration of holes that exists on the n-region side of the junction for an applied voltage equal to V_o. Further, since the n region has a specified hole lifetime, we expect the excess holes to decay with distance such that at $x = W$ the hole concentration equals the equilibrium value p_n, yielding boundary Eq. (5-40). Because nearly all the holes recombine within a distance approximately equal to a diffusion length L_p, Eq. (5-40) presumes that W is much greater than L_p. Substituting the first condition into (5-38), we obtain

$$p_n(\epsilon^{qV_o/kT} - 1) = A + B \tag{5-41}$$

The second boundary condition yields

$$0 = A\epsilon^{W/L_p} + B\epsilon^{-W/L_p} \tag{5-42}$$

Solving Eqs. (5-41) and (5-42) simultaneously and introducing hyperbolic functions for the exponential terms, we obtain

$$A = \frac{-p_n(1 - \epsilon^{qV_o/kT})\epsilon^{-W/L_p}}{2 \sinh (W/L_p)} \tag{5-43}$$

$$B = \frac{p_n(1 - \epsilon^{qV_o/kT})\epsilon^{-W/L_p}}{2 \sinh (W/L_p)} \tag{5-44}$$

Inserting the above expressions into Eq. (5-38) and simplifying, we obtain

$$p = p_n - \frac{p_n(1 - \epsilon^{qV_o/kT})}{2 \sinh (W/L_p)} (\epsilon^{(W-x)/L_p} - \epsilon^{-(W-x)/L_p})$$

$$p = p_n - p_n(1 - \epsilon^{qV_o/kT}) \frac{\sinh [(W - x)/L_p]}{\sinh (W/L_p)} \tag{5-45}$$

Eq. (5-45) gives the concentration of holes as a function of distance in the n region. We know that the hole current density

$$J_p = -qD_p \frac{dp}{dx} \tag{5-46}$$

We therefore differentiate Eq. (5-45) to obtain

$$\frac{dp}{dx} = \frac{+p_n(1 - \epsilon^{qV_o/kT})}{L_p} \frac{\cosh [(W - x)/L_p]}{\sinh (W/L_p)} \tag{5-47}$$

Evaluating $\frac{dp}{dx}$ at $x = 0$ to obtain the hole junction current, we get

$$\left.\frac{dp}{dx}\right|_{x=0} = \frac{p_n}{L_p} (1 - \epsilon^{qV_o/kT}) \frac{1}{\tanh (W/L_p)} \tag{5-48}$$

Substituting this result in Eq. (5-46) and multiplying by the area A to get the total hole current, we obtain, finally,

$$I_p = \frac{qAD_pp_n}{L_p}(\epsilon^{qV_o/kT} - 1)\frac{1}{\tanh(W/L_p)} \tag{5-49}$$

For the condition that $W \gg L_p$, we have

$$I_p = \frac{qAD_pp_n}{L_p}(\epsilon^{qV_o/kT} - 1) \tag{5-50}$$

This is the elementary p-n junction equation which describes the current due to the hole diffusion only. A completely identical analysis for similar boundary conditions in the p region for the electron flow, using

$$\frac{\partial n}{\partial t} = \frac{n_p - n}{\tau_n} + D_n\frac{d^2n}{dx^2} \tag{5-51}$$

and

$$J_n = qD_n\frac{dn}{dx} \tag{5-52}$$

would give

$$I_n = \frac{qAD_nn_p}{L_n}(\epsilon^{qV_o/kT} - 1) \tag{5-53}$$

Thus it is seen that Eq. (5-53) is complementary to Eq. (5-50). To get the total current $I = I_p + I_n$, we combine these equations and we get

$$I = \left(\frac{qAD_pp_n}{L_p} + \frac{qAD_nn_p}{L_n}\right)(\epsilon^{qV_o/kT} - 1) \tag{5-54}$$

This result is the basic voltage-current relationship for the p-n junction. It demonstrates clearly the effect of impurity concentrations in the semiconductor regions, since the diffusion constant, the minority-carrier concentration, and the diffusion length (lifetime) are all related to the impurity level.

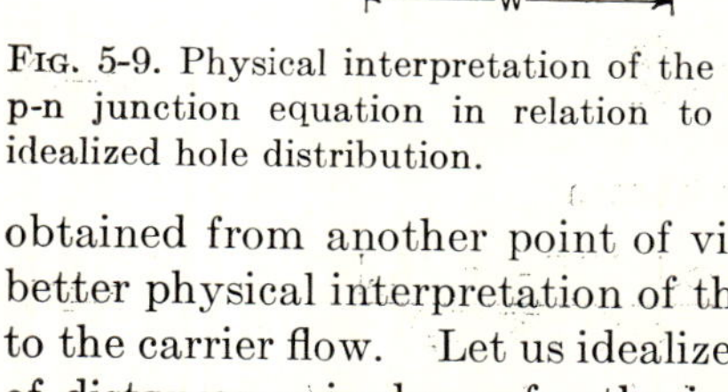

FIG. 5-9. Physical interpretation of the p-n junction equation in relation to idealized hole distribution.

Although the derivation stemmed from a simple differential equation and boundary conditions, it would be helpful to see how the same result is obtained from another point of view. This second approach will give a better physical interpretation of the relation of the p-n junction equation to the carrier flow. Let us idealize the carrier concentration as a function of distance, as is shown for the junction of Fig. 5-9 by the dashed lines. In the p region, the concentration is constant and equal to the equilibrium level p_p. Just at the n-region side of the junction, the injected hole level is given by $p_n\epsilon^{qV_o/kT}$. Now, if we assume that the concentration gradient is linear in the n region and that all the holes recombine at a distance equal to the diffusion length L_p, then

$$\frac{dp}{dx} = -\frac{p_n\epsilon^{qV_o/kT} - p_n}{L_p} \tag{5-55}$$

This is simply the slope of the diffusion concentration gradient drawn in the figure. It is negative since the concentration decreases with distance. Using this result with the equation for the hole diffusion current

$$I_p = -qAD_p\frac{dp}{dx}$$

and substituting for dp/dx, we get

$$I_p = \frac{qAD_pp_n}{L_p}(\epsilon^{qV_o/kT} - 1)$$

which is identical to Eq. (5-50), obtained more formally. The same method is applicable to the determination of the electron current. Thus we see that recombination of the injected carriers establishes a diffusion gradient which in turn yields a flow of current proportional to the *slope*. This is the mechanism basic to carrier diffusion in semiconductor junctions.

Let us now examine the effect of the potential V_o on the junction current I in Eq. (5-54). If the junction is reverse-biased, i.e., if V_o is negative, then the exponential term $\epsilon^{-qV_o/kT}$ quickly reduces to zero, giving

$$I_R = -\left(\frac{qAD_pp_n}{L_p} + \frac{qAD_nn_p}{L_n}\right) \tag{5-56}$$

where the current is designated I_R, the reverse-saturation current. For all practical purposes, we may consider I_R to be constant with voltage for $V_o > 1$ volt. Note that this current is primarily a function of the minority-carrier concentrations p_n and n_p, and would therefore decrease with decreasing resistivity. If the highest possible reverse resistance is desired for a p-n junction, then I_R is reduced by decreasing both the junction area A and the p- and n-region resistivities. In practical junction designs, the p region is usually very heavily doped so that n_p is negligible compared to p_n and the reverse current consists mainly of minority hole flow from the n region. Thus,

$$I_R \approx \frac{-qAD_pp_n}{L_p} \qquad \text{for } \rho_p < \rho_n \tag{5-57}$$

If V_o is made positive so that the p-n junction becomes forward-biased, the exponential term of Eq. (5-54) rapidly becomes greater than 1, giving

$$I_F = I_R\epsilon^{qV_o/kT} \tag{5-58}$$

where I_F is the forward junction current. At room temperature the exponent kT/q equals 0.026 volt. There, V_o may be as small as 0.1 volt for Eq. (5-58) to hold. Because of the positive exponential dependence on voltage, the forward current I_F increases rapidly, yielding a low resistance in the forward direction. The conditions are clearly shown in Fig. 5-10*a* and *b* for reverse- and forward-biased p-n junctions. This is for room temperature and it is assumed that I_R equals unity. In the next chapter, actual electrical characteristics of p-n junctions will be studied in greater detail with respect to the bulk semiconductor parameters.

Thus far, the discussion in this section has been concerned with the case where the diffusion lengths are small compared to the thickness of the

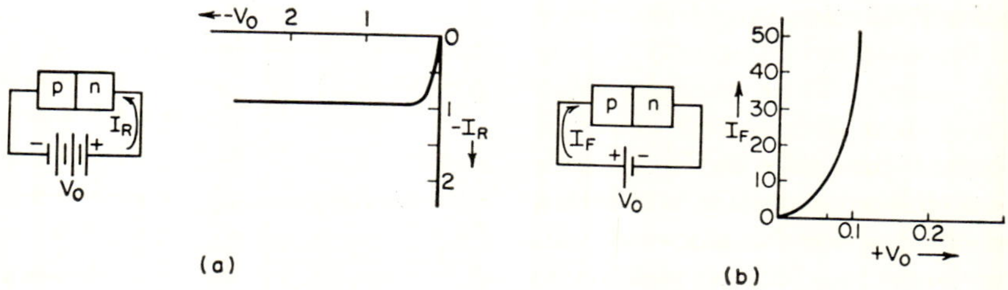

FIG. 5-10. Currents in a biased p-n junction. (*a*) Reverse bias; (*b*) forward bias.

base region. However, as will be seen when junction transistors are studied, another case of interest is when the inverse is true, i.e., when W is much smaller than L_p. This occurs in material with high minority-carrier lifetimes in a relatively thin n region. If $W \ll L_p$, then $1/\tanh (W/L_p) \approx L_p/W$ and Eq. (5-49) becomes

$$I_p = \frac{qAD_p p_n}{W} (\epsilon^{qV_o/kT} - 1) \tag{5-59}$$

The diffusion length L_p has been replaced by the thickness W, making the hole diffusion current independent of lifetime. This may be understood by referring again to Fig. 5-9. If the condition that $W \ll L_p$ is imposed, then all the injected holes will recombine at $x = W$ because of the infinite recombination velocity at the ohmic contact there. Therefore, at $x = W$, the hole concentration must equal the equilibrium level p_n. Consequently, the concentration gradient dp/dx becomes

$$\frac{dp}{dx} = \frac{-p_n(\epsilon^{qV_o/kT} - 1)}{W} \tag{5-60}$$

which would yield the same result as Eq. (5-59).

5-8. Step-junction Capacitance. In Sec. 5-3, where the equilibrium p-n junction was discussed, it was stated that in the transition region from p to n type, there exists a charge dipole region or depletion layer which was created by the carriers diffusing out of the regions and leaving the ionized impurity atoms on either side unneutralized. The distribution of the charge density is such that it produces the equilibrium contact potential V_T as described.[5] This is shown in Fig. 5-11 where the depletion-layer boundaries are $-x_p$ and x_n and the step transition is at $x = 0$, but the potential is increased by the application of reverse bias V_o. The effect of the total potential V_o is to charge the layer by repelling more majority carriers away from $x = 0$ and exposing more impurity ions on both sides. Thus the depletion layer widens with voltage, and it is apparent that this region behaves like a charged capacitance C with a dielectric $\kappa\epsilon_o$ for the material. In mks units, ϵ_o, the permittivity, equals 8.85×10^{-14} farad/cm. In the following analysis of the depletion-layer thickness and capacitance as a function of voltage it will be assumed that the conductivities of the p and n regions are such that all the applied voltage is dropped across the depletion layer.

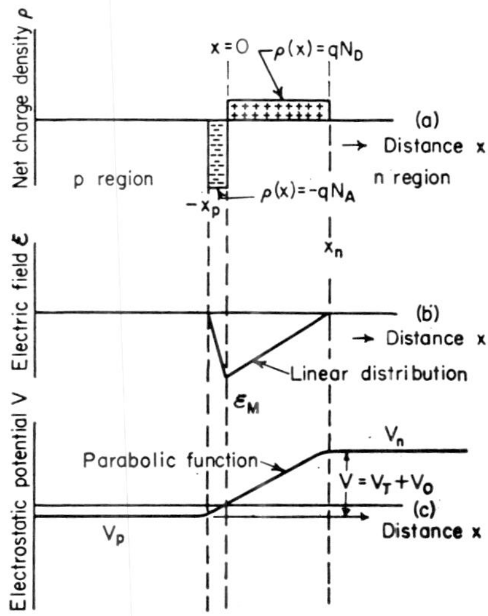

FIG. 5-11. Characteristics of the step-junction depletion layer.

The potential V is related to the charge density ρ by Poisson's equation,

$$\nabla^2 V = \frac{-\rho}{\kappa\epsilon_0}$$

which in one dimension becomes

$$\frac{d^2V}{dx^2} = \frac{-\rho(x)}{\kappa\epsilon_0} \tag{5-61}$$

where κ is the dielectric constant. From Fig. 5-11, it is seen that the charge density $\rho(x)$ is given as

$$\rho(x) = -qN_A \qquad -x_p < x < 0 \tag{5-62}$$
$$\rho(x) = +qN_D \qquad 0 < x < x_n \tag{5-63}$$
$$\rho(x) = 0 \qquad x < -x_p \qquad x > x_n \tag{5-64}$$

where N_A and N_D are net acceptor and donor impurity concentrations, respectively. Substituting Eq. (5-62) and (5-63) into Poisson's Eq.

(5-61), we obtain

$$\frac{d^2V_p}{dx^2} = \frac{+qN_A}{\kappa\epsilon_o} \qquad \text{p region} \tag{5-65}$$

$$\frac{d^2V_n}{dx^2} = \frac{-qN_D}{\kappa\epsilon_o} \qquad \text{n region} \tag{5-66}$$

Integrating these results twice to obtain first the electric field dV/dx and then the potential V, we get

$$\frac{dV_p}{dx} = \frac{qN_Ax}{\kappa\epsilon_o} + A_p \tag{5-67}$$

$$V_p = \frac{qN_Ax^2}{2\kappa\epsilon_o} + A_px + B_p \tag{5-68}$$

$$\left.\right\} \text{p region} \quad -x_p < x < 0$$

$$\frac{dV_n}{dx} = \frac{-qN_Dx}{\kappa\epsilon_o} + A_n \tag{5-69}$$

$$V_n = \frac{-qN_Dx^2}{2\kappa\epsilon_o} + A_nx + B_n \tag{5-70}$$

$$\left.\right\} \text{n region} \quad 0 < x < x_n$$

To evaluate the constants A and B in these equations, it is recognized that at $x = 0$, (5-68) must equal (5-70) since the voltage is continuous, and therefore $B_p = B_n$. Also at $x = 0$, the electric field must be continuous, making (5-67) equal (5-69), and $A_p = A_n$.

Further, at $x = -x_p$ and $x = x_n$, the electric field dV/dx must be zero, since it is continuous and exists only within the layer boundaries. Then, from (5-67) and (5-69) we have

$$A_p = A_n = \frac{qN_Ax_p}{\kappa\epsilon_o} = \frac{qN_Dx_n}{\kappa\epsilon_o} \tag{5-71}$$

From this result we see that

$$N_Ax_p = N_Dx_n \tag{5-72}$$

which is an obvious conservation result, stating merely that the total negative charge on one side must equal the total positive charge on the other side. Introducing the expressions for A_p and A_n into (5-68) and (5-70), we have

$$V_p = \frac{+qN_A}{2\kappa\epsilon_o}x^2 + \frac{qN_Ax_p}{\kappa\epsilon_o}x + B \tag{5-73}$$

$$V_n = \frac{-qN_D}{2\kappa\epsilon_o}x^2 + \frac{qN_Dx_n}{\kappa\epsilon_o}x + B \tag{5-74}$$

To obtain the potential difference from $-x_p$ to x_n, we subtract $V_p(-x_p)$ from $V_n(x_n)$, or

$$V = V_o + V_T = V_n - V_p = \frac{q}{2\kappa\epsilon_o}(N_Dx_n{}^2 + N_Ax_p{}^2) \tag{5-75}$$

Since $N_A = N_D(x_n/x_p)$ from Eq. (5-72),

$$V = \frac{qN_D}{2\kappa\epsilon_o} x_n(x_n + x_p) \tag{5-76}$$

In most good p-n junctions $N_A \gg N_D$, and therefore $x_n \gg x_p$ meaning that the charge density in the p region is so great that practically all the layer widening or spreading takes place in the n region. For this approximation, letting $x_m = x_n + x_p \approx x_n$, the total depletion-layer width, we have, finally,

$$x_m = \sqrt{\frac{2\kappa\epsilon_o V}{qN_D}} \quad \text{cm} \quad \text{n type} \tag{5-77}$$

$$x_m = \sqrt{\frac{2\kappa\epsilon_o V}{qN_A}} \quad \text{cm} \quad \text{p type} \tag{5-78}$$

Equation (5-77) gives the depletion-layer width as a function of voltage and impurity concentration for the case where the n region is the high-resistivity side of the junction. For the case of an n-p junction where the p region is the high-resistivity side, N_D in Eq. (5-77) is replaced by N_A as shown by Eq. (5-78).

In order to obtain the expression for step-junction capacitance, we may treat the depletion layer as a parallel-plate capacitance with the charges separated a distance x_m in a dielectric κ, or

$$C_T = \frac{\kappa\epsilon_o}{x_m} \quad \text{farads/cm}^2 \tag{5-79}$$

where C_T is the capacitance per unit area. Equation (5-79) is the identical result that is obtained if we consider the true definition of capacitance, viz., dQ/dV, where Q is the junction charge density. If Eqs. (5-77) and (5-78) are substituted into (5-79), we have

$$C_T = \sqrt{\frac{q\kappa\epsilon_o N_D}{2V}} \quad \text{farads/cm}^2 \quad \text{n type} \tag{5-80}$$

$$C_T = \sqrt{\frac{q\kappa\epsilon_o N_A}{2V}} \quad \text{farads/cm}^2 \quad \text{p type} \tag{5-81}$$

Thus we see that for a step junction, the capacitance per unit area decreases as the square root of voltage. Further, if the appropriate resistivity relationships are substituted for the impurity concentration N, we see that the capacitance also decreases as the square root of resistivity. In Fig. 5-12 are graphical plots of Eqs. (5-77) or (5-78) and (5-80) or (5-81) as functions of the ratio V/N, where V is the voltage and N is the impurity concentration, for both germanium and silicon. The curves differ only due to the differences in dielectric constant for the semiconductors.

The expressions for the step-junction capacitances that were just presented are for the case of the p-n junction under reverse bias. For this condition, the reverse current I_R flowing through the depletion region is usually very small and therefore does not disturb the charge distribution in the region. On the other hand, when the junction is biased in the forward direction, the situation is somewhat different. Firstly, it follows from physical reasoning that forward bias would cause the depletion layer to narrow, since the forward potential V_o reduces the net junction

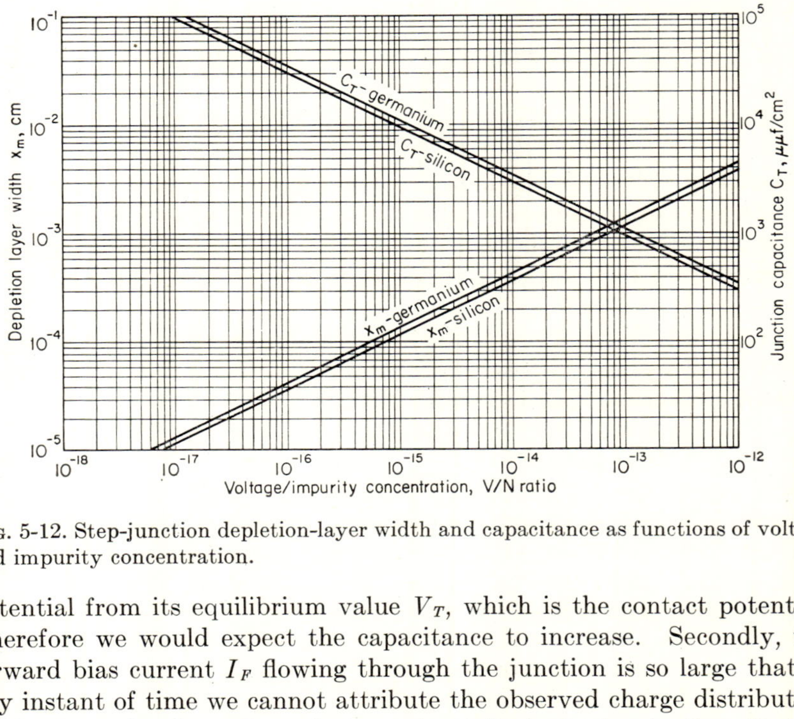

FIG. 5-12. Step-junction depletion-layer width and capacitance as functions of voltage and impurity concentration.

potential from its equilibrium value V_T, which is the contact potential. Therefore we would expect the capacitance to increase. Secondly, the forward bias current I_F flowing through the junction is so large that at any instant of time we cannot attribute the observed charge distribution to the impurity ions alone, but must attribute it also to the effective number of carriers there. Since the charge is increased, the capacitance will increase proportionately. The forward-bias transistor capacitance may be in the order of two times as great as the capacitance at equilibrium alone. Nevertheless, the equilibrium capacitance is determined by the relation (capacitance per unit area)

$$C_T = \sqrt{\frac{q\kappa\epsilon_o N}{2V_T}} \tag{5-82}$$

where V_T is the contact potential given earlier by Eq. (5-15).

5-9. Graded-junction Capacitance. The model of the step junction is most applicable to the case of the p-n junction which is fabricated by the alloy technique, wherein one obtains a rather abrupt transition from one impurity type to the other. In Chap. 1, however, it was shown that many junction structures are formed by the diffusion method. This results in what is referred to as a *graded junction*, since the diffused region establishes a gradual transition from one impurity type to the other. This is illustrated by the p-n-junction impurity profiles in Fig. 5-13.

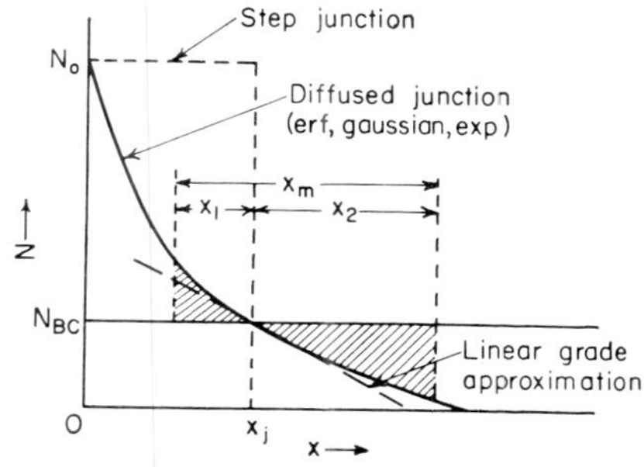

FIG. 5-13. Impurity profile and depletion-layer characteristics for diffused junctions.

Diffused junctions are characterized by several types of impurity distributions depending upon the method employed for carrying out the diffusion. For the case of diffusion from the gaseous vapor where the surface concentration is always constant, we have an impurity distribution governed by the *error function*, or

$$N(x) = N_o\left(1 - \text{erf}\,\frac{x}{\sqrt{4Dt}}\right) \tag{5-83}$$

where N_o = surface concentration
D = impurity diffusion constant
t = diffusion time

For the case of the diffusion which proceeds from a fixed concentration at the surface, we obtain a *gaussian* impurity distribution given by

$$N(x) = N_o\epsilon^{-x^2/4Dt} \tag{5-84}$$

In addition to the error-function and gaussian diffusion distribution, one may also obtain other variations which result in exponential or $\int$erfc profiles. In any event, it is interesting to note that in all cases of diffused graded junctions one can readily approximate the crossover point by an impurity grade which is *linear*. This is an obvious result since in a diffused layer the impurity concentrations decrease rapidly from the surface and approach a somewhat shallow, constant slope at the point of transition.

We shall now study the characteristics of the depletion layer for the model of the p-n junction having a linear grade. In Fig. 5-14*a* is the charge distribution that exists when a reverse potential is applied to a graded junction in which the impurity concentration is linear through $x = 0$, or

$$N(x) = ax \tag{5-85}$$

where a is the grade constant in atoms/cm^4. The ionized donors in the n region establish a positive charge which is neutralized by the negative charge of the ionized acceptors in the p region. Because of the linear grade of the impurities, the net charge density will also be a linear function, or

$$\rho(x) = -qax \tag{5-86}$$

where q is the electron charge and a is the grade constant. It should be evident that for a distance x on either side of $x = 0$, the areas are equal, which indicates that the depletion layer spreads equally in both directions for a linear-grade junction. This is in contrast to the step junction in which the layer spreads more into the higher-resistivity region. Thus, if the total barrier width is x_m, the spreading in any one region is $x_m/2$. In order to relate the distance x_m to voltage, we begin with Poisson's equation for one dimension, viz.,

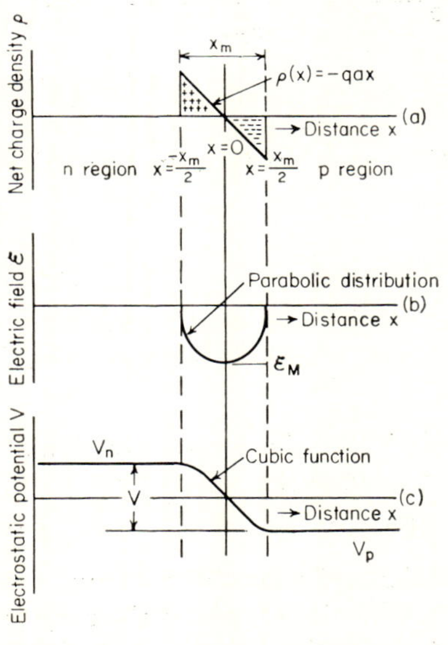

FIG. 5-14. Characteristics of the linear graded-junction depletion layer.

$$\frac{d^2V}{dx^2} = \frac{-\rho(x)}{\kappa\epsilon_o} \tag{5-87}$$

Substituting (5-86) into (5-87), we have

$$\frac{d^2V}{dx^2} = \frac{qax}{\kappa\epsilon_o} \tag{5-88}$$

Integrating (5-88), we obtain dV/dx, the electric field E.

$$\frac{dV}{dx} = \frac{qax^2}{2\kappa\epsilon_o} + K_1 \tag{5-89}$$

Integrating again, we obtain the voltage expression.

$$V = \frac{qax^3}{6\kappa\epsilon_o} + K_1x + K_2 \tag{5-90}$$

To evaluate the constant K_1, we make use of the assumption that all the applied voltage is dropped across the depletion region. In other words, the electric field is zero outside the junction. Therefore, with the boundary conditions that $\mathcal{E} = 0$ at $x = x_m/2$ and at $x = -x_m/2$ inserted into (5-89), K_1 becomes

$$K_1 = \frac{-qax_m^2}{8\kappa\epsilon_o}$$

or

$$\mathcal{E} = \frac{qax^2}{2\kappa\epsilon_o} - \frac{qax_m^2}{8\kappa\epsilon_o} \qquad \frac{-x_m}{2} < x < \frac{x_m}{2} \tag{5-91}$$

Eq. (5-91) represents the way the electric field varies with distance through the depletion layer, and is shown by the parabolic function in Fig. 5-14*b*. Note that the field is maximum at $x = 0$.

$$\mathcal{E}_M = \frac{-qax_m^2}{8\kappa\epsilon_o} \qquad \text{volts/cm} \tag{5-92}$$

The applied voltage V is equal to the potential V_n at $-x_m/2$ less the potential V_p at $x_m/2$.

$$V = V_n - V_p \tag{5-93}$$

If Eq. (5-90) is evaluated at the appropriate outer boundaries to obtain V_n and V_p and the results are inserted into (5-93), the constants K_1 and K_2 cancel out, and we get, finally,

$$x_m = \left(\frac{12\kappa\epsilon_o V}{qa}\right)^{1/3} \tag{5-94}$$

We see that for a linear-graded junction, the depletion thickness will vary as the cube root of the voltage. For the step junction, x_m varied as $V^{1/2}$. For this same linear geometry, the capacitance per unit area is derived by utilizing

$$C = \frac{\kappa\epsilon_o}{x_m} \tag{5-95}$$

which is the expression for the parallel-plate equivalent capacitance. Substituting Eq. (5-94) into (5-95), we obtain, after simplifying,

$$C_T = \left[\frac{(\kappa\epsilon_o)^2 q}{12}\right]^{1/3}\left(\frac{a}{V}\right)^{1/3} \qquad \text{farads/cm}^2 \tag{5-96}$$

Here, too, the capacitance for a linear-grade junction varies as $V^{-1/3}$, whereas for the step junction it varies as $V^{-1/2}$.

Although the linear grade was established as an approximation for the gradient of a diffused junction, actual experimental measurements of the variation of capacitance with voltage in diffused junctions indicate that at low voltages one observes cube-root dependence.[6] This indicates that the approximation is quite good. As voltage is increased, however, the cube-root dependence gradually changes to a square-root dependence and the diffused junction starts to behave like a step junction. The range between the $V^{-1/3}$ and $V^{-1/2}$ behavior is the most important range of interest in device design and was analyzed by Lawrence and Warner.[7] The results are plotted in Figs. 5-15 to 5-30. All the symbols in these design curves are as defined in Fig. 5-13. By entering the curves with the appropriate value of V/N_{BC} and x_j (where x_j denotes that distance from the surface at which the diffused impurity concentration equals the background concentration N_{BC}, one obtains values for the junction capacitance, the total depletion-layer thickness x_m, and the distance the

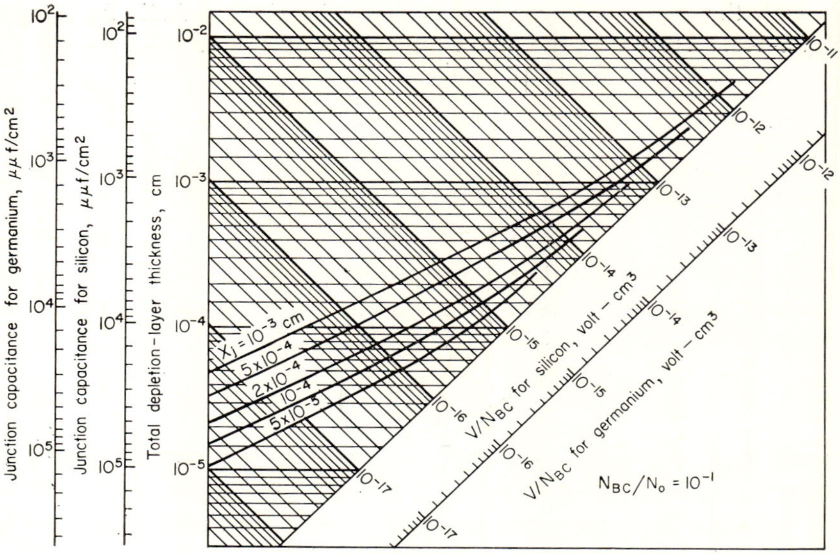

FIG. 5-15. Graded-junction curves for N_{BC}/N_o range of 3×10^{-2} to 3×10^{-1}. (*After H. Lawrence and R. M. Warner, Jr.*)

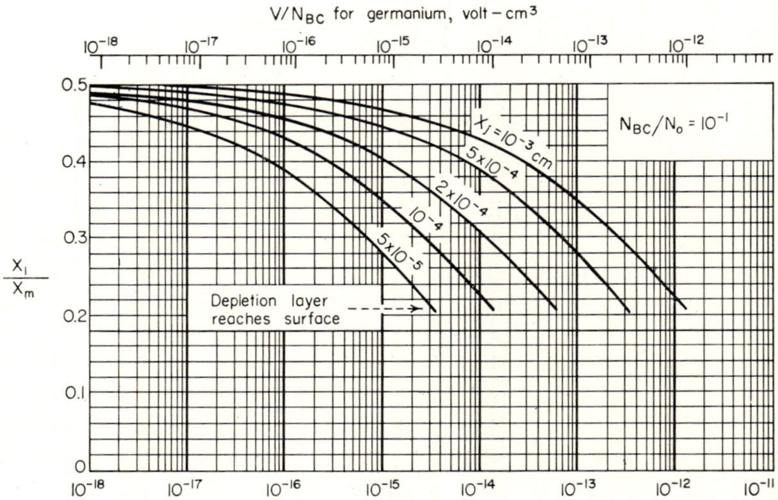

FIG. 5-16. Graded-junction curves for N_{BC}/N_o range of 3×10^{-2} to 3×10^{-1}. (*After H. Lawrence and R. M. Warner, Jr.*)

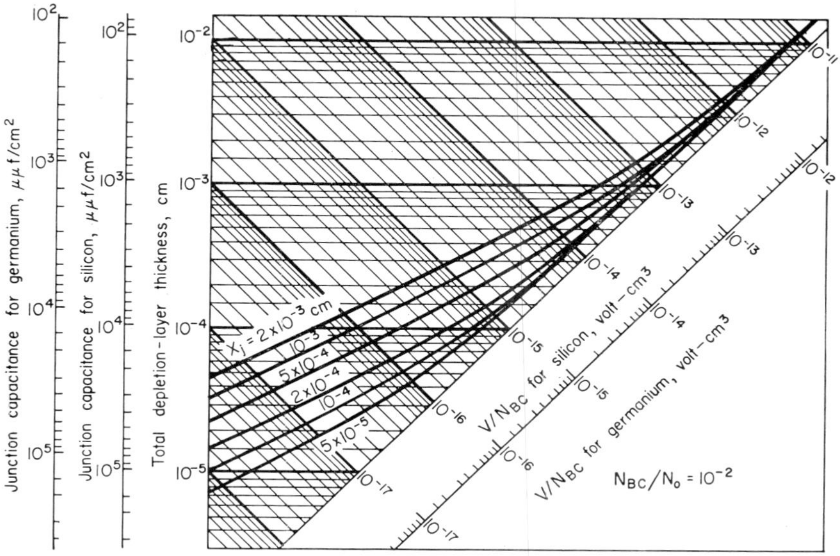

FIG. 5-17. Graded-junction curves for N_{BC}/N_o range of 3×10^{-3} to 3×10^{-2}. (*After H. Lawrence and R. M. Warner, Jr.*)

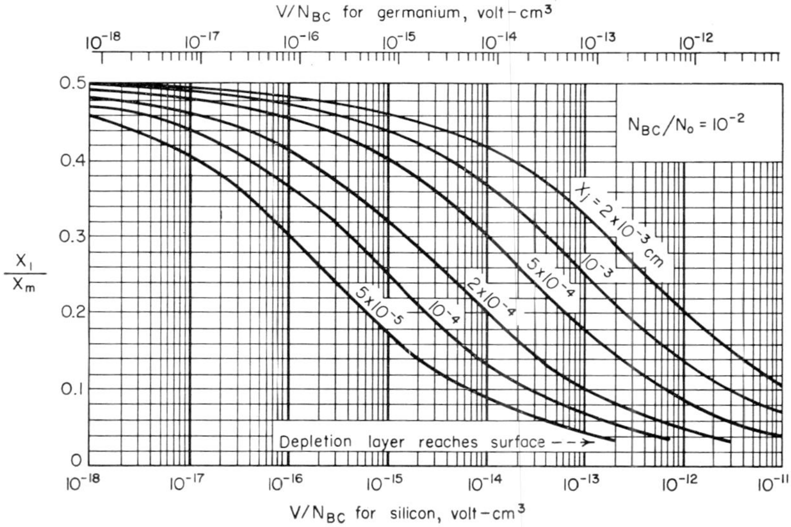

FIG. 5-18. Graded-junction curves for N_{BC}/N_o range of 3×10^{-3} to 3×10^{-2}. (*After H. Lawrence and R. M. Warner, Jr.*)

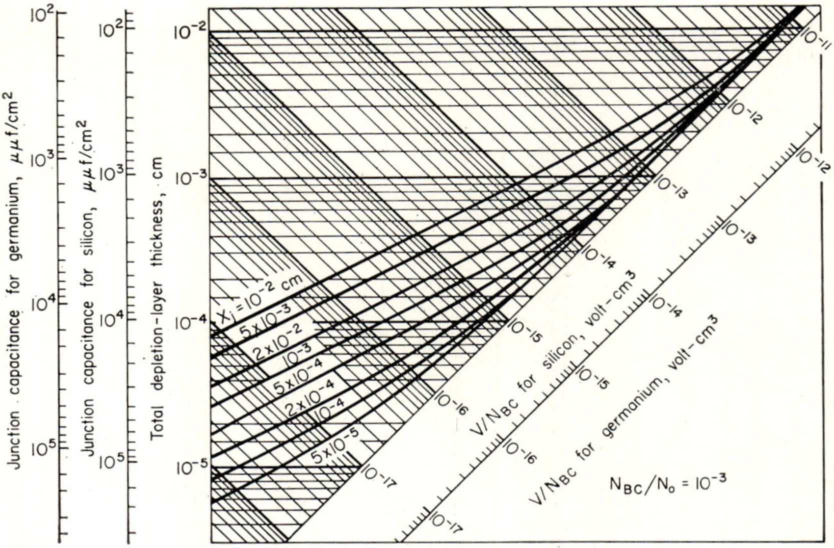

FIG. 5-19. Graded-junction curves for N_{BC}/N_o range of 3×10^{-4} to 3×10^{-3}. (*After H. Lawrence and R. M. Warner, Jr.*)

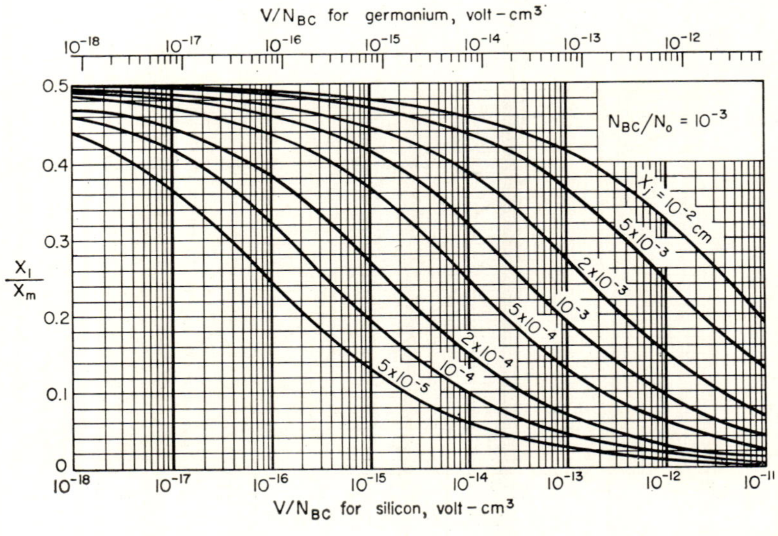

FIG. 5-20. Graded-junction curves for N_{BC}/N_o range of 3×10^{-4} to 3×10^{-3}. (*After H. Lawrence and R. M. Warner, Jr.*)

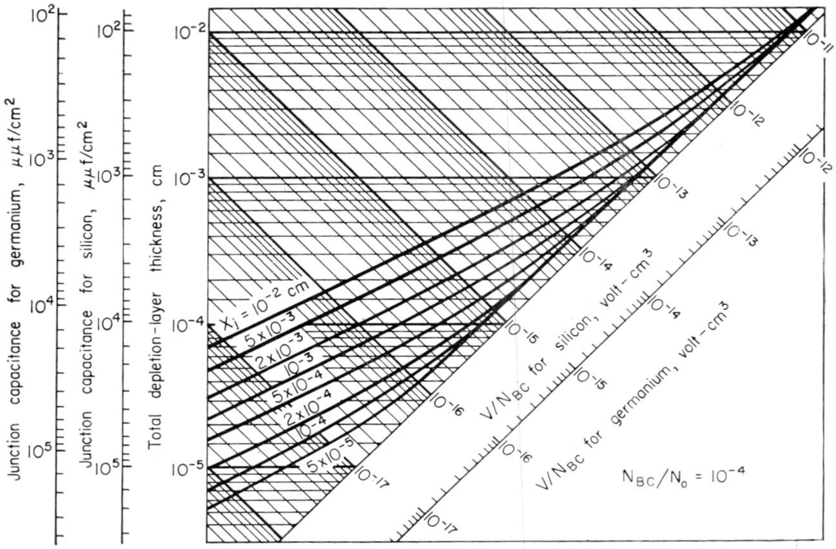

FIG. 5-21. Graded-junction curves for N_{BC}/N_o range of 3×10^{-5} to 3×10^{-4}. (*After H. Lawrence and R. M. Warner, Jr.*)

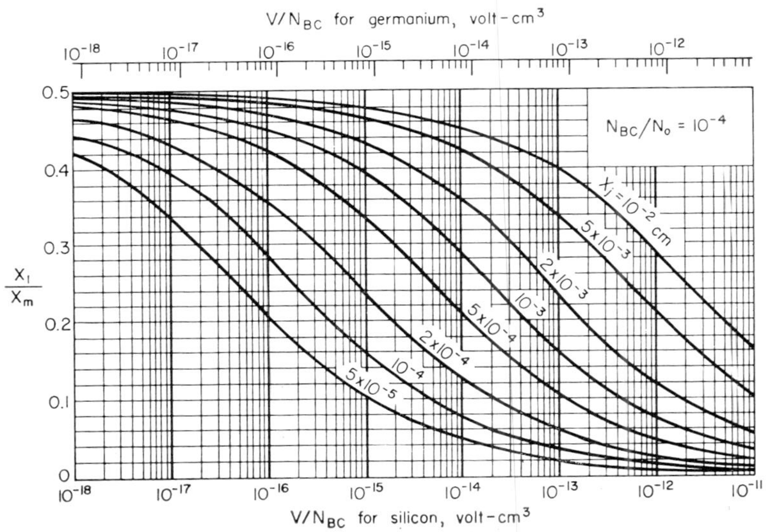

FIG. 5-22. Graded-junction curves for N_{BC}/N_o range of 3×10^{-5} to 3×10^{-4}. (*After H. Lawrence and R. M. Warner, Jr.*)

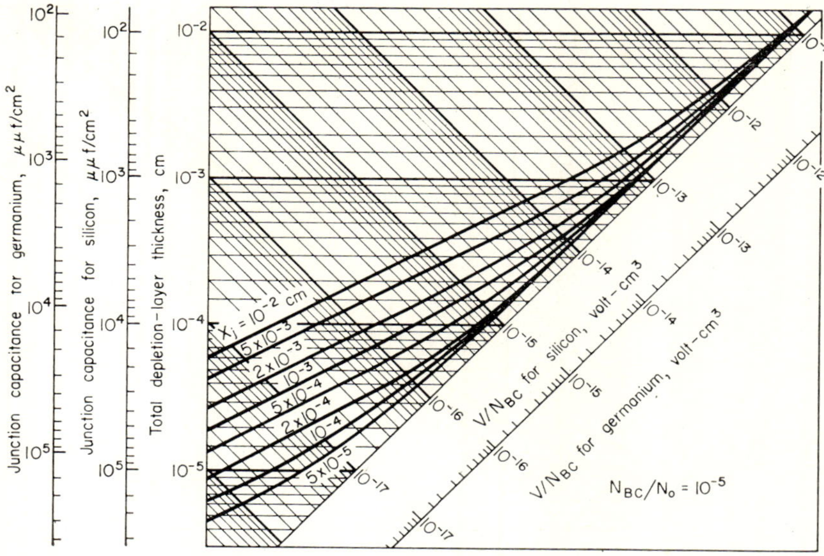

FIG. 5-23. Graded-junction curves for N_{BC}/N_o range of 3×10^{-6} to 3×10^{-5}. (*After H. Lawrence and R. M. Warner, Jr.*)

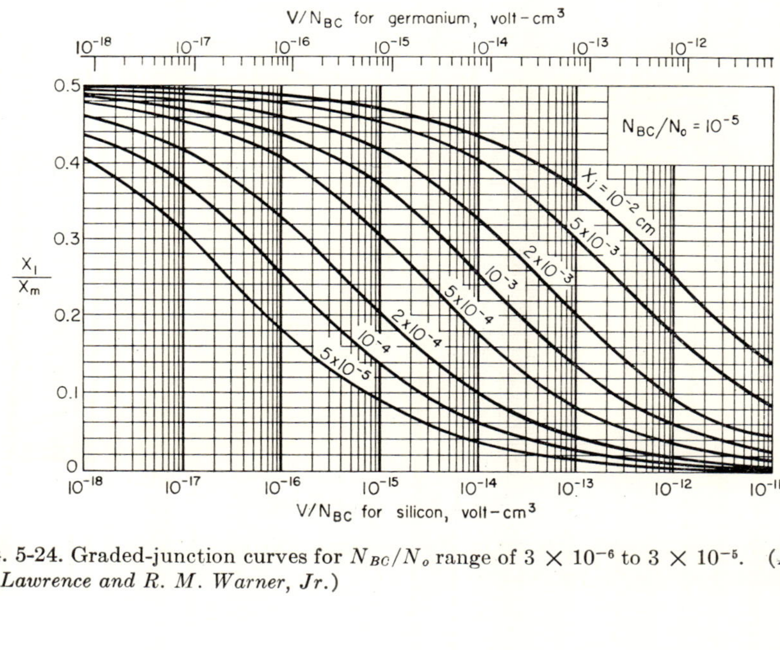

FIG. 5-24. Graded-junction curves for N_{BC}/N_o range of 3×10^{-6} to 3×10^{-5}. (*After H. Lawrence and R. M. Warner, Jr.*)

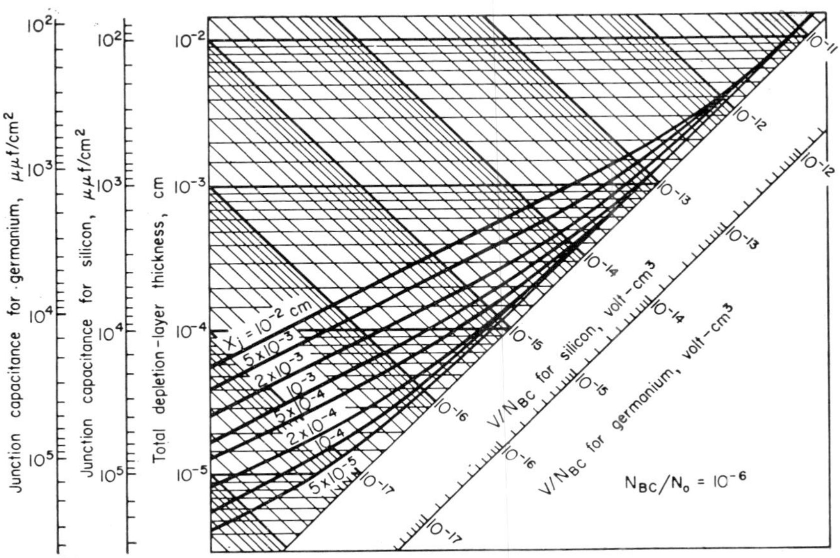

FIG. 5-25. Graded-junction curves for N_{BC}/N_o range of 3×10^{-7} to 3×10^{-6}. (*After H. Lawrence and R. M. Warner, Jr.*)

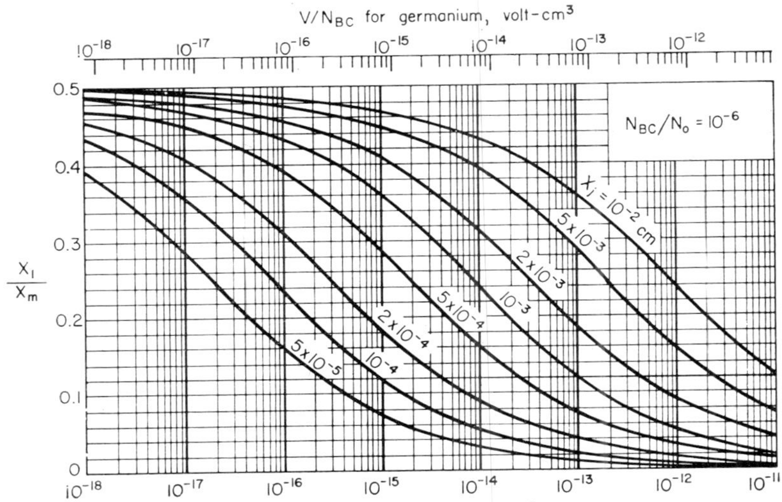

FIG. 5-26. Graded-junction curves for N_{BC}/N_o range of 3×10^{-7} to 3×10^{-6}. (*After H. Lawrence and R. M. Warner, Jr.*)

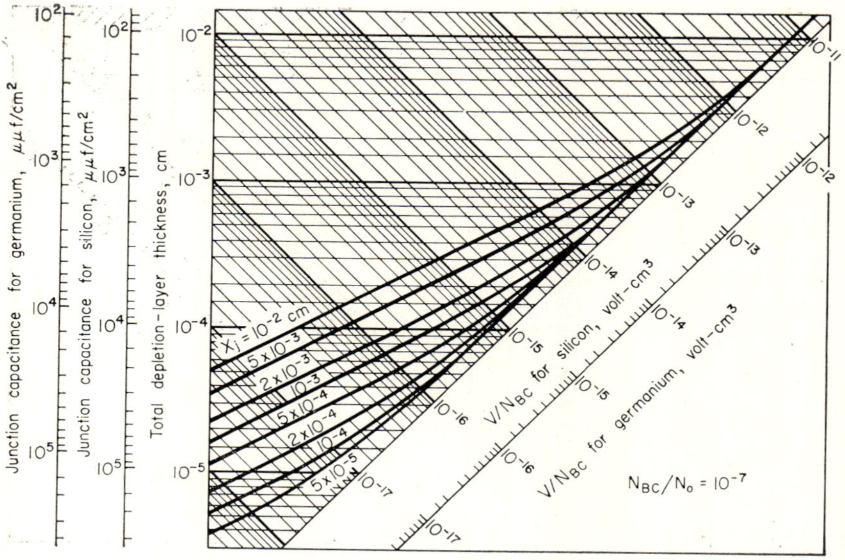

FIG. 5-27. Graded-junction curves for N_{BC}/N_o range of 3×10^{-8} to 3×10^{-7}. (*After H. Lawrence and R. M. Warner, Jr.*)

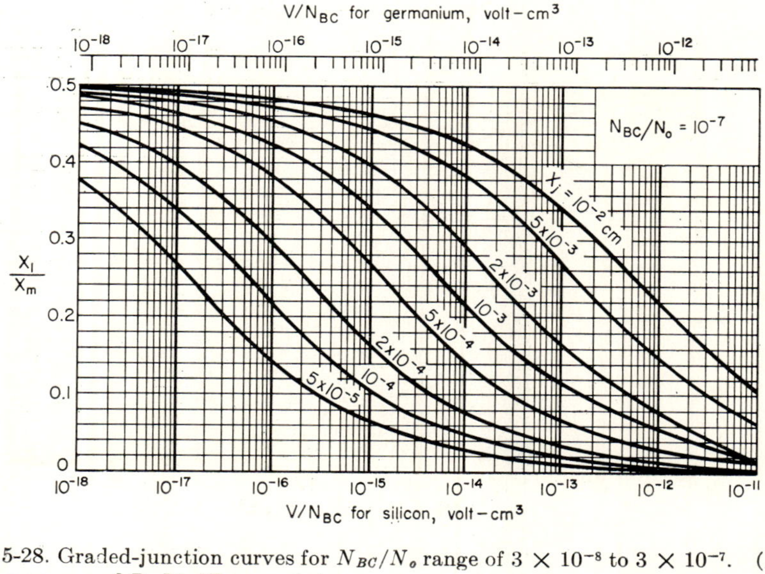

FIG. 5-28. Graded-junction curves for N_{BC}/N_o range of 3×10^{-8} to 3×10^{-7}. (*After H. Lawrence and R. M. Warner, Jr.*)

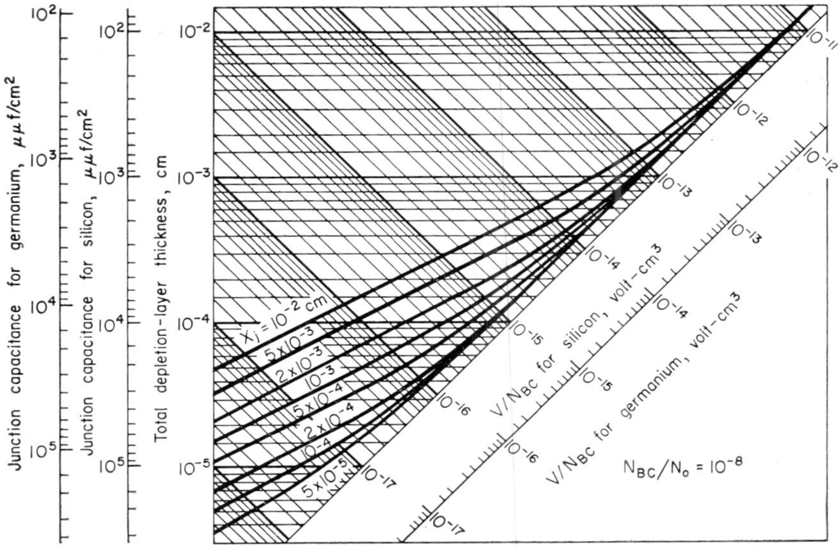

FIG. 5-29. Graded-junction curves for N_{BC}/N_o range of 3×10^{-9} to 3×10^{-8}. (*After H. Lawrence and R. M. Warner, Jr.*)

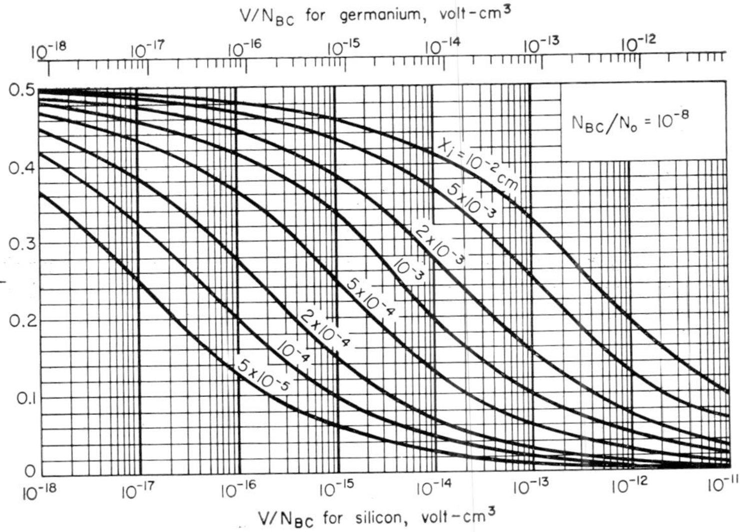

FIG. 5-30. Graded-junction curves for N_{BC}/N_o range of 3×10^{-9} to 3×10^{-8}. (*After H. Lawrence and R. M. Warner, Jr.*)

depletion layer spreads into the diffused region x_1. Further, there is a pair of curves for each range of the ratio N_{BC}/N_o, where N_o denotes the magnitude of the surface concentration. These curves may be used for either error-function or gaussian impurity distributions since the difference between the two is extremely small. The reader will note, using Fig. 5-15 as an illustrative example, that the curves start out with a slope corresponding to cube-root dependence (linear behavior) and gradually converge to a slope corresponding to square-root dependence (step behavior).

PROBLEMS

5-1. A p-n junction is made with a 3-ohm-cm resistivity n-type germanium pellet by alloying indium on one face of the pellet such that the p region is formed having a net impurity density of 3×10^{18} acceptors/cm³. The area of the junction is 2,500 mil² and the n-region thickness is 100 mils. The hole lifetime τ_p in the n region is 20 μsec. At room temperature determine the following:

a. The resistivity of the p region
b. The contact potential of the junction at equilibrium
c. The reverse saturation current in microamperes for a reverse bias $V_o = -5$ volts
d. The forward current in milliamperes for a forward bias $V_o = +0.25$ volt
e. The capacitance at 5 volts reverse bias
f. The reverse voltage necessary to spread the space-charge layer a distance of 0.001 in. into the n region

5-2. A silicon p-n junction is made by alloying aluminum into an n-type pellet of 3-ohm-cm resistivity. The resistivity of the p region is 0.02 ohm-cm. The area is 2,500 mil², the n-region thickness is 1 mil, and the hole lifetime in the n region is 150 μsec. Determine the following room-temperature parameters:

a. The reverse saturation current at −5 volts bias
b. The equilibrium junction capacitance

5-3. For a p-n junction at equilibrium, the hole concentration just at the n region may be given as

$$p = K\epsilon^{-qV_T/kT}$$

where K is a constant and V_T is the equilibrium contact potential. Using this result and the fact that at equilibrium the hole diffusion-current density equals the drift-current density such that the net hole-current density is zero, derive the Einstein relationship:

$$D = \frac{kT}{q}\,\mu$$

5-4. For a step junction in which the depletion layers spreading into both sides are comparable, show that the junction capacitance is given by

$$C_T = \sqrt{\frac{q\kappa\epsilon_0}{2V}\,\frac{N_AN_D}{N_A + N_D}}$$

5-5. Antimony is diffused into p-type germanium having a resistivity of 0.18 ohm-cm. The surface concentration of antimony is 10^{17} atoms/cm³ and the diffusion depth is 1.5 μ. Determine the capacitance per unit area and the distance the depletion layer spreads into the diffused region at a reverse voltage of 20 volts.

REFERENCES

1. Shockley, W.: "Electrons and Holes in Semiconductors," pp. 461–464, D. Van Nostrand Company, Inc., Princeton, N.J., 1953.
2. Shockley, W.: Transistor Electronics: Imperfections Unipolar and Analog Transistors, *Proc. IRE*, vol. 40, pp. 1304–1305, November, 1952.
3. Shockley, W.: The Theory of P-N Junctions in Semiconductors and P-N Junction Transistors, *Bell System Tech. J.*, vol. 28, pp. 435–489, July, 1949.
4. Middlebrook, R. D.: "An Introduction to Junction Transistor Theory," chap. 7, pp. 115–130, John Wiley & Sons, Inc., New York, 1957.
5. Saby, J. S., and W. C. Dunlap: Impurity Diffusion and Space Charge Layers in Fused-impurity p-n Junctions, *Phys. Rev.*, vol. 90, May, 1953.
6. Pritchard, R. L.: Transition Capacitance of p-n Junctions, G.E. Research Laboratory Report No. 57-RL-1817, October, 1957.
7. Lawrence, H., and R. M. Warner, Jr.: Diffused Junction Depletion Layer Calculations, *Bell System Tech. J.*, vol. 39, pp. 389–404, March, 1960.

6

Characteristics of p-n Junctions

6-1. Introduction. For a typical semiconductor p-n junction where the p-region resistivity is very much lower than that of the n region, we have seen that the current flow is due primarily to the hole diffusion component. In the forward direction, the current consists of the holes that are injected as minority carriers into the n region, where they recombine in accordance with the effective lifetime. In the reverse direction, the current consists of those minority holes in the n region within a diffusion length of the junction that diffuse to the depletion layer and then get swept across by the electric field. For both cases, we assumed that the electron components are negligible. From the theory of the previous chapter, the junction current is governed by the basic equation

$$I = \frac{qAD_p p_n}{W} (\epsilon^{qV/kT} - 1) \tag{6-1}$$

with the assumption that the diffusion length of the n region is much greater than its thickness W. We see that in the forward direction the current increases exponentially with voltage, whereas the reverse current rapidly saturates to a constant value. This nonlinear relationship for the p-n junction makes it an excellent rectifying device.

In this chapter, we shall examine the characteristics of p-n junctions as rectifiers. Although the principles developed are applicable to rectifier design, the intent is to establish an understanding of the electrical properties of the p-n junction as an integral part of the transistor. Both the reverse and forward directions will be studied in greater detail, and comparisons will be made between germanium and silicon. We shall also consider limiting conditions of voltage and current in the reverse and forward directions, respectively. Lastly, the frequency response of the junction to pulses will be studied.

6-2. Reverse-current Characteristics. From Eq. (6-1), if the voltage V is negative and greater than 0.1 volt, the reverse saturation current of the junction is given by

$$I_R = \frac{qAD_p p_n}{W} \tag{6-2}$$

If we normalize Eq. (6-1) by letting I_R equal unity, a plot as a function of reverse voltage would appear as shown in Fig. 6-1. The voltage range greater than 0.1 volt is referred to as the *voltage-saturation range,* since the current rapidly becomes almost constant with voltage. If I_R is small, on the order of microamperes, it is apparent that the resistance of a reverse-biased junction is very large, on the order of megohms. For example, if the reverse current measured at 5 volts is 1 μa, the reverse (or back) resistance equals 5 megohms. For all practical purposes, in most circuit applications this approaches an open circuit. To evaluate the resistance to alternating current, the slope of the V-I characteristic must be considered. In Fig. 6-1, the slope is practically zero, corresponding to a theoretical dynamic resistance of thousands of megohms. Later it will be shown that typical values are much lower.

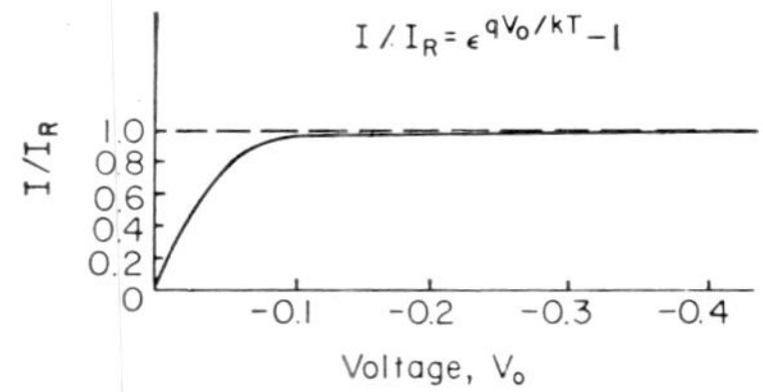

FIG. 6-1. Normalized reverse V-I characteristic.

The parameters of Eq. (6-2) tell us what the requirements are to keep the reverse saturation current I_R small. Firstly, I_R is reduced by making the junction area small. Secondly, an increase in the n-region thickness W decreases I_R, provided that the lifetime τ_p increases to maintain the assumption that $L_p \gg W$. Thirdly, a decrease in the minority hole concentration p_n would decrease I_R. In effect, this means lowering the resistivity of the n region. In other words, the reverse current increases with resistivity. As was studied in Chap. 4, p_n is very much lower for silicon than for germanium at the same resistivity. Consequently, other things being equal, typical reverse currents for germanium would be in the order of microamperes while, for silicon, I_R might be in the order of micromicroamperes. This infinitesimally small current for silicon is only a theoretical value. We shall soon see that surface effects and charge-generation phenomena bring the value up to the millimicroampere range, as is observed experimentally for silicon junctions.

Actually, Eq. (6-2) must be corrected to take into account the effect of the spreading of the depletion layer with reverse voltage. We saw in the previous chapter that the depletion-layer thickness x_m increases as the square root to cube root of applied voltage, depending on the nature of the junction. If the equilibrium thickness is W_o, the effective thickness W is written as

$$W = W_o - x_m \tag{6-3}$$

or

$$W = W_o - K^{1/n}\sqrt{V} \tag{6-4}$$

where K is an arbitrary constant. This decrease in W with voltage has the effect of increasing I_R accordingly, as shown in Fig. 6-2. Thus the V-I curve acquires a finite slope, yielding a considerably lower a-c dynamic resistance than that stated earlier. Also, it is seen that its d-c resistance decreases with voltage.

In practical junction devices it is quite difficult to achieve an absolutely theoretical reverse-current characteristic. The chief reason for this is the dominant role the semiconductor surface plays in contributing to excess leakage current. The surface leakage current across the p-n junction will appear as an additive component of the bulk reverse current. This is understandable if we view the surface as a conduction path in parallel with the p-n junction. Thus the total observed current is the

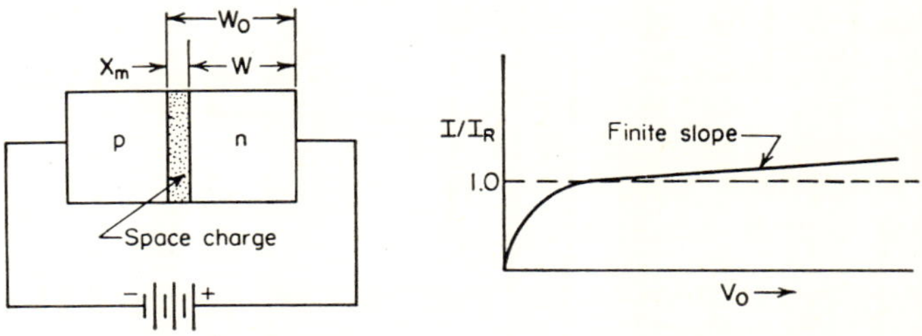

FIG. 6-2. Effect of depletion layer spreading on the reverse-current characteristic.

sum of the surface and bulk currents. One of the most formidable problems in semiconductor technology is to obtain proper treatment of the surface in order to minimize leakage effects. It is known that both moisture and ionic contamination on the surface will increase the reverse current. However, it is not too clear just what is happening physically. Presently it is believed that leakage conduction occurs either as ionic current in the oxide film[1,*] or as conduction in the inversion layers discussed in Chap. 4. In the case of the latter, the inversion layer might appear in the form of a *channel*, as shown in Fig. 6-3. The channel is considered to be an extension of the p region about the n-region surface and its properties would be determined by the nature of the surrounding ambient. It corresponds to an increase in the effective area of the junction and electrically it is an excellent model for explaining the behavior of the reverse current. A typical reverse-current characteristic showing the effects of surface leakage is also drawn in Fig. 6-3. At high voltages the back current may rise appreciably, representing a decrease in resistance and an increase in reverse power dissipation. As we shall see later on, this can limit the useful power rating for transistors.

* References, indicated in the text by superscript figures, are listed at the end of the chapter.

6-3. Temperature Dependence of Reverse Current. For a semiconductor device, whether it be a rectifier or a transistor, it is usual practice to base the maximum power rating on the theoretical allowable temperature the junction can attain. For germanium this is approximately 125°C and for silicon it is about 250°C. Beyond these temperatures, the conduction becomes predominantly intrinsic, and the junction loses its rectification properties. More conservative junction-temperature ratings used practically are 100 and 200°C for germanium and silicon, respectively. These junction temperatures are reached by the combination of the electrical power dissipated and the temperature of the surrounding

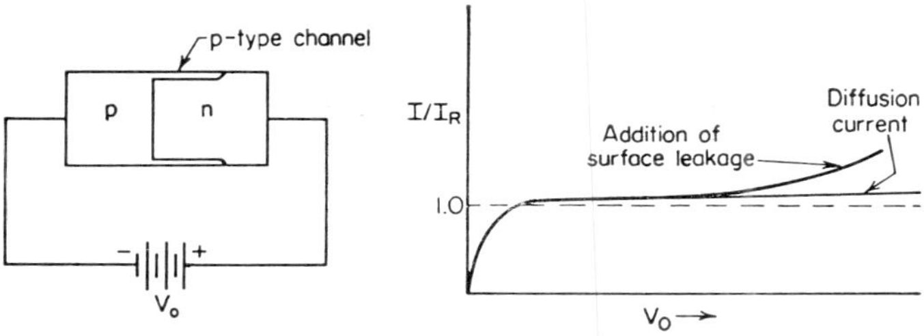

FIG. 6-3. A p-n junction channel and typical reverse-current characteristic, illustrating the effects of surface leakage.

environment. It should be evident, therefore, that if a device is to be operated at high ambient temperature, the power dissipation must be decreased to maintain the constant junction-temperature rating.

For these reasons, it is important that the reverse power dissipation of a junction be as small as possible, since it represents useless power consumption. Unfortunately, as temperature increases, the reverse current for a p-n junction increases too, thereby increasing the power dissipated for a fixed voltage V. Neglecting surface effects, we shall now examine the manner in which I_R varies with temperature. From Eq. (6-2), it is seen that only the diffusion constant D_p and the minority-carrier concentration p_n are temperature-dependent terms. If we make the substitutions $D_p = (kT/q)\mu_p$ and $p_n \approx n_i^2/N_D$, we have

$$I_R = \frac{Ak}{WN_D} T\mu_p n_i^2 \tag{6-5}$$

In Chap. 3 we saw that n_i^2 is proportional to

$$n_i^2 \propto T^3 \epsilon^{-E_{GO}/kT} \tag{6-6}$$

where E_{GO} is the band gap for the semiconductor at 0°K. Equations (4-8) and (4-9) indicate that the lattice mobility for germanium varies with temperature as $T^{-1.66}$ for electrons and $T^{-2.33}$ for holes. In order to arrive at a single general expression for the temperature variation of I_R

for germanium, we let the carrier mobility vary approximately as T^{-2}. Therefore, combining all the temperature dependences with Eq. (6-5), we get, finally,

$$I_R \propto T^2\epsilon^{-E_{GO}/kT} \qquad \text{germanium} \tag{6-7}$$

Let us now apply this result to the case of germanium for which $E_{GO} = 0.785$ ev. A normalized plot of Eq. (6-7) is given in Fig. 6-4, referenced to a room-temperature value of 27°C. One observes that the

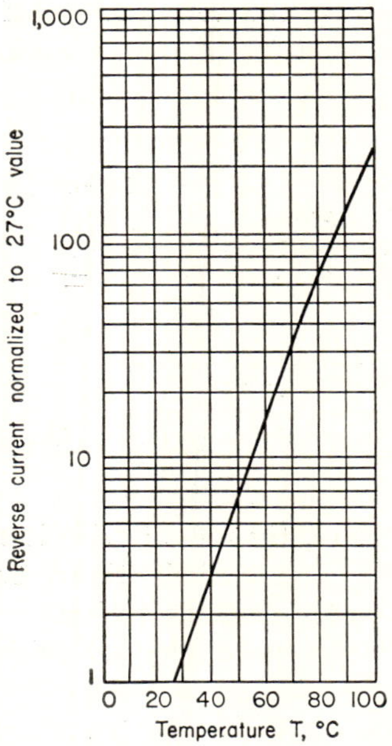

FIG. 6-4. Theoretical variation of reverse current with temperature for a germanium p-n junction. $I = KT^2\,\epsilon^{-9100/T}$

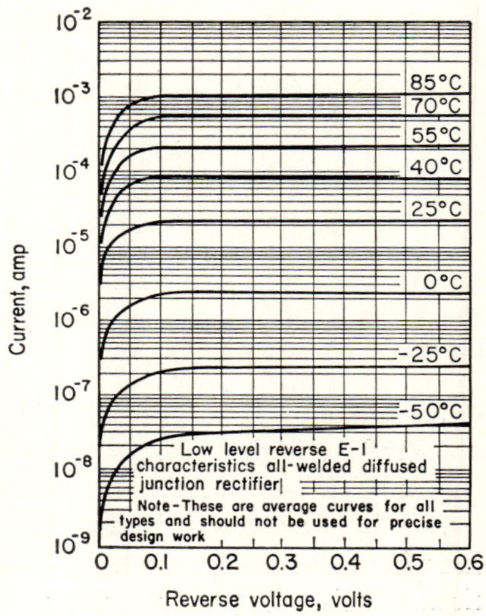

FIG. 6-5. Typical reverse-current temperature dependence for the IN93 germanium rectifier. (*Courtesy of General Electric Co.*)

reverse current increases an order of magnitude over a temperature change of about 26 to 32°C. Actual measurements of commercial germanium rectifiers, such as the curves of Fig. 6-5 for the GE IN93, correlate reasonably well with the theory. The reverse currents are measured at low voltages close to the saturation value so as to eliminate the effects of surface leakage. Generally, at ordinary temperatures the bulk current may be swamped by the surface component, which may tend to be somewhat independent of temperature. However, at the higher temperatures, the bulk diffusion current becomes very large, making the leakage current negligible by comparison. Thus, the true temperature dependence may be obtained by extrapolating down from high-temperature measurements.

For silicon, the mobility variation behaves approximately as $T^{-2.6}$ and Eq. (6-7) must be modified to

$$I_R \propto T^{1.4}\epsilon^{-E_{GO}/kT} \qquad \text{silicon} \tag{6-8}$$

However, for silicon p-n junctions at room temperature and higher, this current is so extremely small that it does not account for the observed reverse currents. In silicon, another mechanism, to be described in the next section, predominates.

6-4. Depletion-layer Charge Generation.[2] We recall from the explanation of the Shockley-Read-Hall recombination theory discussed in Chap. 4 that the lifetime for a semiconductor is determined from consideration of the net rate of recombination of the carriers, in other words, the net rate of recombination and the total emission or generation. In the appendix this is given as

$$R = \frac{np - n_i^2}{\tau_{pr}(n + n_r) + \tau_{nr}(p + p_r)} \tag{6-9}$$

where n and p correspond to the total free carriers. In Eq. (6-9), $np > n_i^2$, yielding a net rate of recombination. However, when a p-n junction is reverse-biased we know that a depletion layer is formed, representing a depletion of majority carriers in the region just at the junction. Also, the thickness of the layer increases as the square root to cube root of the reverse voltage. We would expect, therefore, that in the barrier region the net rate of recombination is negative, corresponding to a net rate of carrier emission or *generation*. If we transpose the numerator of Eq. (6-9) to account for the reduction of n and p in the barrier region, it becomes

$$e = \frac{n_i^2 - np}{\tau_{pr}(n + n_r) + \tau_{nr}(p + p_r)} \tag{6-10}$$

where e is the net rate of generation of carriers. It must be noted that although all the majority carriers are exhausted from the depletion region, n and p in (6-10) are those minority carriers that diffuse into the depletion region from the adjoining p and n regions under conditions of applied reverse bias. In essence, this is the diffusion current considered earlier, which flows through the depletion layer.

Particularly for silicon, where n and p are very small at ordinary temperatures, Eq. (6-10) may be approximated as

$$e \approx \frac{n_i^2}{\tau_{pr}n_r + \tau_{nr}p_r} \tag{6-11}$$

To recapitulate for a moment, τ_{pr} and τ_{nr} are the limiting lifetimes determined by the density of impurity recombination centers and their capture probabilities. Also, n_r and p_r are defined as the equilibrium concentrations that would exist if the Fermi level coincided with the energy of the centers. To obtain a qualitative indication of the significance of (6-11), we may assume, for the sake of illustration, that the recombination centers lie at the middle of the band gap. Thus, $n_r = p_r = n_i$ and Eq. (6-11) becomes

$$e \approx \frac{n_i}{\tau_{pr} + \tau_{nr}} \qquad \text{for } E_R \approx \frac{E_G}{2} \tag{6-12}$$

If the thickness of the depletion layer is x_m, the junction area is A, and the electron charge is q, Pell has shown that the reverse current due to charge generation in the depletion layer is[2]

$$I_G = Kqex_mA \tag{6-13}$$

where K is a proportionality constant which approaches unity for reverse voltages greater than 1 volt. Because x_m varies as the square root to cube root of voltage, the charge generation current changes accordingly.

From the foregoing we can now conclude that the total reverse current for a p-n junction is the sum of three components, namely

$$I_R = I_D + I_G + I_S$$

where I_D = bulk diffusion current
I_G = charge generation current
I_S = surface leakage

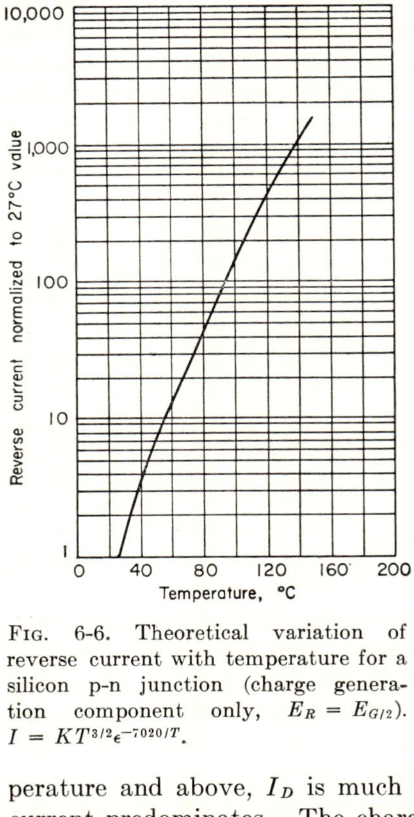

FIG. 6-6. Theoretical variation of reverse current with temperature for a silicon p-n junction (charge generation component only, $E_R = E_{G/2}$). $I = KT^{3/2}\epsilon^{-7020/T}$.

Neglecting I_S, we can make a comparison of germanium with silicon. For germanium, experimental measurements reveal that at room temperature and above, I_D is much greater than I_G, so that the diffusion current predominates. The charge generation component first becomes significant at low temperatures on the order of 200°K or about −73°C. On the other hand, for silicon at room temperature and above, I_D is so small that the generation current I_G predominates. In fact, the diffusion current does not become appreciable until a temperature of about 175°C is reached. The denominator of the generation expression, Eq. (6-12), would indicate that e may be large for silicon, since the sum of the limiting lifetimes $(\tau_{pr} + \tau_{nr})$ may be as low as 0.1 to 1 μsec for most crystals. In germanium, however, $(\tau_{pr} + \tau_{nr})$ might be of the order of 100 μsec. Therefore, using Eq. (6-12) in (6-13), we see that, for silicon, the temperature variation of the reverse current is dependent primarily on n_i, which is given by Fig. 3-10. A normalized plot of $I_R \approx I_G$ as a function of temperature, for silicon, is shown in Fig. 6-6.

6-5. Reverse-voltage Avalanche Breakdown. Thus far, in the discussion of the reverse characteristic, no mention has been made of any upper limits that may be imposed on the maximum applied voltage. A plot of current as a function of voltage for the p-n junction equation developed herein would indicate complete saturation with voltage ad infinitum. In actual semiconductor junctions, however, this is not the case. It is observed that when the reverse voltage reaches some critical value, the reverse current increases very rapidly to extremely large values, thereby deviating strongly from the saturation condition. This condition is referred to as *reverse-voltage breakdown*. A typical example of this is shown in Fig. 6-7. The breakdown voltage V_B is defined as that voltage at which the current increases at almost an infinite rate. However, in the small range just before V_B is reached, it is seen in the figure that the current commences to increase gradually. This is called the *multiplication range*. Thus we see that the effect of voltage breakdown is to set an upper limit for the maximum reverse voltage that may be applied to a p-n junction. It should be noted that in the breakdown range the currents may be quite large, since the junction resistance under these conditions is somewhat equivalent to a forward-biased junction.

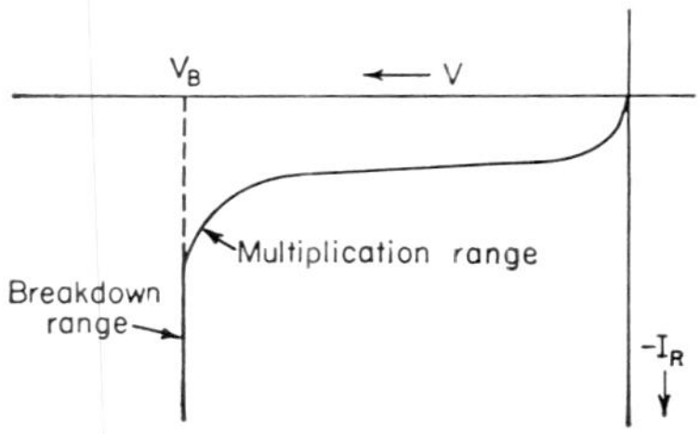

FIG. 6-7. Reverse-voltage breakdown.

McKay and McAfee offered the first explanation of reverse-voltage breakdown phenomena.[3] In their theory, they postulated that the breakdown process in a p-n junction was very much like the Townsend avalanche breakdown observed in gases. In a gas, it is known that an electron, in the presence of a strong electric field, can acquire sufficient energy to ionize a gas atom upon collision. The liberated electron in turn will ionize other atoms, and so on, resulting in a rapid multiplicative process leading to a complete avalanche breakdown of the gas. By analogy, the same avalanche process can occur in the space-charge region of a reverse-biased junction. As we have seen, a strong electric field exists in the space-charge region, since the applied voltage is essentially dropped across the thickness x_m. The field becomes stronger with increasing voltage, until a point is reached for which the electrons and holes (which comprise the reverse current flowing through the space-charge region) acquire sufficient energy to break additional valence bonds upon collision. This results in further generation of electron-hole pairs, causing the reverse current to multiply. If the voltage is increased further to the threshold value V_B, the process becomes so cumulative that an avalanche occurs and the junction "breaks down" completely. It is important to note here that breakdown is not a permanent effect and that the junction will recover when the voltage is reduced again. The collision of a carrier with a valence-bond electron is equivalent to ionization, since an electron is freed, leaving a hole behind.

The avalanche process in a step p-n junction may be analyzed by considering Fig. 6-8. It is assumed here that the n-region resistivity is much higher than the p-region resistivity, so that the depletion layer spreads mostly into the n region. Also, the reverse current consists mainly of minority holes p_o that diffuse from the n region. If current multiplication occurs in the space-charge region such that at any point x the total number of holes flowing out is p, McKay has shown[4] that the number of holes produced in an incremental length dx is

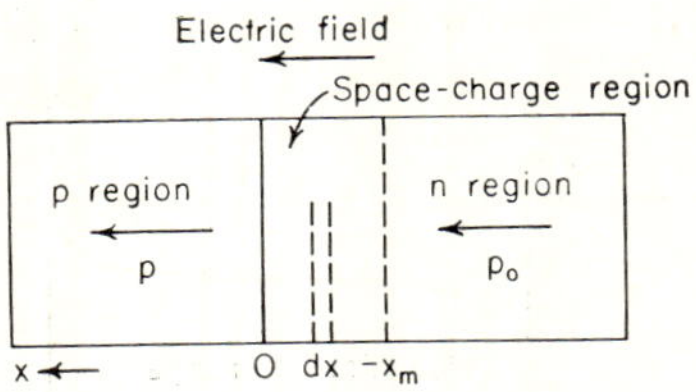

FIG. 6-8. Geometry for calculating avalanche breakdown. (*After K. G. McKay and K. B. McAfee.*)

$$dp_1 = p\alpha_i\,dx \tag{6-14}$$

In Eq. (6-14), α_i is the ionization rate for holes and dp_1 is the number of holes produced. To obtain the total number of holes produced, we integrate Eq. (6-14) over the distance x_m, or

$$\int_0^{p-p_0} dp_1 = \int_{-x_m}^{0} p\alpha_i\,dx \tag{6-15}$$

Note that at $x = -x_m$, the number of holes produced is zero, whereas at $x = 0$, the number of holes produced is equal to $(p - p_o)$. Completing the integration and simplifying, we have

$$1 - \frac{p_o}{p} = \int_0^{x_m} \alpha_i\,dx \tag{6-16}$$

If we define the multiplication factor $M = p/p_o$, we get, finally,

$$1 - \frac{1}{M} = \int_0^{x_m} \alpha_i\,dx \tag{6-17}$$

Examination of this result indicates that breakdown occurs when M approaches ∞ or when the integral equals unity. Equation (6-17) will also apply to electron multiplication if we assume that the ionization rate is the same for both electrons and holes. The ionization rate α_i is not a constant but a function of the electric field $\mathcal{E}$. Experimental plots of

$\alpha_i(\mathcal{E})$ for both silicon[4] and germanium[5] are given in Fig. 6-9. The curve for germanium is an interpolation which holds approximately for either holes or electrons.

In order to evaluate Eq. (6-17) to determine the avalanche-breakdown voltage, it is necessary to know the manner in which the electric field varies over the distance x_m. Equations (5-69) and (5-71) show that, for a step p-n junction, the electric field across the depletion layer is

$$\mathcal{E} = \frac{dV_n}{dx} = \frac{qN_D}{\kappa\epsilon_o}(x_m - x) \quad (6\text{-}18)$$

where x_n is replaced by x_m, since it is assumed that the depletion layer extends mostly into the n region. Eq. (6-18) is maximum for $x = 0$, which means that the electric field is maximum at the junction.

$$\mathcal{E}_M = \frac{qN_D}{\kappa\epsilon_o}x_m \quad (6\text{-}19)$$

Substituting this result into (6-18), we get the desired field function for the p-n junction.

$$\mathcal{E} = \mathcal{E}_M\left(1 - \frac{x}{x_m}\right) \quad (6\text{-}20)$$

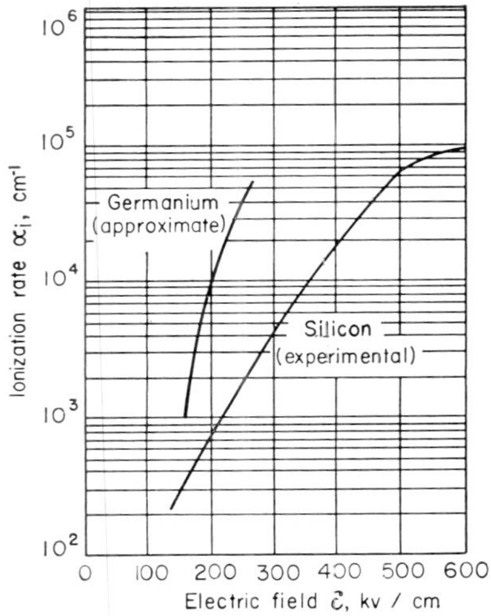

FIG. 6-9. Ionization rates for electrons and holes in reverse-biased semiconductor junctions.

Thus we see that the electric field is maximum at $x = 0$ and decreases linearly to zero at $x = x_m$, as shown in Fig. 5-11*b*. Differentiating (6-20) in order to obtain new integration limits for (6-17), we have $d\mathcal{E} = (\mathcal{E}_M/x_m)\,dx$, or from (6-19),

$$dx = \frac{\kappa\epsilon_o}{qN_D}\,d\mathcal{E} \quad (6\text{-}21)$$

Inserting this into Eq. (6-17), we get, finally,

$$1 - \frac{1}{M} = \frac{\kappa\epsilon_o}{qN_D}\int_0^{\mathcal{E}_M}\alpha_i(\mathcal{E})\,d\mathcal{E} \quad (6\text{-}22)$$

Thus the avalanche-breakdown voltage is reached at that value of $\mathcal{E}_M$ at which the right-hand side of Eq. (6-22) equals unity or $M = \infty$. The actual value of V_B is obtained from the $\mathcal{E}_M$ expression which is given by (6-19). If we substitute the relation for x_m as a function of voltage given by Eq. (5-77) and solve for V_B, we have

$$V_B = \frac{\kappa\epsilon_o}{2qN_D}\mathcal{E}_{MB}{}^2 \quad (6\text{-}23)$$

where $\mathcal{E}_{MB}$ is the critical electric field at breakdown. Since V_B is inversely proportional to the impurity concentration, it increases with increasing resistivity. The dependence of avalanche breakdown upon resistivity is an important transistor design relationship. Using Eqs. (6-22) and (6-23) in conjunction with the ionization rates of Fig. 6-9 and integrating graphically, one would obtain a curve of V_B as a function of impurity concentration. This is shown in Fig. 6-10 for both germanium[5] and silicon. The silicon curve was calculated from McKay's data,[4] using the resistivity data of Fig. 4-10. If the resistivity of the high-resistivity side of a p-n junction is known, the avalanche-breakdown voltage is determined by

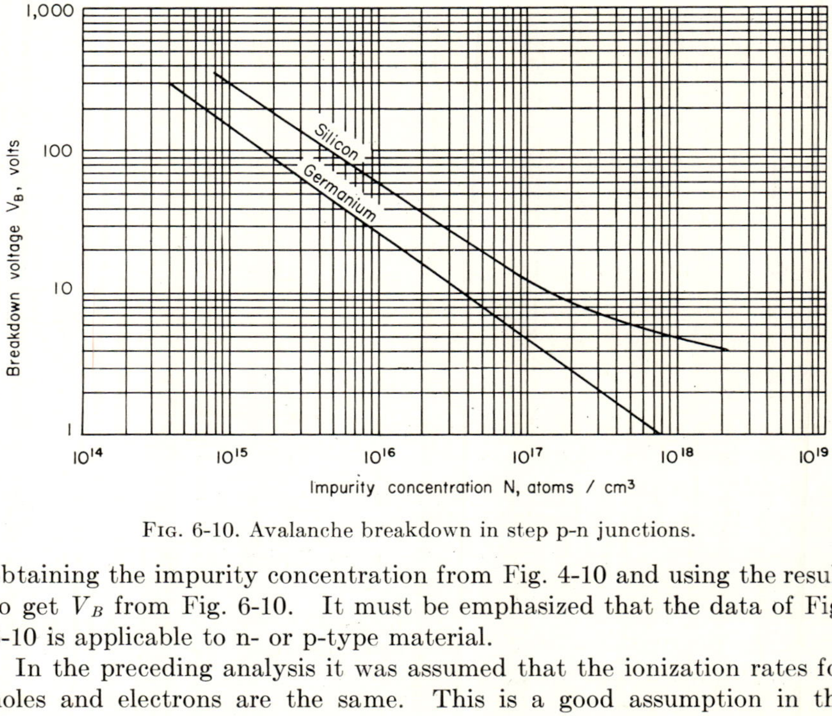

Fig. 6-10. Avalanche breakdown in step p-n junctions.

obtaining the impurity concentration from Fig. 4-10 and using the result to get V_B from Fig. 6-10. It must be emphasized that the data of Fig. 6-10 is applicable to n- or p-type material.

In the preceding analysis it was assumed that the ionization rates for holes and electrons are the same. This is a good assumption in the region close to breakdown. However, Miller has pointed out that in the multiplication range for a reverse-biased junction, the values of M will differ depending on whether the high-resistivity side is n or p type.[5,6] This indicates that the ionization rates are not quite equal. From measurements of M at voltages approaching breakdown, Miller has given the following empirical relation:

$$M = \frac{1}{1 - (V/V_B)^n} \tag{6-24}$$

where n has the values of Table 6-1. These empirical relationships are presented in Fig. 6-11, where M is plotted as a function of the parameter V/V_B. Of course when $V = V_B$, $M = \infty$. The increase of current in the multiplication range will be useful in understanding the current-gain characteristics of junction transistors.

TABLE 6-1. VALUES OF n FOR EQUATION (6-24)

Semiconductor	n type	p type
Germanium	3	6
Silicon	4	2

6-6. Avalanche Breakdown in Graded Junctions. From the avalanche-breakdown theory presented in the previous section for the step junction, we saw that voltage breakdown was due to a carrier ionization mechanism, governed by the relation

$$1 - \frac{1}{M} = \int_0^W \alpha_i(\mathcal{E})\, dx \qquad (6\text{-}25)$$

where $\alpha_i(\mathcal{E})$ is the carrier ionization rate as a function of electric field in the junction and M is the current multiplication factor. Breakdown occurs when $M = \infty$ or when the value of the integral equals unity.

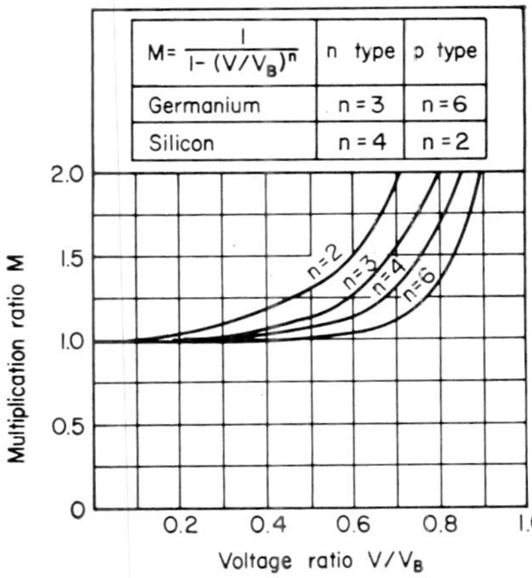

FIG. 6-11. Avalanche multiplication in p-n junctions.

For a graded p-n junction, which is usually formed by a diffusion process, the step-junction results are not applicable. This is attributed to the fact that because of the graded error-function or gaussian impurity distribution, the junction can withstand a higher voltage before $\mathcal{E}_{MB}$ is reached. In other words, for a given applied voltage, the depletion-layer thickness of a graded junction would be greater than that of a step junction. A representative profile of a graded junction was given in Fig. 5-13. The avalanche breakdown for such a structure may be obtained by utilizing Poisson's equation for the error-function or gaussian charge distribution and obtaining the appropriate function for the electric field, then using that in conjunction with the carrier ionization rates and Eq. (6-25) to obtain the desired result.[7] This is a rather complex approach and may be considerably simplified by using the linear-grade approximation for the diffused junction. In this case, the grade constant a for the junction is equal to the slope of the impurity gradient at $x = x_j$ (see Fig. 5-13). Further, to express a in terms of the surface concentration N_o, the background concentration N_{BC}, and the junction depth x_j, we can use an exponential impurity distribution as our model in order to simplify the analysis. Differentiating the exponential function and evaluating the slope at $x = x_j$, we have

$$a = \frac{N_{BC}}{x_j} \ln \frac{N_o}{N_{BC}} \tag{6-26}$$

Equation (6-26) is an approximate expression for the grade constant of a diffused junction. It remains now to determine how the avalanche breakdown varies with the grade constant a, using a linear model as the basis for analysis.

To evaluate Eq. (6-25) for a linear-grade junction, it is necessary to express the limits of integration in terms of the electric field $\mathcal{E}$. Modifying Eqs. (5-91) and (5-92), we have

$$\mathcal{E} = \mathcal{E}_M \left[1 - \left(\frac{2x}{x_m} \right)^2 \right] \tag{6-27}$$

where

$$\mathcal{E}_M = \frac{qax_m{}^2}{8\kappa\epsilon_o} \tag{6-28}$$

This is a parabolic function as was shown in Fig. 5-14*b*. Equation (6-27) may be differentiated with respect to x and solved for dx. The expression for x is obtained by solving (6-27) directly. Finally, the new integration limits are obtained by evaluating (6-27) for the electric field at $x = 0$ and $x = x_m/2$. Substituting these results into (6-25), we have, for the linear grade,

$$1 - \frac{1}{M} = \int_0^{\mathcal{E}_M} \frac{x_m \alpha_i(\mathcal{E})}{2\mathcal{E}_M{}^{1/2}(\mathcal{E}_M - \mathcal{E})^{1/2}} \, d\mathcal{E} \tag{6-29}$$

From (6-28), we can substitute the terms for $x_m/\mathcal{E}_M{}^{1/2}$ into (6-29) and we have, finally,

$$1 - \frac{1}{M} = \frac{1}{2} \left(\frac{8\kappa\epsilon_o}{qa} \right)^{1/2} \int_0^{\mathcal{E}_M} \frac{\alpha_i(\mathcal{E})}{(\mathcal{E}_M - \mathcal{E})^{1/2}} \, d\mathcal{E} \tag{6-30}$$

where a is the junction grade constant.

Since breakdown occurs when the right-hand side of (6-30) equals unity, it is necessary to evaluate the integral term. From Fig. 6-9, which is a plot of $\alpha_i(\mathcal{E})$ for both germanium and silicon, we can fit the following approximate empirical relations:

$$\alpha_i \approx 6.25 \times 10^{-34} \mathcal{E}^7 \qquad \text{germanium} \tag{6-31}$$

$$\alpha_i \approx 1.65 \times 10^{-24} \mathcal{E}^5 \qquad \text{silicon} \tag{6-32}$$

These relations presume that the ionization rates for holes and electrons are equal, which is a fair assumption. Inserting (6-31) and (6-32) into

(6-30) and performing the integration, we get

$$1 - \frac{1}{M} = \frac{1}{2}\left(\frac{8\kappa\epsilon_o}{qa}\right)^{1/2} (6.25 \times 10^{-34})(0.638\mathcal{E}_M{}^{7.5}) \qquad \text{germanium} \qquad (6\text{-}33)$$

$$1 - \frac{1}{M} = \frac{1}{2}\left(\frac{8\kappa\epsilon_o}{qa}\right)^{1/2} (1.65 \times 10^{-24})(0.739\mathcal{E}_M{}^{5.5}) \qquad \text{silicon} \qquad (6\text{-}34)$$

If these equations are set equal to unity, $\mathcal{E}_M$ becomes the maximum electric field in the junction at the breakdown voltage V_B. $\mathcal{E}_{MB}$ may be expressed

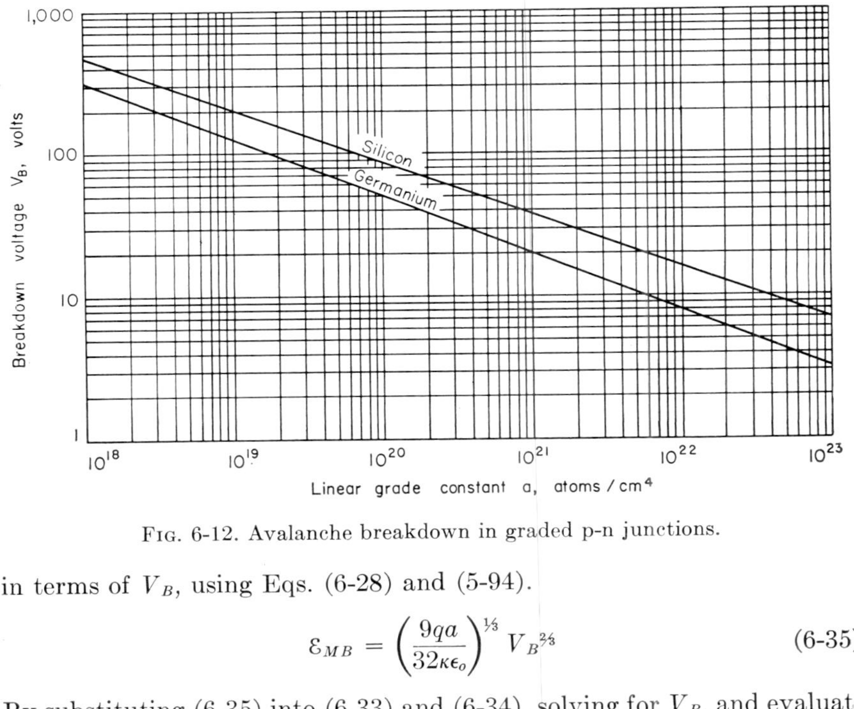

FIG. 6-12. Avalanche breakdown in graded p-n junctions.

in terms of V_B, using Eqs. (6-28) and (5-94).

$$\mathcal{E}_{MB} = \left(\frac{9qa}{32\kappa\epsilon_o}\right)^{1/3} V_B{}^{2/3} \qquad (6\text{-}35)$$

By substituting (6-35) into (6-33) and (6-34), solving for V_B, and evaluating all the constants for both germanium and silicon, we obtain the desired result, namely,

$$V_B = 5.05 \times 10^9 a^{-0.400} \qquad \text{germanium} \qquad (6\text{-}36)$$

$$V_B = 1.71 \times 10^9 a^{-0.364} \qquad \text{silicon} \qquad (6\text{-}37)$$

where V_B is the avalanche-breakdown voltage in volts and a is the grade constant in atoms per centimeter[4]. These results are plotted in Fig. 6-12. It is seen that as a becomes larger corresponding to a steeper grade, the breakdown voltage decreases. In summary, Fig. 6-12 is used in conjunction with Eq. (6-26) to determine the avalanche-breakdown voltage for a graded junction.

6-7. Forward-current Characteristics. We may now focus our attention on the forward-current characteristic of the p-n junction. In Eq. (6-1), when the impressed voltage V is positive and greater than 0.1 volt, the exponential term becomes large compared to unity, and the forward current I_F is expressed as

$$I_F = \frac{qAD_p p_n}{W} \epsilon^{qV/kT} \tag{6-38}$$

or

$$I_F = I_R \epsilon^{qV/kT} \tag{6-39}$$

It is evident that the exponential factor increases very rapidly for small values of V, since kT/q equals 0.026 volts at room temperature, 300°K. Consequently, large currents will flow for relatively small values of voltage, corresponding to a very low junction resistance in the forward direction. A plot of $\epsilon^{qV/kT}$ as a function of V is given in Fig. 6-13 to illustrate the degree of variation. For example, if I_R is approximately 1 μa and $V = 0.3$ volt, the forward current is about 100 ma, corresponding to a resistance of 3 ohms. Because of the exponential dependence, the forward resistance will decrease quite rapidly with applied voltage.

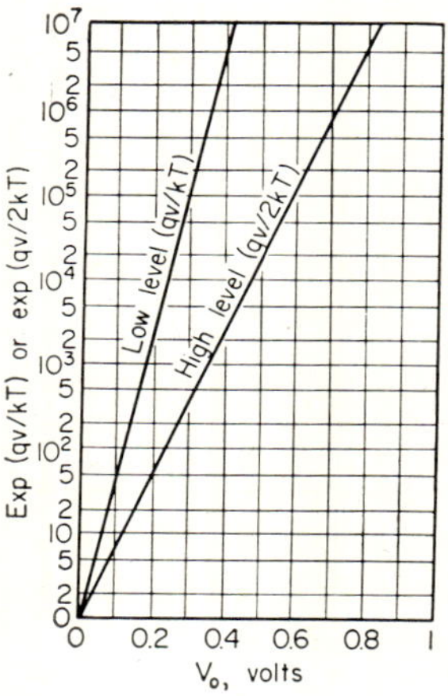

FIG. 6-13. Low-level and high-level injection factors for junction rectifiers at 300°K. (*After J. S. Saby.*)

A comparison of germanium with silicon will best illustrate the voltage-current characteristic of the forward-biased junction. In Eq. (6-38), if we make the substitution $p_n = n_i^2/N_D$, then

$$I_F = \frac{qAD_p}{WN_D} n_i^2 \epsilon^{qV/kT} \tag{6-40}$$

In Chap. 3 we saw that n_i^2 is proportional to $\epsilon^{-E_G/kT}$, where E_G is the band gap. If we assume that D_p is the same for both germanium and silicon and that the other parameters are equal, then

$$I_F = K\epsilon^{-E_G/kT}\epsilon^{qV/kT} \tag{6-41}$$

Since E_G is the band-gap energy expressed in electron-volts, it may also be written as qV_G, where V_G is the band-gap potential in volts. Therefore the forward-current expression becomes

$$I_F = K\epsilon^{q(V-V_G)/kT} \tag{6-42}$$

At room temperature (300°K), V_G for germanium is 0.72 volt and for silicon V_G is 1.1 volts. These differences in band gap clearly point out the differences in the forward characteristics between germanium and silicon. Other things being the same, for equal forward currents the voltage across a silicon p-n junction will be about 0.38 volt greater than that for an equivalent germanium junction. If we take into consideration the fact that D_p is somewhat lower for silicon than for germanium, the voltage difference may be as much as 0.4 volt. Figure 6-14 shows a normalized plot of the forward-current characteristic for both germanium and silicon p-n junctions, illustrating the dependence on band-gap potential. This result is easy to understand when it is realized that for the same impurity concentration in the n region, the minority hole concentration is considerably smaller for silicon than for germanium. Therefore it takes a larger voltage for silicon to inject the same amount of current as germanium. For certain applications, this property of silicon has decided advantages, particularly in computer switching applications where it may be necessary to have one transistor turn off several other transistors.

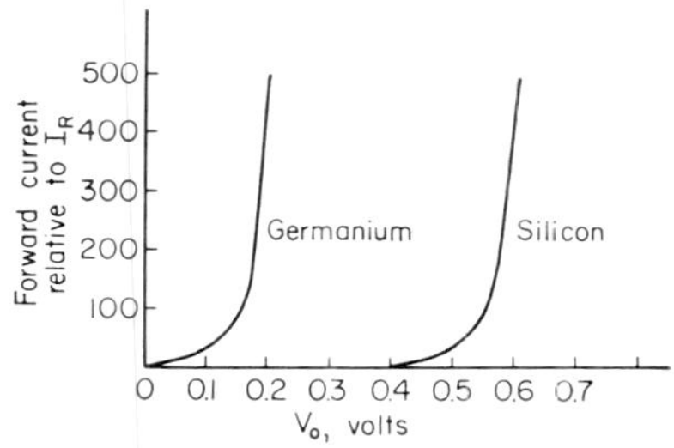

FIG. 6-14. Effect of band-gap potential on the forward-current characteristic.

The temperature dependence of the forward-current characteristic is also of interest here. We see from Eq. (6-7) that the reverse saturation current varied as $T^2\epsilon^{-E_G/kT}$, where I_R consisted mainly of the diffusion current for germanium. Since charge generation would not occur for conditions of forward bias, the temperature relationship given by Eq. (6-8) should hold for silicon. Substituting into Eq. (6-39), we obtain

$$I_F \propto T^{2.0}\epsilon^{q(V-V_G)/kT} \qquad \text{germanium} \tag{6-43}$$

$$I_F \propto T^{1.4}\epsilon^{q(V-V_G)/kT} \qquad \text{silicon} \tag{6-44}$$

where $E_G = qV_G$. The important conclusion to be noted here is that for any given value of forward voltage (less than the band-gap potential V_G), the forward current will increase with increasing temperature. In other words, the d-c resistance of a forward-biased p-n junction decreases with temperature.

6-8. High Forward-current Effects. As was indicated in Chap. 5, the analysis of the p-n junction is based on the assumption that the density of holes injected into the n region is small compared to the density of electrons (majority carriers) in that region. For this condition, there is a negligible disturbance of the charge neutrality in the n region, such that the holes move only by diffusion towards the ohmic contact. In other words, all of the applied voltage is dropped across the junction such that there are no voltage drops in the adjacent p or n regions. This means that the injected holes in the n region cannot move by drift effects, since the electric field is zero there. Therefore, in the forward direction, the p-n junction equation holds only for small forward currents, where the injected hole density $p_n\epsilon^{qV/kT}$ is small compared with the electron concentration $n \approx N_D$. At high forward currents this assumption is no longer valid and Eq. (6-1) is not applicable. As will be seen in this section, the correct expression is obtained if we consider that at high currents the voltage drop in the n region becomes comparable with the junction drop such that it reduces the effective applied voltage. This effect manifests itself in the form of a simple resistance that appears in series with the junction at high currents.

Let us now examine in detail exactly what happens when the injected hole concentration becomes large. Again we shall use the model where

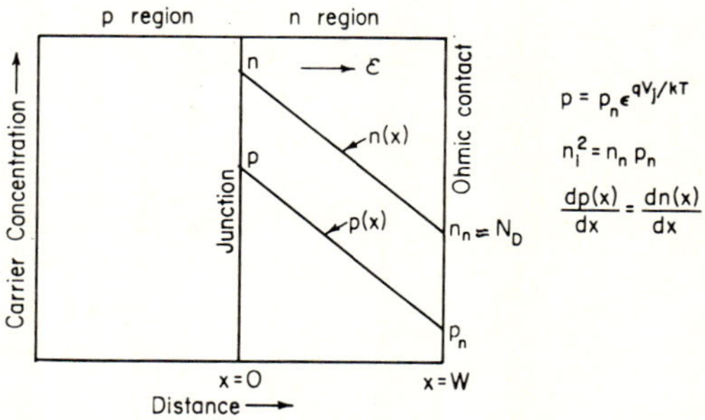

FIG. 6-15. High forward-current effects in a p-n junction.

$\rho_n \gg \rho_p$, making the electron current injected into the p region negligible by comparison. Also, we shall consider the case where the hole lifetime is sufficiently large to make L_p much larger than the n-region thickness W. These conditions are illustrated in Fig. 6-15. At the ohmic contact ($x = W$) the carrier concentrations are at their equilibrium values, since the recombination rate is presumed to be infinite. Thus, the hole concentration equals p_n and the majority-electron concentration equals the net donor density N_D. If the injected hole concentration at $x = 0$ is p, then the concentration as a function of distance is given by the line $p(x)$. Since we are neglecting recombination in the n region, $p(x)$ is a straight line, and diffusion of the holes is proportional to the slope. However, when p becomes appreciably large, there is no longer any charge neutrality in the vicinity of $x = 0$. In order that charge neutrality be maintained, the electron concentration in the n region must have the same concentration gradient as the holes. This is shown in Fig. 6-15 by $n(x)$, for which the slope is equal to that of $p(x)$. In addition to the hole diffusion current, the electron-concentration gradient will try to establish electron diffusion in the same direction. However, as soon as the electrons try to move away, charge neutrality is violated and an electric field $\mathcal{E}$ is created which acts to keep the electrons in place against their concentration gradient. Thus the net electron current is zero, since the electric field acts to generate an electron drift current which is equal and opposite to the electron diffusion current. The essential point, however, is that the electric field is also in a direction which aids the flow of holes to the ohmic contact. Since the field arises from a gradient similar to that of the holes, the net effect is to double the hole diffusion current. Finally, the existence of an electric field in the n region requires that there be a voltage drop also.

From Eqs. (5-31) and (5-32) we see that the total one-dimensional current-density flow in a semiconductor is given by

$$J_p = q\mu_p p\mathcal{E} - qD_p \frac{dp}{dx} \tag{6-45}$$

$$J_n = q\mu_n n\mathcal{E} + qD_n \frac{dn}{dx} \tag{6-46}$$

These are equations which represent the combined drift and diffusion currents in one dimension for holes and electrons, respectively. From the assumptions of the foregoing paragraph we can immediately write that

$$\frac{dp}{dx} = \frac{dn}{dx} \tag{6-47}$$

and that J_n, the electron current density, equals zero. Making these substitutions into Eq. (6-46) and solving for the field, we obtain

$$\mathcal{E} = \frac{-D_n}{\mu_n} \frac{1}{n} \frac{dp}{dx} \tag{6-48}$$

From the Einstein relationship $D = \frac{kT}{q} \mu$, (6-48) becomes

$$\mathcal{E} = \frac{-kT}{q} \frac{1}{n} \frac{dp}{dx} \tag{6-49}$$

If we insert this result into Eq. (6-45), the hole current density becomes

$$J_p = q\mu_p p \frac{-kT}{q} \frac{1}{n} \frac{dp}{dx} - qD_p \frac{dp}{dx} \tag{6-50}$$

Since $(kT/q)\mu_p$ equals D_p, (6-50) becomes

$$J_p = -qD_p \frac{dp}{dx} \left(1 + \frac{p}{n}\right) \tag{6-51}$$

In the limit, at high forward currents, $p \approx n$, and

$$J_p = -q2D_p \frac{dp}{dx} \tag{6-52}$$

This is the high forward hole current density, which is seen to be twice the low-level value. In other words, we may say that for high-level injection in a p-n junction, the diffusion constant doubles in the limit. To obtain the actual current I_p, the derivative in (6-52) must be evaluated. From Fig. 6-15, if the injected concentration p is $p_n\epsilon^{qV_j/kT}$ at $x = 0$ and then drops linearly to p_n at $x = W$, the slope becomes

$$\frac{dp}{dx} = \frac{-p_n}{W}(\epsilon^{qV_j/kT} - 1) \tag{6-53}$$

Insertion of this result into (6-52) and multiplication by the junction area A give the current expression.

$$I_p = \frac{qA2D_pp_n}{W}(\epsilon^{qV_j/kT} - 1) \tag{6-54}$$

It should be noted that in Eq. (6-54), V_j is not the total applied voltage V_o, but only the voltage drop across the junction. Therefore, Eq. (6-54) is not the true high-level V-I equation, as we have not as yet taken into account the series potential drop V_s in the n region.[8] Considering that

$$V_o = V_j + V_s \tag{6-55}$$

we must now evaluate V_s. At high levels, since the hole and electron concentrations and the gradients are equal, we can write

$$\mathcal{E} = \frac{-kT}{q}\frac{1}{n}\frac{dn}{dx} \tag{6-56}$$

Therefore

$$V_s = \int_0^W \mathcal{E}\,dx = \frac{-kT}{q}\int_n^{n_n}\frac{dn}{n} \tag{6-57}$$

or

$$V_s = \frac{kT}{q}\ln\frac{n}{n_n} \tag{6-58}$$

Since $n \approx p$ and $n_n = n_i^2/p_n$, (6-58) becomes

$$V_s = \frac{kT}{q}\ln\frac{pp_n}{n_i^2} \tag{6-59}$$

Also, $p = p_n\epsilon^{qV_j/kT}$; taking the natural logarithm, we have

$$V_s = \frac{kT}{q}\left[\ln\left(\frac{p_n}{n_i}\right)^2 + \frac{qV_j}{kT}\right] \tag{6-60}$$

$$V_s = \frac{kT}{q}\ln\left(\frac{p_n}{n_i}\right)^2 + V_j \tag{6-61}$$

Letting $V_s = V_o - V_j$ and rearranging terms, we have, finally,

$$V_j = \frac{V_o}{2} - \frac{kT}{q}\ln\frac{p_n}{n_i} \tag{6-62}$$

This result gives the junction voltage in terms of the applied voltage V_o

and the minority-carrier concentration p_n in the n region. If (6-62) is substituted for V_j in Eq. (6-54), we have

$$I_p = \frac{qA2D_p p_n}{W} (\epsilon^{qV_o/2kT - \ln(p_n/n_i)} - 1) \tag{6-63}$$

or more directly

$$I_F = \frac{qA2D_p n_i}{W} \epsilon^{qV_o/2kT} \tag{6-64}$$

The -1 term was dropped because it is negligible compared to the exponential for forward-bias conditions.

Thus, Eq. (6-64) describes the V-I characteristic of a p-n junction for the high-level case. Comparison of this result with the low-level case given by Eq. (6-38) shows several differences. Firstly, the current becomes proportional to twice the diffusion constant, as stated earlier. Secondly, the high-level current becomes independent of resistivity and simply a function of n_i, which is constant for a particular semiconductor. Thirdly and most important, the magnitude of the exponent is reduced by a factor of 2. On a semilog plot of current with forward voltage, we would therefore expect a two-to-one change in slope. This is shown in Fig. 6-16, which represents a typical over-all forward characteristic. At low current levels, the slope is q/kT as given by Eq. (6-38). As the current increases, the slope goes through a gradual transition range. This is the range where the injected hole concentration is becoming comparable to the majority-carrier concentration n_n. It should be apparent that the transition range would be dependent upon the resistivity of the n region. Since resistivity decreases as n_n increases, we would expect the transition range to occur at higher voltages as resistivity decreases. Finally, in the high-level range the slope becomes $q/2kT$ as indicated by Eq. (6-64). The magnitude of the parameter $\epsilon^{qV/2kT}$ is also plotted in Fig. 6-13. Experimental measurements of actual germanium and silicon rectifiers have shown excellent agreement with the high-level theory presented here.[8,9]

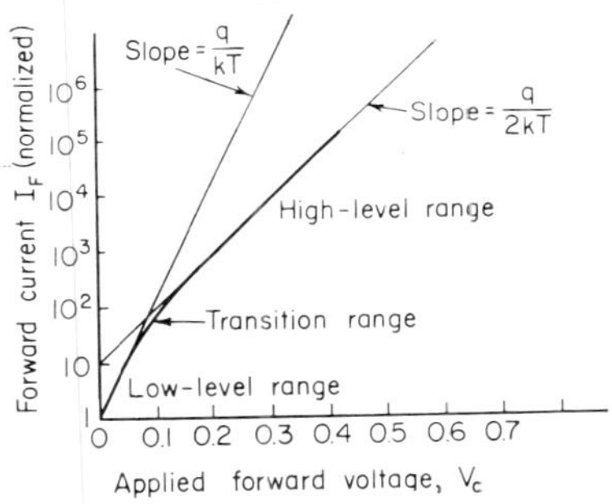

FIG. 6-16. Effects of high forward currents on V-I characteristics.

6-9. Minority-carrier Storage in p-n Junctions. Solutions for forward and reverse V-I characteristics of the p-n junction that have been obtained so far are for steady-state d-c conditions and therefore do not describe the transient behavior of the junction in response to pulses. The pulse response of the junction is a very important consideration if we are to predict the behavior of junction transistors used in switching circuits. The speed at which the junction responds to an applied signal becomes a critical factor if it is required that the output signal be an almost exact replica of the input. We shall examine here qualitatively some of the factors that limit the pulse response of p-n junctions.

Consider, for example, the p-n junction employed in the single square-wave circuit of Fig. 6-17, where the input signal is a true square wave of voltage. If the junction characteristics are completely resistive, we have

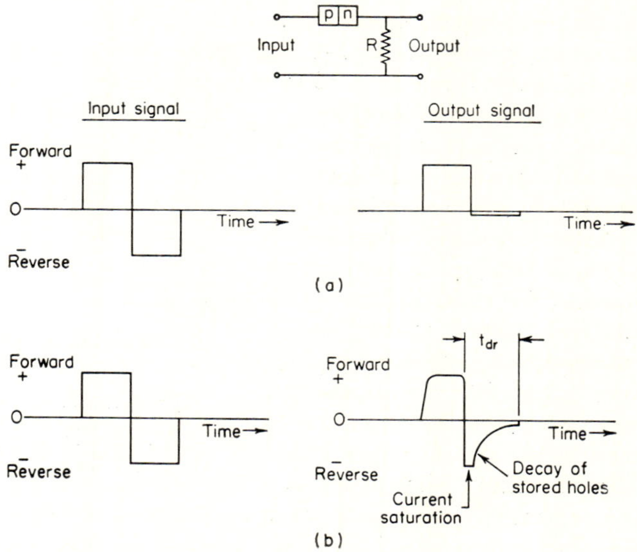

FIG. 6-17. Pulse response of p-n junctions illustrating carrier-storage effects. (*a*) Ideal response; (*b*) actual response.

an ideal response as shown by the output signal in Fig. 6-17*a*. During the positive portion of the input signal the function is forward-biased. Since the voltage drop across the junction is small, almost all the applied voltage appears across the output resistor R. When the input signal switches negative, the junction is reverse-biased. During this time, the reverse saturation current flowing around the circuit is very small, and a slightly negative voltage drop appears across R. This demonstrates how the p-n junction operates as a half-wave rectifier. If the input signal is sinusoidal, the output is a half sine wave. The pulse response in Fig. 6-17*a* is shown to be ideal, since the output waveform rises and falls at an infinite rate.

In an actual p-n junction the output response under the same conditions is far from ideal, as shown in Fig. 6-17*b*. The observed deviations from the ideal response are not surprising when we consider some of the capacitive-reactive effects that must be taken into account. For instance, there exists the junction-barrier capacitance C_T, which is voltage dependent and therefore must be charged and discharged in accordance with the applied voltage signal. For the high-frequency components contained in the square-wave signal, the junction capacitance presents a bypassing effect. Consequently, the leading and trailing edges of the output pulse would appear with a slight slope and somewhat rounded off at the corners, as shown. Therefore, the effect of the capacitance limits the speed of response of the junction to fast signals. Transient response, in terms of rise and fall times, will be studied quantitatively in Chap. 16, in reference to junction-transistor switches.

The discussion of the preceding paragraph was concerned primarily with the forward-conducting portion of the output signal. As shown in Fig. 6-17*b*, the reverse portion of the cycle deviates markedly from the ideal response of (*a*). This observed increase in reverse current is explained by the phenomena of *minority-carrier storage* in the n region. Consider again the model of the thickness-limited junction, i.e., $L_p \gg W$. In the forward direction, a large number of holes are injected into the n region and are stored there in the form of a concentration gradient (Fig. 6-18*a*). This charge distribution is maintained as long as the junction is conducting. If the applied voltage is suddenly switched to a reverse bias, as in the case of the input square wave, the hole (or charge) concentration at $x = 0$ must drop to zero, since the reverse-biased gradient appears as in Fig. 6-18*b*. This means that the total charge represented by the shaded area in Fig. 6-18 must be swept out of the n region. It is apparent that this cannot occur instantly, and therefore, at the first instant after switching, we would expect a large reverse current to flow. Thus, as the current flows, the stored charge is reduced gradually, as shown by the dashed lines in Fig. 6-18, until the equilibrium point is reached, corresponding to the normal reverse saturation current. It is this minority-carrier storage effect that accounts for the reverse-voltage tail shown in Fig. 6-17*b*. The flat portion represents a saturation condition for which the initial reverse current is limited by the series load resistance *R*.

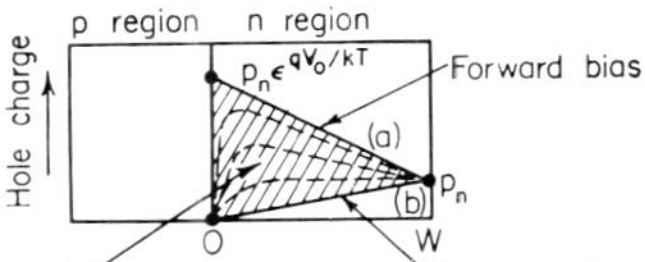

FIG. 6-18. Hole-storage effects during switching.

It becomes evident that the recovery time from carrier storage is greater for large forward currents before switching since there is a larger concentration of injected carriers. Also, the storage time may be reduced by decreasing the magnitude of the minority-carrier lifetime in the region where the charge is stored. The fact that very low lifetime is required for low storage times is understood when one realizes that the injected minority carriers are removed only by recombination with the majority carriers. Theoretical analysis and experimental data on the subject of minority-carrier storage[10] indicate that the recovery or storage time of a junction is related to the lifetime as follows:

$$t_{dr} = 0.9\tau \qquad \text{step junction} \tag{6-65}$$

$$t_{dr} = 0.5\tau \qquad \text{graded junction} \tag{6-66}$$

where τ is the minority-carrier lifetime. Very low recovery times can be obtained simply by adding large concentrations of deep-lying impurity recombination centers such as nickel or gold into the junction. In particular the latter is common practice in the fabrication of very-high-speed diodes for computer applications.

6-10. Commercial Junction Rectifiers. It should be clear now that the semiconductor p-n junction by itself is an excellent rectifying device. It has all the properties of an ideal rectifier, namely, a very low forward resistance approaching a short circuit and an extremely high reverse resistance approaching an open circuit. Because of these reasons, germanium and silicon p-n junctions are in widespread use throughout the semiconductor industry as the basic elements of commercial junction rectifiers. In the manufacturing process, it is usual practice to form the junction by either alloying or diffusion. Indium metal is used in alloyed germanium junctions, whereas aluminum is used with silicon. Rectifiers made with silicon offer the advantages of having very low reverse currents and of operating at high ambient temperatures. Germanium rectifiers, on the other hand, offer the advantage of higher efficiency, since the forward voltage drop is lower at high currents.

In Fig. 6-19 are shown some typical commercial junction rectifiers. The smaller unit is a germanium device capable of operating at 300 volts and 0.3 amp. The larger unit is a silicon rectifier designed for delivering 50 amp at 200 volts. The high current rating is obtained by utilizing a very large junction area. Because of the excessive power dissipation at this rating, a large copper stud is used as a heat sink for the junction. Removal of the heat from the junction is a major problem in rectifier design. Depending on the actual ratings desired, the package design may vary anywhere from small hermetically sealed housings to large cells mounted on radiating fins.

In addition to selection of the correct package for thermal purposes, rectifier design also includes the problem of establishing the necessary specifications for the n-type pellet. Firstly, the resistivity must be determined such that the avalanche-breakdown voltage is well above the maximum operating voltage that will be applied to the rectifier. Secondly, a compromise must be reached between junction area and pellet thickness such that both the forward voltage drop and the reverse saturation current are within the objective specifications. This also requires that the minority-carrier lifetime of the pellet be high. Thirdly, the temperature requirements of the final device will determine whether germanium or silicon is used. All of these design factors stem from the basic p-n junction equations that have been discussed in this chapter, for both the low-level and the high-level cases. By appropriately relating parameters such as A, W, and p_n, any number of design monographs may be prepared to enable one to select the correct pellet characteristics consistent with particular reverse-voltage and forward-current ratings.

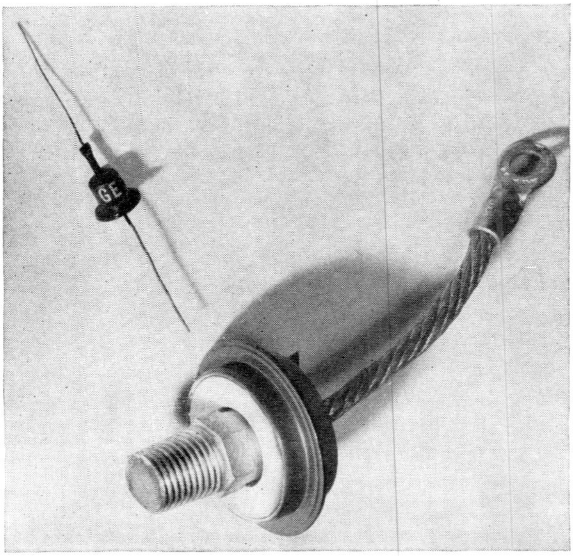

FIG. 6-19. Commercial junction rectifiers. The small unit is a germanium rectifier rated at 0.3 amp. The large unit is a silicon rectifier designed for 50 amp operation. (*Courtesy of General Electric Co.*)

It is not the intent of this text to cover the manifold ramifications of rectifier design, but many of the principles will be established in later chapters in reference to the design philosophy of junction transistors.

PROBLEMS

6-1. The reverse current at $V = -5$ volts is measured for both a germanium and a silicon p-n junction at room temperature. For germanium, the current is 1 μa and is largely a diffusion component. For silicon, the current is 1 mμa and is determined to be primarily due to charge generation. Assuming negligible surface leakage, what is the back resistance for each rectifier at 100°C and $V = -5$ volts?

6-2. An indium dot is alloyed into an n-type pellet of germanium, forming a recrystallized p region of 0.001 ohm-cm resistivity. The junction area is 200 mil^2, the effective n-region thickness is 1 mil, and the pellet resistivity is 2.5 ohm-cm. Recombination in the n region is negligible ($L_p \gg W$). Using the low-level theory determine the forward current at which the injected hole concentration equals the electron concentration in the pellet at room temperature.

6-3. A silicon n-p junction is prepared by growing a p-type crystal and abruptly changing the type by adding a large concentration of n-type impurities. The net impurity concentration in the n region is 10^{19} atoms/cm^3, whereas in the p region it is 5×10^{15} atoms/cm^3. The lifetime in the p region is 0.2 μsec and is presumed to be equal to the sum of the limiting lifetime due to recombination centers lying at the middle of the band gap. For a junction area of 225 mils^2, determine the following at room temperature:

a. The reverse current at 5 volts due to charge generation
b. The avalanche-breakdown voltage
c. The reverse voltage at which the reverse current is multiplied twice. Assume that M takes into account x_m variations.

6-4. Design a germanium rectifier capable of handling 5 amp forward current and 200 volts reverse voltage. At these ratings, the forward voltage drop shall not exceed 0.55 volt and the reverse current at -5 volts shall not exceed 4 μa at room temperature (neglect surface effects). Assume that, for good quality germanium, lifetime varies 50 μsec/ohm-cm and that the pellet thickness W shall be no greater than $0.1L_p$. For the design, specify the following n-type pellet characteristics:

a. Resistivity
b. Lifetime
c. Effective pellet thickness (mils)
d. Active junction area (mils^2)

Consider the p region to be very heavily doped with indium.

6-5. Determine the avalanche-breakdown voltage for a diffused silicon junction in which boron is diffused into a 2 ohm-cm n-type wafer to a junction depth of 3 μ. The boron surface concentration is 8×10^{17} atoms/cm^3.

REFERENCES

1. Kingston, R. H.: Review of Germanium Surface Phenomena, *J. Appl. Phys.*, vol. 27, pp. 101–114, February, 1956.
2. Pell, E. M.: Reverse Current and Carrier Lifetime as a Function of Temperature in Germanium Junction Diodes, *J. Appl. Phys.*, vol. 26, pp. 658–665, June, 1955.
3. McKay, K. G., and K. B. McAfee: Electron Multiplication in Silicon and Germanium, *Phys. Rev.*, vol. 91, pp. 1079–1084, September, 1953.
4. McKay, K. G.: Avalanche Breakdown in Silicon, *Phys. Rev.*, vol. 94, pp. 877–884, May, 1954.
5. Miller, S. L.: Avalanche Breakdown in Germanium, *Phys. Rev.*, vol. 99, pp. 1234–1241, August, 1955.
6. Miller, S. L.: Effects of Avalanche Multiplication in Silicon Transistors, presented at AIEE-IRE Semiconductor Device Research Conference, Purdue University, June, 1956.
7. Root, C. D., D. P. Lieb, and B. Jackson: Avalanche Breakdown Voltages of Diffused Silicon and Germanium Diodes, *IRE Trans.*, vol. ED-7, pp. 257–262, October, 1960.
8. Saby, J. S.: Junction Rectifier Theory, unpublished.
9. Hall, R. N.: Power Rectifiers and Transistors, *Proc. IRE*, vol. 40, pp. 1512–1518, November, 1952.
10. Bakanowski, A. E., and J. H. Forster: Electrical Properties of Gold-doped Diffused Silicon Computer Diodes, *Bell System Tech. J.*, vol. 39, pp. 87–104, January, 1960.

7

Introduction to Transistors

7-1. Introduction. In the forthcoming chapters, the various carrier mechanisms involved in the junction transistor will be analyzed quantitatively, beginning with the simplest model and systematically adding more effects until the complete theoretical model of the device is formulated. Design relationships will be established along the way. However, in order to present an effective guide through these theoretical developments, it would be advantageous to have a broad, general description of the transistor and its significant principles of operation and characteristics, and a familiarity with the new terminology. Therefore, this chapter will present a qualitative nonmathematical survey of the junction transistor as an introduction to the chapters to follow.

7-2. The Ideal Amplifier. In almost all electronic systems, active elements or devices are necessary to amplify small electrical signals to higher levels of power (or voltage). Obviously, such devices must be as efficient as possible, giving high amplification gains, minimum internal power loss, and faithful, undistorted reproductions of the applied input signals over the widest range of frequencies. Any device offering the highest degree of these performance characteristics would certainly be ideal.

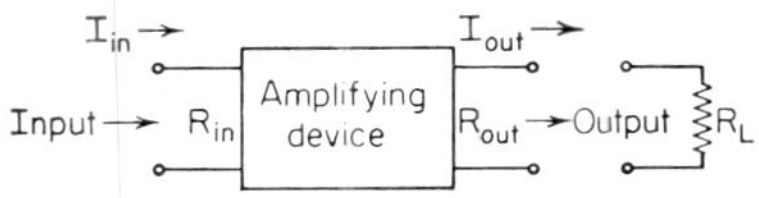

FIG. 7-1. Amplifier in a "black box."

For the sake of illustration, we shall suppose that an amplifying device of unknown structure is available and is completely contained within a "black box," such that only pairs of terminals are evident at the input and output, as in Fig. 7-1. If such an amplifier is an ideal one, we can make definite statements regarding what we can expect for the characteristics of the input and output terminals. Suppose that the input signal to be amplified is in the form of the current I_{IN}. In order for the amplifier to have high gain, the first requirement is that the necessary input power consumption be as small as possible. This would mean that the resistance R_{IN} which the current I_{IN} sees at the input terminals must be as small as possible, approaching zero for the ideal amplifier. At the output terminals, however, we would expect a different characteristic. It would be desirable that the current I_{IN} pass through the amplifier at least undiminished, so that it equals the output current I_{OUT}, as in Fig. 7-1. In order to get power gain, then, it is necessary that the output resistance R_{OUT} be as large as possible, approaching infinity for the ideal amplifier. In effect, this is merely stating that the output current of the amplifier flows out from a *constant current source* (very high resistance)

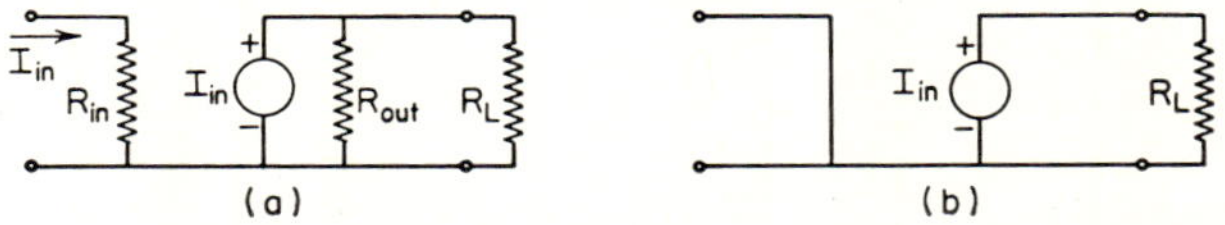

FIG. 7-2. Hypothetical amplifier equivalent circuits. (*a*) Practical amplifier; (*b*) ideal amplifier.

such that this current is independent of any load resistor R_L placed across the output terminals. If the amplifier output resistance R_{OUT} is very high (approaching infinity), we may also use very large values of R_L for the load, giving high power gains for the amplifier. This may be seen by comparing the input and output powers in terms of I^2R. If the resistance on the input side is R_{IN} (very small), then the input power is $I_{IN}{}^2R_{IN}$. At the load, the output power is $I_{OUT}{}^2R_L$ where R_L is very large. Taking the ratio for the power gain, we have

$$G = \frac{I_{OUT}{}^2R_L}{I_{IN}{}^2R_{IN}} \qquad R_{OUT} \gg R_L \gg R_{IN} \tag{7-1}$$

For the condition that $I_{IN} = I_{OUT}$,

$$G = \frac{R_L}{R_{IN}} \tag{7-2}$$

Thus it is seen that a device capable of permitting a current to enter at low resistance, go through undiminished, and come out at high resistance would certainly approach an ideal amplifier. The equivalent circuit of such a device might appear as shown in Fig. 7-2*a*, where the generator is a constant-current generator delivering a current equal to the input current. For the ideal case, the input would approach short circuit and the output open circuit as shown in Fig. 7-2*b*.

7-3. Amplifier Synthesis with p-n Junctions. In order to obtain an amplifying device on the basis of the criteria we have just discussed, viz., (1) low resistance input, (2) high resistance output, and (3) unattenuated current transfer, we may take advantage of the electrical characteristics of p-n junctions as a function of voltage bias. By suitably employing p-n junctions, we can properly synthesize an amplifying device.

As was discussed in Chap. 5, when a p-n junction is biased in the forward direction, very large currents will flow, consisting primarily of holes, crossing the junction, which diffuse into the n region where they recombine with the electrons there. This, of course, occurs for the case where the p-region resistivity is very much lower than the n-region

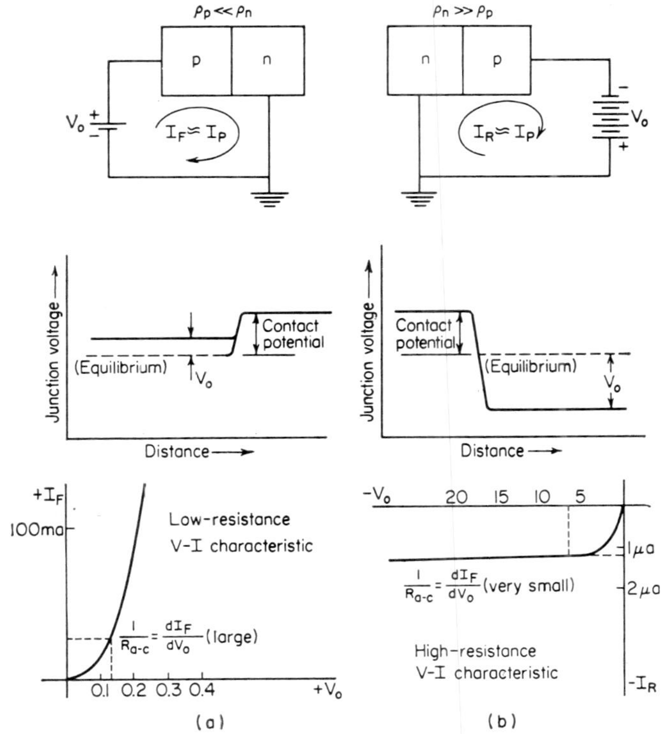

FIG. 7-3. Characteristics of p-n junctions. (*a*) Forward bias; (*b*) reverse bias.

resistivity such that $p_p \gg n_n$. For these conditions, the current increases rapidly and exponentially with voltage, giving rise to very small forward resistances. These effects are illustrated in Fig. 7-3*a*. For purposes of discussion, the potential of the n region is held fixed at ground potential, so that the application of forward bias raises the potential of the p region by V_o volts. The *V-I* characteristic, as drawn, is strictly arbitrary, signifying only the general order of magnitude of the currents involved. At a particular bias point, shown by the dotted lines, the small-signal a-c conductance ($1/R_{AC}$) which is equal to the slope of the curve (dI_F/dV_o) evaluated at that point, is rather large, yielding a small a-c resistance.

Thus, the forward-biased p-n junction could satisfy the first amplifier criterion as far as the input is concerned.

In Fig. 7-3*b*, the p-n junction is shown in the reverse-biased condition with the application of a negative voltage to the p region. Here again, for the sake of reference, the n region is maintained at ground potential. The current that flows in the circuit is the reverse saturation current which, for the chosen resistivities, consists primarily of the holes (minority-carrier concentration) in the n region that diffuse to the junction and

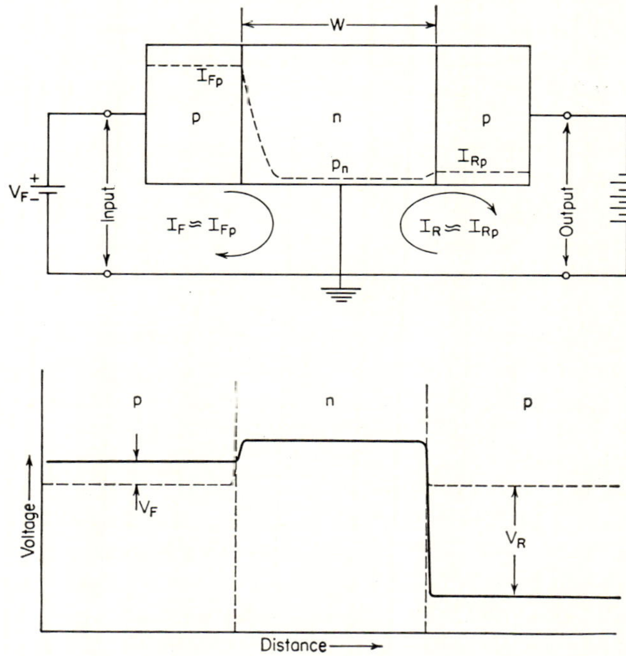

FIG. 7-4. p-n junctions back-to-back.

drop into the p region. Since p_n is usually quite small, this current is quite small and is also practically constant with voltage as shown in the approximate sketch of the *V-I* characteristic. At a particular reverse bias the slope of the curve is almost zero, giving rise to an extremely large a-c resistance which satisfies the second criterion for the ideal amplifier.

To construct the amplifier, the obvious procedure would be to combine the p-n junctions as described into a homogeneous structure, in order to obtain the desired characteristics at the input and output terminals. Such a structure would appear as shown in Fig. 7-4, where the p-n junctions are placed back to back, with the n regions common and at ground potential. The bias voltages have been designated V_F and V_R for the forward and reverse biases respectively. In this model it is assumed that the p regions are of equal and very low resistivity compared to the n region, such that the currents are predominantly due to hole flow. If the thickness W of the n region is very much greater than the diffusion length L_p for holes in that region, the hole current distribution as a function of distance appears as shown by the dashed curve in Fig. 7-4. For the forward-biased junction (at the left), the hole current is shown recombining with electrons in the n region, constituting the first loop current I_F. The second loop current I_R arises from the concentration of holes in the n region which diffuse to the reverse-biased junction (at the right). The important result to note is that since these diffusion mechanisms are remote from each other within the n region, no interactions occur and the structure behaves just as though the junctions were biased independently.

We may conclude, then, that the structure satisfies the first and second criteria, but not the third, which requires that the input current I_F appear at the output terminals. In Fig. 7-4, the output current I_R is independent of and several orders of magnitude smaller than the input I_F. The next section will treat the modifications which are necessary in order to obtain good amplifier characteristics.

7-4. The p-n-p Junction Transistor. If the structure of Fig. 7-4 is modified by making the thickness W of the n region much smaller than the hole diffusion length, an effective means is provided to transmit the input current to the output. Such a structure is appropriately called a transistor, which means "carry across." In the new structure, shown in Fig. 7-5, where $W \ll L_p$, a concentration of holes will enter the n region from the forward-biased junction and, because of the large concentration gradient in the base region, will diffuse across to the reverse-biased junction. Since the distance is much less than a diffusion length, only a very small fraction of the holes will recombine with electrons in the n region, and most of the holes will reach the reverse-biased junction. The arriving holes are then easily swept across this junction, making the hole concentration equal to zero there. The hole current flowing out from the reverse-biased junction is only slightly less than the input value. Thus we have satisfied the final condition for power gain, i.e., the output current is a constant current from a high-resistance source.

In Fig. 7-5, the forward-biased junction is called the *emitter,* since it emits or injects carriers into the n region. In transistor nomenclature, the n region is called the *base*, shown at ground potential in the diagram. The reverse-biased junction is called the *collector,* for reasons that are obvious. Finally, the input and output voltages and currents are called the *emitter* and *collector voltages* and *currents,* respectively (V_E, I_E and V_C, I_C). The current flowing out of the base region due to those holes that recombine with electrons there is called the *base current* I_B.

The action of the p-n-p junction transistor may alternatively be understood by referring to the voltage-profile diagram. All of the voltages are shown to be constant in the emitter, base, and collector regions; this is so for the assumption that the conductivities of these regions are sufficiently high so that the applied potentials are completely dropped across the junctions only. With the base region fixed at ground, the emitter voltage V_E effectively decreases the base-to-emitter potential hill, permitting large numbers of holes to diffuse into the base. Although some recombination will occur within the base, most of the holes will reach the

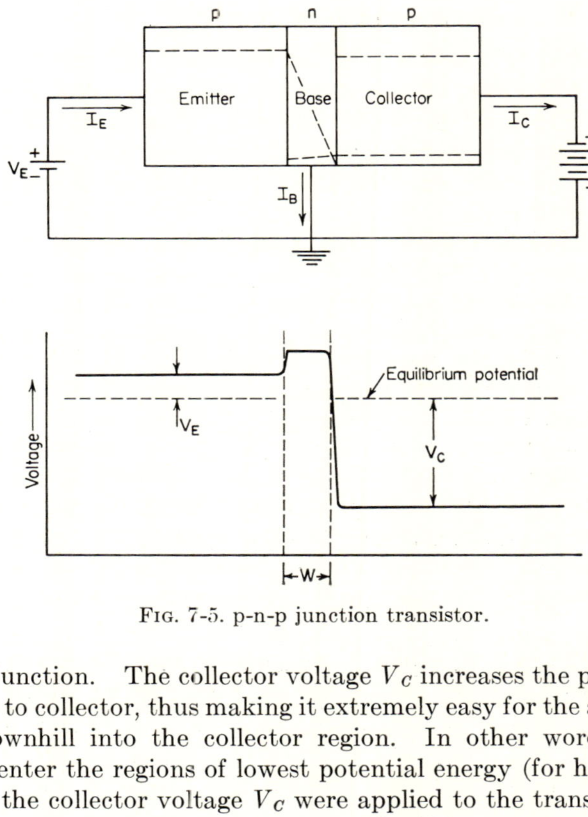

FIG. 7-5. p-n-p junction transistor.

collector junction. The collector voltage V_C increases the potential drop from base to collector, thus making it extremely easy for the arriving holes to fall downhill into the collector region. In other words, the holes prefer to enter the regions of lowest potential energy (for holes).

If only the collector voltage V_C were applied to the transistor, so that there was no emitter current I_E, we would still expect a small current to flow in the output. This would be the reverse saturation current of the collector junction; however, it would not be the same as in connection with Fig. 7-4, due to the fact that the base-region thickness is now much less than the hole diffusion length. This means that this current would consist primarily of the equilibrium concentration of holes in the n region, nearly all of which would easily diffuse to the collector junction. The equilibrium floating potential of the open-circuited emitter junction would also contribute to the lowering of the saturation current. Denoted as I_{CBO}, the collector current for open-circuited emitters, it is seen that this current would flow whether or not there was an input signal, and therefore represents useless power dissipation in the collector circuit of the transistor. If emitter current was fed into the transistor, the output collector current would be equal to the sum of the I_{CBO} saturation current and some fraction (close to unity for good transistors) of the emitter current I_E, or

$$I_C = I_{CBO} + \alpha I_E \tag{7-3}$$

where α (alpha) is a measure of the efficiency of transport of carriers through the transistor. Thus, if no recombination took place, α would equal unity. In good junction transistors, I_{CBO} is usually very small compared to normal operating values of I_E, so that Eq. (7-3) simply becomes $I_C = \alpha I_E$, meaning that alpha is the ratio of the collector current to the emitter current.

7-5. Current-gain Theory. The parameter α represents one of the most important of the parameters that characterize the operation of a junction transistor. As was just stated, alpha is the ratio of the collector current to the emitter current, or, in more general terms applicable to a-c signals,

$$\alpha \equiv \left.\frac{\partial I_C}{\partial I_E}\right|_{V_C = K} \tag{7-4}$$

which is the small-signal current gain about a particular d-c operating bias. From the standpoint of the current criterion for good power gain, it is desirable that alpha be as close as possible to unity. It was described previously that the recombination of injected minority carriers with the majority carriers of the base region limits alpha to values slightly less than unity.

Actually, the over-all alpha of a transistor may be designated as the product of three separate carrier mechanisms, written as follows:

$$\alpha = \gamma\beta^*\alpha^* \tag{7-5}$$

where α = alpha (over-all current gain)
γ = emitter efficiency
β^* = base-transport factor
α^* = collector multiplication ratio

It is apparent that if each of these mechanisms is equal to unity, then the over-all current gain is unity. Each will be explained in the following paragraphs.

Emitter efficiency γ is the ratio of the number of minority carriers injected into the base region from the emitter to the total number of carriers crossing the emitter junction. In other words, γ is the ratio of the injected current to the total emitter current. For a p-n-p transistor, the emitter efficiency is

$$\gamma = \frac{I_p}{I_p + I_n} \tag{7-6}$$

where I_p is the current due to holes injected from the emitter p-n junction. The total current crossing the junction consists of the component I_p and a current component I_n due to the electrons injected into the emitter from the base. Although this forward-bias component was usually neglected in the simple p-n junction theory, it cannot be so neglected in terms of alpha, where we are concerned with small deviations from unity. Since only the hole current I_p contributes to transistor action, it is desirable that I_n be kept as small as possible, to maintain an emitter efficiency as close to unity as possible. It is apparent that from a design standpoint this can be attained by making the ratio of the base region resistivity ρ_b to the emitter resistivity ρ_e as large as possible, thereby making the emitter hole concentration p_{pe} much greater than the base electron concentration n_{nb}.

The second right-hand term of Eq. (7-5), the base-transport factor, is a significant term which represents the ratio of the number of holes arriving at the collector junction to the number of holes injected into the base region from the emitter. The latter corresponds to the hole current I_p in the emitter efficiency Eq. (7-6). As the holes diffuse through the base region, some of the holes will recombine with electrons, constituting a small current which is not collected and which flows out of the base region. Since it is required that this internal current loss be a minimum, it is necessary to keep the amount of recombination in the base region as small as possible. The base width W and the hole diffusion length L_{pb} therefore appear in the expression for the base-transport factor. It should be recalled that the diffusion length represents an arbitrary average distance a carrier will diffuse before it recombines, and is equal to $L_{pb} = \sqrt{D_{pb}\tau_{pb}}$. In the case of the transistor, τ_{pb} is the lifetime for holes in the base. A high-lifetime base region whose thickness W is much smaller than L_{pb} then gives a value for the base-transport factor very close to unity. In other words, the carriers reach the collector junction in a distance far less than their diffusion length and therefore have a high probability of not recombining with electrons. In junction-transistor design, the thickness of the base region becomes a critical factor, not only for current gain but, as will be seen later, for other parameters as well.

The last term of Eq. (7-5), the collector multiplication ratio α^*, is a ratio of the total current crossing the collector junction to the hole current (for a p-n-p transistor) arriving at the junction. Under normal conditions α^* is usually equal to unity. For certain conditions, however, it may exceed unity. This arises from the fact that the holes entering the collector provoke a flow of electrons from the p-type collector region into the base. This electron current is not a diffusion current but a drift current, established by the electric field created by the hole current in the collector. For a given hole concentration (or current) in the collector region, the electron drift current will increase as the resistivity increases, because the electron concentration (minority carriers) becomes greater. Thus it is expected that α^* becomes greater than 1 in transistors having relatively high collector resistivities, particularly at higher temperatures where the minority-carrier concentration increases due to thermal

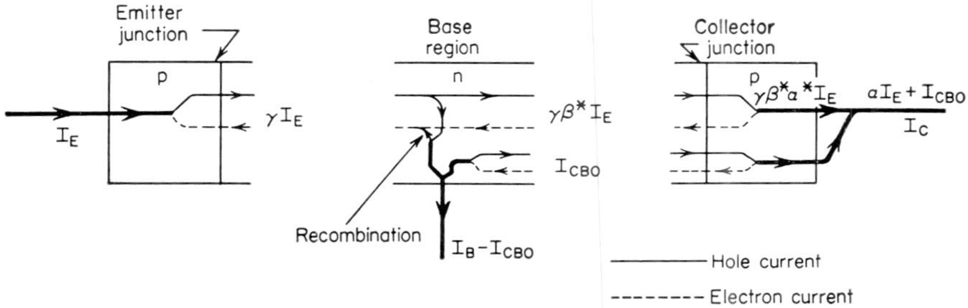

FIG. 7-6. Current schematic for composite transistor alpha.

generation. Mesa transistors made by the diffusion process have collector resistivities higher than that of the base or emitter. However, alloy junction types have very low collector resistivities, and therefore collector multiplication may be considered negligible. It should be emphasized that α^* is mostly a function of the relationship of the magnitude of the injected-carrier concentration to the majority-carrier concentration in the collector and is therefore a function of current injection level.

In summary, we see that through judicious choice of resistivities and lifetimes for the emitter, base, and collector regions of the junction transistor, both the base-transport factor β^* and the collector multiplication ratio α^* can be set approximately equal to unity, thereby making the over-all transistor current gain α dependent primarily on the emitter efficiency γ. The contributions of the three mechanisms γ, β^*, and α^* to over-all alpha may be readily pictured by the simple current schematic of Fig. 7-6. The solid lines depict the hole components, the dashed lines are the electron currents, and the heavy lines represent the total currents. The three transistor regions are shown separated for easier visualization.

The components of the I_{CBO} current are also drawn in. This is the current that flows in the collector for zero emitter current, i.e., open-circuit emitter. These components arise from the diffusion of the thermally generated minority carriers.

7-6. Grounded-base Transistor Characteristics. Having established the current-gain mechanism in the junction transistor, we can now describe the voltage-current characteristics of the output collector circuit. A typical family of curves is drawn in Fig. 7-7, which is a plot of collector current I_C versus collector voltage V_C for different values of emitter current. In certain regions the curves have been exaggerated for the purpose of illustrating special effects. Also, the curves are those which would be obtained for the circuit as shown, in which the base region of the transistor is grounded.* With emitter current equal to zero, the first

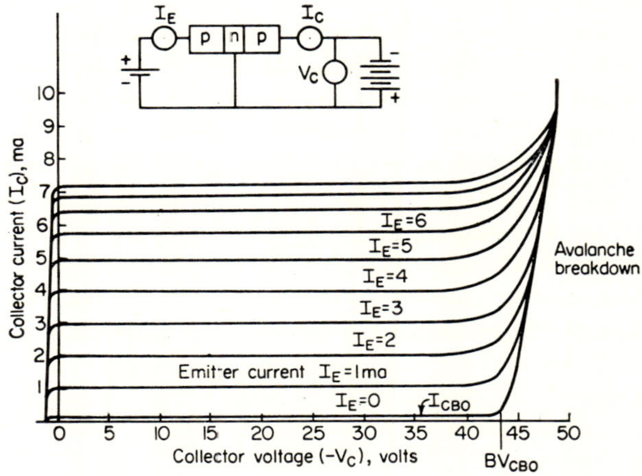

FIG. 7-7. V_C-I_C characteristic for grounded-base junction transistor.

V_C-I_C curve corresponds to the I_{CBO} current as a function of voltage and is similar to the reverse saturation-current characteristics of a p-n junction. Since I_{CBO} is a function of p_{nb} in the base (also n_{pc} in the collector, but to a far lesser degree if $\rho_{nb} > \rho_{pc}$), it would be expected that I_{CBO} has a temperature dependence similar to that of p_{nb}. As the temperature increases, I_{CBO} also rises, and in so doing shifts the whole family of curves upwards by the same amount. The V_C-I_{CBO} power dissipation represents useless junction heating and therefore limits the power capabilities of the device by its contribution to the transistor's operating temperature. It is important, therefore, in good transistor design, to keep the I_{CBO} current

* It should be pointed out here that although the discussion in this chapter refers to the p-n-p transistor structure, the same theory holds for the n-p-n structure. In the n-p-n case, the active minority carriers are electrons which are injected into a p-type base region. Furthermore, the required bias voltages are reversed; i.e., a negative voltage is necessary to forward-bias the emitter n-p junction and a positive reverse bias is necessary for the collector. Lastly, all hole terms and subscripts for the p-n-p case become electron terms and subscripts for the n-p-n case, and vice versa.

as small as possible. In actual transistor devices, the inherent I_{CBO} currents are very small, on the order of microamperes or much less for silicon. However, surface leakage plays a dominant role in contributing to the total current.

When the emitter current I_E is equal to 1 ma, the collector current I_C is equal to αI_E, or almost I_E for α close to unity. Thus we have the first curve of I_C versus V_C for $I_E = 1$. As the emitter current is increased in equal increments of 1 ma, the collector current also increases in increments of αI_E, giving a direct linear relationship as shown. However, at the higher emitter currents it is seen that the curves begin to crowd together. This effect is due to the fact that alpha falls off gradually as the emitter current (density) is increased, and is related primarily to a decrease in emitter efficiency due to an effective decrease of the base resistivity by the large injected current. In circuit applications in which the emitter current swings over a wide range, the alpha-crowding at the higher levels results in highly nonlinear operation, distorting the collector output signal. It is good design practice, therefore, to keep current gain as a function of I_E as constant as possible over the allowable range of emitter currents. The upper emitter-current limit is set by the maximum permissible junction temperature, which is a function of the power dissipation and the ambient temperature. Transistors are temperature limited because of the degradation of semiconductor bulk parameters, particularly as resistivities become intrinsic due to thermal generation. It is for these reasons that silicon transistors can operate at higher temperatures than germanium.

Although not shown in Fig. 7-7, it is also possible for alpha to decrease for very small emitter currents, owing to the extensive recombination of injected carriers with either surface carriers or impurity recombination centers. This subject will be treated in detail later. For those circuits requiring amplification of very small a-c signals, it is desirable that alpha peak rapidly with emitter current, in order that the smallest possible d-c emitter bias current be used, for minimum power-supply drain.

In Fig. 7-7, as the collector reverse-bias voltage is increased, it is seen that the curves begin to slope upward slightly. This effect is due to the widening of the depletion layer of the reverse-biased collector junction. We recall that the thickness x_m of the depletion layer of a p-n junction is proportional to $\sqrt[n]{V}$ and for a step junction will spread almost completely into the higher resistivity region. Thus, for the p-n-p transistor, the spreading with voltage will effectively reduce the thickness W of the n region. This will result in two significant effects: (1) the base transport factor will increase, thereby increasing alpha slightly, and (2) the I_{CBO} current will increase since it is inversely proportional to W. Therefore we have the change of slope of the collector current, as shown.

If the base-region resistivity is high and the thickness W quite small, a point is reached at which the collector voltage spreads the depletion layer far into the n region, such that it reaches the emitter junction. At this critical point, a phenomenon called *voltage punchthrough* occurs and large currents flow which can literally burn out the transistor if there is not adequate series resistance in the circuit. Essentially, at the punch-through condition the emitter charge dipole is disturbed by the spreading, so that many more holes are swept out from the emitter region in order to maintain charge neutrality at the emitter junction. It becomes apparent from the theory that, for a given base thickness W, punchthrough will occur at higher voltages for lower resistivity base regions. The V_C-I_C characteristics, however, show the collector junction "breaking down" by the avalanche mechanism described in Chap. 6. It is assumed, in this case, that for the resistivities involved, avalanche breakdown is occurring before punchthrough. From the McKay theory, avalanche breakdown is approximately directly proportional to resistivity. Thus, the maximum allowable collector voltage of the transistor is limited by the lower of the two reverse mechanisms, avalanche breakdown and punchthrough.

Referring to Fig. 7-7 again, we see that as the voltage approaches the breakdown value BV_{CBO}, the collector currents begin to bend up rather rapidly. This effect is actually a current multiplication brought about by the approach to breakdown. In this region, the carrier ionization rate is increasing with electric field, rendering an increase to the over-all transistor alpha, since the collector current is becoming greater. If we designate the multiplication factor by M, then I_C is given as

$$I_C = \alpha M I_E = \gamma\beta^*\alpha^* M I_E \quad (7\text{-}7)$$

It is possible then for alpha (also called collector-to-emitter current gain) to be greater than unity, since $M > 1$ near BV_{CBO}. In fact, at breakdown alpha approaches infinity.

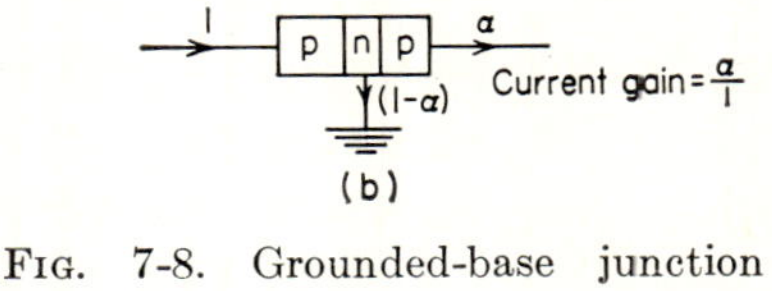

FIG. 7-8. Grounded-base junction transistor.

7-7. Grounded-emitter Transistor Characteristics. When a junction transistor is operated in a grounded-base circuit, power gain is achieved by driving the low-resistance input from a constant current source, such that the input current is multiplied by alpha and the output power is developed across the load resistor in the collector circuit. Such a circuit would appear as shown in Fig. 7-8*a*, where R_L is the load resistor and R_E is much greater than the d-c resistance of the emitter junction, so that the emitter bias current is made constant. For an alternating input current I_{AC}, the voltage output would obviously be $\alpha I_{AC}R_L$. From the standpoint of the currents, if the input current is referenced as 1, then by definition the output current is α. The difference between the two, viz., the base current, must therefore be equal to $(1 - \alpha)$, as shown schematically in Fig. 7-8*b*. For example, if $\alpha = 0.98$, then for unity input the base current is $(1 - \alpha)$, or 0.02.

The fact that high-alpha transistors have extremely small base currents makes it quite conceivable to drive the device using the base as the input and the emitter at ground potential, as far as alternating currents are concerned. Such an arrangement is called *grounded-emitter* operation; a typical circuit for a p-n-p transistor is drawn in Fig. 7-9*a*. Note that as far as the direct emitter current and collector voltage are concerned, the junctions are still biased in the same manner as for the grounded-base case. The resistance R_E is contained within the emitter-base circuit to maintain a constant-current emitter bias. The significant aspect of this configuration is shown in Fig. 7-9*b*. If the same current relationships are retained, then it is seen that as far as input and output are concerned, the current gain is

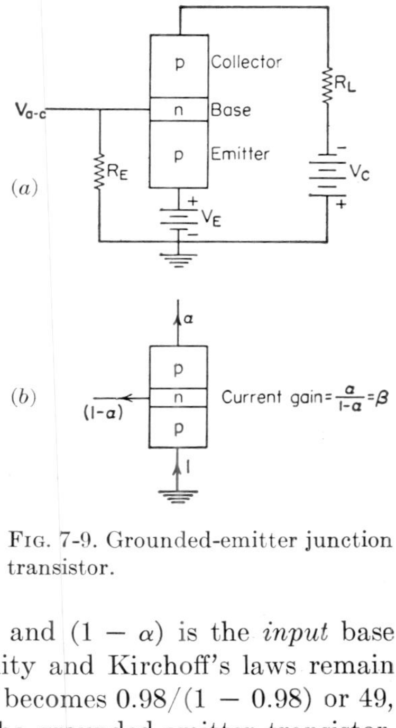

FIG. 7-9. Grounded-emitter junction transistor.

$$\beta = \frac{\alpha}{1 - \alpha} \tag{7-8}$$

where α is the *output* collector current and $(1 - \alpha)$ is the *input* base current. The emitter current is still unity and Kirchoff's laws remain unviolated. If alpha is again 0.98, beta becomes $0.98/(1 - 0.98)$ or 49, which is therefore the *current gain* for the grounded-emitter transistor. Under these conditions, a simple analysis will show that the input resistance for grounded emitter becomes equal to

$$R_{\text{grounded emitter}} = R_{\text{grounded base}} \times \frac{1}{1 - \alpha} \tag{7-9}$$

From Eq. (7-9), the higher the alpha, the smaller is the base current and the higher are both the beta and the input resistance. This higher input resistance permits the grounded-emitter stage to be driven from a low-resistance voltage source V_{AC}, making it more suitable than the grounded-base circuit for most amplifying circuits. It should be noted that beta, per Eq. (7-8), increases very rapidly as α approaches 1.

The V_C-I_C collector characteristics must now be modified for the grounded-emitter case. The new family of curves is drawn in Fig. 7-10 for the same collector current and voltage scales, but the varying parameter is the base current I_B. The first important result to observe is that the collector saturation current for $I_B = 0$ is equal to βI_{CBO}. For good transistors, where beta may be about 10 and I_{CBO} about 1 μa, βI_{CBO} for grounded emitter is 10 μa at room temperature. At 85°C, for example, thermal generation increases βI_{CO} to about 1 ma. Obviously, this cuts heavily into the useful range of operation for the transistor. It is this multiplication of I_{CBO} that makes it so essential that I_{CBO} be as small as

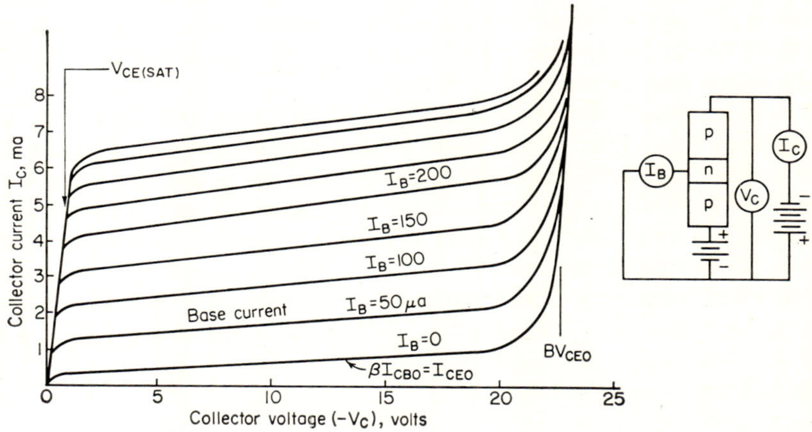

FIG. 7-10. V_C-I_C characteristic for grounded-emitter junction transistor.

possible. In silicon transistors, the room temperature I_{CBO} is on the order of millimicroamperes so that the range of operation may be extended to higher temperatures than germanium. By definition, the βI_{CBO} current is written as I_{CEO}.

Since the effective base thickness varies as the collector depletion layer widens with voltage, we expect a small increase in alpha to occur. Beta, being quite sensitive to small changes in alpha, increases at a much faster rate. It is for this reason that the curves of Fig. 7-10 slope upward considerably more than for the grounded-base transistor. Similarly, the sensitivity of beta to changes in alpha also creates a more rapid falloff of beta (collector-to-base current gain) with emitter current density. This accounts for the more pronounced crowding of the V_C-I_C characteristic at the higher input base currents.

Maximum collector voltage for grounded emitter is also limited by either punchthrough or avalanche breakdown, whichever is lower, depending on the base resistivity and thickness. For the case of avalanche, intensive current multiplication will also occur near the breakdown region. In Eq. (7-8) it may be seen that a voltage can be reached where the multiplication factor M makes the over-all alpha equal to 1; the magnitude of beta becomes infinite, thereby breaking down the collector junction at a voltage less than BV_{CBO}. This common-emitter breakdown voltage is denoted as BV_{CEO}. Another characteristic, evident from Fig. 7-10, is the grounded-emitter saturation voltage $V_{CE(SAT)}$. This parameter is especially important in grounded-emitter switching applications. As I_{CBO} is the saturation current as collector voltage is increased, $V_{CE(SAT)}$ is the saturation voltage as collector current is increased. Physically speaking, $V_{CE(SAT)}$ is comprised of the emitter and collector junction potentials and the IR drops in the series resistance of the collector region. The latter are particularly significant in transistors made by the single-ended impurity-contact processes.

Two additional parameters, both related to the emitter junction, are BV_{EBO} and V_{BE}. The former is the avalanche-breakdown voltage of the emitter junction and is primarily dependent on the resistivity of the base region. The latter is the base-to-emitter forward voltage of the emitter junction and is simply the junction voltage necessary to maintain the forward-bias emitter current.

7-8. High-frequency Characteristics. Measurements and theoretical analyses of the junction transistor reveal that a-c parameters such as current gain and input and output impedances will vary as a function of frequency in a rather complex manner. At this point it will suffice to say that diffusion capacitances and transit-time effects will appear in the a-c terms as we consider higher frequencies of operation. Nevertheless, it may be said that the most significant parameter variation concerns the variation of current gain with frequency. Theory and experiment indicate that in well-designed transistors the grounded-base current gain remains fairly constant with frequency up to a point and then begins to fall off in a manner similar to the frequency response of a simple R-C network. By definition, the frequency at which alpha decreases to 0.707 (or $1/\sqrt{2}$) of its low-frequency value is called the *alpha-cutoff frequency* $f_{\alpha b}$. A typical variation of alpha with frequency is plotted in Fig. 7-11, where α_o is the low-frequency value and α_{co} is equal to 0.707 α_o at the alpha-cutoff frequency $f_{\alpha b}$.

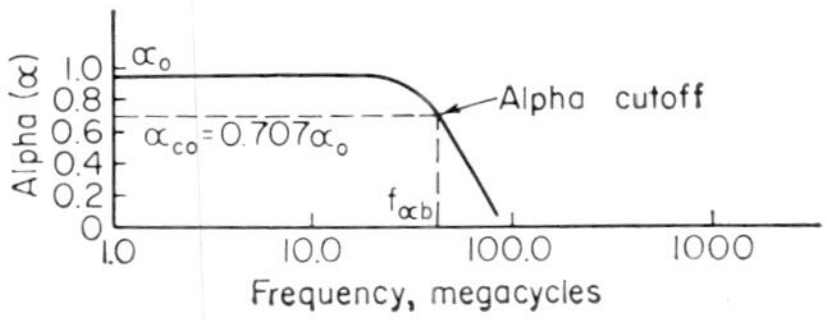

FIG. 7-11. Variation of alpha with frequency.

The alpha-cutoff mechanism is associated with the finite transit time which the minority carriers require to cross the base width. Since the carriers have a certain mobility determined by their effective masses and velocities, a point is reached where the carriers are not able to respond as rapidly to the impressed signal frequency. This creates a phase lag or capacitance effect which enhances recombination, thereby decreasing alpha. On this basis, it would be expected that electrons have a better frequency response than holes, since their mobilities are greater. In this respect, n-p-n transistors are superior to p-n-p transistors, other things being equal. The theory will show that alpha cutoff is inversely proportional to the square of the base width and directly proportional to the minority-carrier mobility. In fact, alpha cutoff will increase with increasing collector bias, since the base width gets narrower. It should be evident that in terms of good amplification performance at high frequencies, $f_{\alpha b}$ should be as high as possible.

For very-high-frequency transistors in which the base widths are extremely thin, the observed frequency response of current gain is considerably less than that predicted by the aforementioned transit time through the base region. This indicates that there are additional high-frequency characteristics which contribute to the reduction of $f_{\alpha b}$. One of these is obviously the existence of the collector junction capacitance C_{Tc}, which appears across the output of the transistor. This capacitance must be charged up through any series resistance in the collector or other regions of the transistor. In this case, we have the collector series resistance r_{SC} [this accounts for the observed $V_{CE(SAT)}$ discussed in Sec. 7-7], which establishes the $r_{SC}C_{Tc}$ time constant.

A second characteristic which limits frequency response is the existence of the emitter junction capacitance, which is usually quite large because the depletion layer is quite narrow under forward-bias conditions. This emitter capacitance C_{Te} shunts the input resistance of the transistor r_e, thereby establishing the time constant r_eC_{Te}. This time constant may become quite large at low operating currents, since r_e is inversely proportional to I_E.

The final high-frequency characteristic of importance is the transit time of the carriers through the collector depletion layer. We saw in Chap. 4 that in the presence of electric fields, carriers will travel with a finite drift velocity. If the depletion-layer thickness is appreciable, one cannot neglect the finite transit time necessary to cross it.

In summary, we see that the over-all frequency response of a transistor is not restricted to the base-transit time, but must include additional time delays such as the emitter time constant r_eC_{Te}, the collector time constant $r_{SC}C_{Tc}$, and the transit time through the collector depletion layer. The theory of transistor frequency response will be studied in greater detail in Chaps. 13 and 14.

If the transistor is operated in grounded-emitter configuration, the effect of the frequency variation of alpha on beta is quite severe. We recall that beta is given by the magnitude of $\alpha/(1 - \alpha)$. As the frequency is increased, the denominator $(1 - \alpha)$ becomes very sensitive to the phase shift of alpha, increasing rapidly and decreasing β equally so. In grounded base, alpha is the ratio of the absolute magnitudes of I_C and I_E, regardless of any phase lag between the two currents. However, in grounded emitter, I_B [or $(1 - \alpha)$] is the vector difference $I_E - I_C$, as illustrated in Fig. 7-12. As the phase angle θ increases with frequency, $(1 - \alpha)$ or I_B also increases, and beta falls off quite rapidly at lower frequencies. From the theory, the frequency of beta cutoff is given as $f_{\alpha e} = (1 - \alpha)f_{\alpha b}$. This is the frequency at which beta is 0.707 of the low-frequency value.

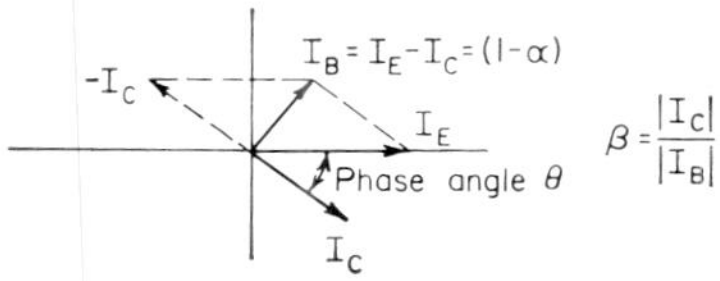

FIG. 7-12. Effect of phase shift on magnitude of $(1 - \alpha)$.

The relationship of $f_{\alpha e}$ to $f_{\alpha b}$ is clearly shown in the frequency-response diagram of Fig. 7-13. The vertical scale is exaggerated in order to permit the plot of both alpha and beta on the same diagram. The frequency at which beta is equal to 1 is defined as f_T, the gain bandwidth of the transistor, which is shown to be slightly higher than the alpha-cutoff frequency $f_{\alpha b}$. In most high-frequency transistors having diffused bases, the excess phase shift causes f_T to be lower than $f_{\alpha b}$. Although beta is unity at f_T, there may be appreciable power gain at this frequency because of the impedance ratios of the transistor. Actually, the maximum frequency of operation for a transistor is the maximum frequency of oscillation $f_{\max}$, which is defined as the frequency at which the power gain is equal to 1.

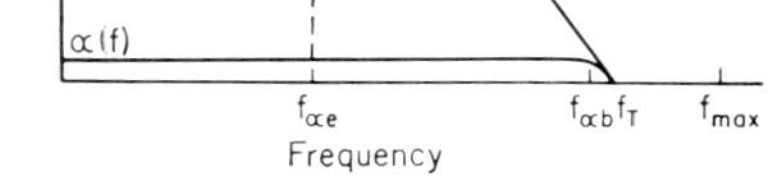

FIG. 7-13. Relationship of grounded-emitter frequency response to grounded-base response.

Both power gain and the maximum frequency of oscillation are inversely related to an additional time constant $r'_B C_{Tc}$, which is associated with the resistance of the base region r'_B. This base resistance is related to the base current flowing out of the base region. Regardless of any particular geometry for the base region of the transistor, the base current I_B must traverse this region to an ohmic contact. Since the base layer has finite thickness and resistivity, we can establish a finite base resistance r'_B. Thus we have the most important time constant $r'_B C_{Tc}$, which must be minimized to achieve maximum high-frequency response and power gain.

7-9. A-C Equivalent Circuit. In applying a junction transistor to an electronic circuit, the designer often finds it convenient to represent the device by its equivalent electric circuit, in order to facilitate the analysis of its performance. In studying the physics of transistors and by analyzing the behavior of the holes and electrons we can readily (as will be done in later chapters) arrive at voltage-current expressions which are related to the physical parameters of the semiconductor materials. These equations give rise to general admittances for both the input and output terminals and enable us to write down an equivalent circuit for the transistor.

From the discussion of the ideal amplifier in the beginning of this chapter and the qualitative descriptions of transistor mechanisms, we are in a position to establish a simple equivalent circuit which approximates

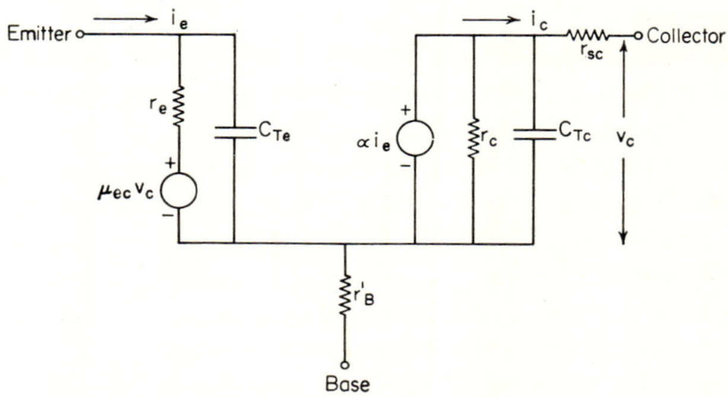

FIG. 7-14. Transistor equivalent circuit.

the model of the transistor described thus far. Such an arrangement is drawn in Fig. 7-14. Equivalent circuits are usually drawn only for the a-c impedances of the device. The signals, therefore, are only the alternating currents and/or voltages, measured at a particular d-c bias. Although the d-c biases are not shown, the circuit holds only for the operating points in question. In transistors, the a-c parameters will change with d-c operating bias.

The input resistance r_e represents the slope of the forward-biased emitter junction; for an emitter current of 1 ma, r_e would be about 26 ohms at room temperature. The output resistance r_c corresponds to the slope of the reverse-biased collector junction at a particular collector voltage. This resistance is considerably lower than that predicted by saturation conditions because, under small-signal operation, the a-c variation of collector voltage will impart an a-c variation to alpha, due to depletion-layer widening effects. This dependence of alpha on collector voltage will increase the slope of the V_C-I_C curves, thereby lowering the collector resistance to a value which may be equal to about 1 megohm. This same a-c variation of the base thickness will also vary the emitter resistance, since less emitter voltage is needed to maintain the same emitter current as the base region becomes thinner. This feedback effect on the emitter by the a-c collector voltage is represented by a small a-c voltage generator placed in series with the emitter resistance r_e, and is given as $u_{ec}v_c$, where u_{ec} is the feedback factor. To depict the current-gain mechanism, a constant-current generator equal to αI_e, where I_e is the input emitter signal current, is placed across the output terminals, with the polarity as shown in Fig. 7-14. The a-c equivalent circuit is then completed with the addition of the emitter capacitance C_{Te} across the input, the collector capacitance C_{Tc} across the output, and the base resistance r'_B in series with the base terminal. Thus we see that this equivalent circuit satisfies the requirements for gain and adequately characterizes the physics of the junction transistor.

7-10. Optimum Transistor Parameter Design. As previously stated, the purpose of this chapter is to introduce the basic mechanism of the

TABLE 7-1. KEY TRANSISTOR PARAMETERS

Symbol	*Parameter*
α	Grounded-base current gain
BV_{CBO}	Collector-to-base breakdown voltage
BV_{EBO}	Emitter-to-base breakdown voltage
I_{CBO}	Collector-to-base leakage current
β	Grounded-emitter current gain
BV_{CEO}	Collector-to-emitter breakdown voltage (open base)
I_{CEO}	Collector-to-emitter leakage current
$V_{CE(SAT)}$	Grounded-emitter saturation voltage
V_{BE}	Base-to-emitter forward-bias voltage
C_{Te}	Emitter junction capacitance
C_{Tc}	Collector junction capacitance
r_e	A-c resistance of forward-biased emitter
r_c	A-c resistance of reverse-biased collector
r'_B	Base-region resistance
$f_{\alpha b}$	Alpha-cutoff frequency
$f_{\alpha e}$	Beta-cutoff frequency
f_T	Gain-bandwidth frequency
f_{max}	Maximum oscillating frequency

junction transistor in a qualitative manner. Also, it serves to highlight the significance of the more pertinent electrical parameters and characteristics of the transistor. In order to properly integrate all the parameters into an optimum device, it would be desirable at this point to examine each one independently in relation to what might be required for ideal junction-transistor design. The key parameters discussed thus far are summarized in Table 7-1.

Exclusive of any limitations on the maximum temperature of operation, the most important criteria for an ideal transistor are gain, frequency response, and power capability. The combination of these three factors dictates what is expected for the individual parameters inherent to the device. For maximum gain, the current gain α should be as close to unity as possible. This yields a high beta in the grounded-emitter circuit configuration, corresponding to large collector-to-base current gains. Further, for other things equal, the transistor alpha should be constant over a wide range of emitter current, falling off as little as possible at the higher currents. This allows for the highest possible emitter current rating at which the transistor still provides gain, thereby extending the power dissipation rating. Also, the transistor base width should not vary with collector voltage; this minimizes any internal feedback effects due to changes of alpha with voltage. The constancy of alpha over the widest range of emitter current and collector voltage not only provides for low input and high output impedances, but also minimizes output distortion for large signal variations.

In addition to minimum bias variations, the current gain must ideally be constant with frequency to as high a frequency as possible. Therefore, for maximum transistor frequency response, the base-transit time should be small. However, this by itself is not sufficient for frequency performance. Additionally, both the input and output capacitances must be as small as possible to eliminate high-frequency bypassing effects. Also, at the higher frequencies, the base resistance r'_B results in unwanted signal-power loss, thereby reducing the over-all power gain of the device. Thus, r'_B should be designed to be as small as possible.

From the standpoint of power rating, we have already mentioned the importance of high current gains at the larger emitter currents. To permit the transistor to operate at some high output power level, it is necessary to be able to apply as large a collector voltage as possible. Therefore, for the ideal transistor, the collector breakdown voltage ratings must be as large as possible. The combination of I_E and V_C will increase the operating temperature of the device so the maximum temperature that the semiconductor material can withstand before losing its extrinsic properties becomes an important design consideration. Also, as explained earlier, the I_{CBO} current represents useless power loss and should be designed to be as small as possible at high temperatures.

In applications where junction transistors are used as switches, most of the foregoing requirements are also applicable. In certain cases, as we will see later, it is necessary to restrict beta to low values in order to minimize minority-carrier storage after the switching pulse is removed. Additionally, for a good switching transistor it is required that the series resistance of the emitter and collector regions be as low as possible to provide a low $V_{CE(\mathrm{SAT})}$.

The transistor designer is faced with the problem of trying to optimize these parameters to attain performance as close to the ideal as possible. Unfortunately, this is not easy. Since each specific parameter is related to the properties and geometry of the actual transistor structure, it is not possible to optimize without getting into conflicting requirements. Thus, compromises must be made, consistent with the objective electrical specifications. Furthermore, although the design theory may show that it is conceivable to obtain a junction transistor approximating the ideal, it may not be too practical to properly fabricate

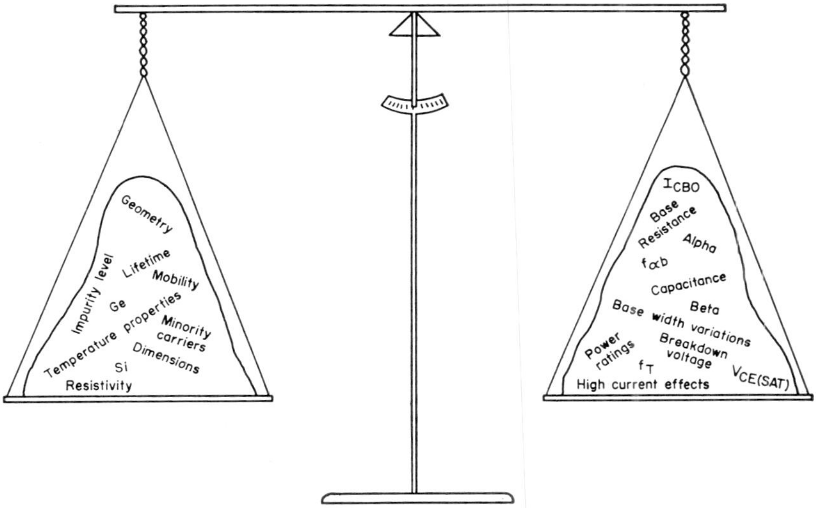

FIG. 7-15. Junction-transistor design.

such a device within the limitations of the state of the transistor art. As was pointed out in Chap. 1, there exist today several basic processes for making junction transistors, each yielding different impurity structures. For each type, however, certain electrical parameters must be sacrificed for the sake of others. Therefore, for some circuit applications, one process type may be more suitable than another. In the chapters to follow, the design expressions for the parameters we have discussed will be developed and related to such factors as physical dimensions, resistivity, lifetime, impurity concentration, and geometry. The transistor design problem, therefore, becomes a problem of *balancing*, as illustrated pictorially in Fig. 7-15. The foregoing is exclusive of the problems of processing, surface treatment, reliability, thermal transfer, packaging, etc., which are not covered in this text.

PROBLEMS

7-1. A germanium n-p-n transistor is connected into the following circuit at room temperature:

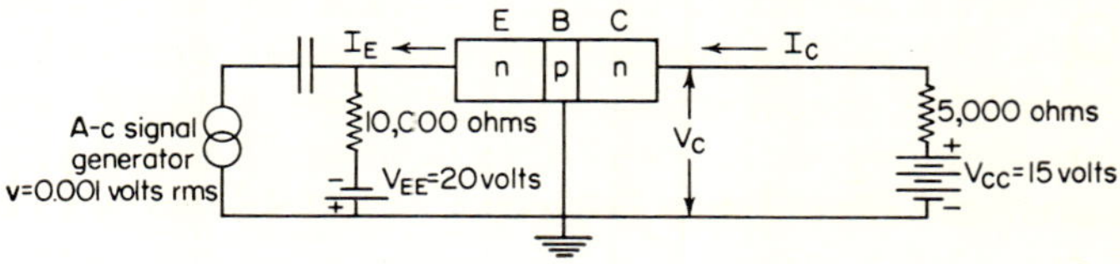

For $\alpha = 0.985$, determine the direct emitter current I_E, the direct collector current I_C, and the d-c collector voltage V_C. What is the power gain of the circuit in decibels at very low frequencies? [Note: Power gain in decibels is defined as $G = 10 \log (P_{\text{OUT}}/P_{\text{IN}})$. Also, $r_e = kT/qI_E$.]

7-2. A silicon n-p-n transistor is made by the epitaxial process so that step junctions are obtained. The following resistivity data are known: $\rho_{ne} = 0.001$ ohm-cm, $\rho_{pb} = 0.5$ ohm-cm, and $\rho_{nc} = 8$ ohm-cm. For an area equal to 100 mils^2, what are the collector capacitance at $V_C = +10$ volts and the emitter capacitance at $V_E = +1$ volt?

7-3. A germanium p-n-p alloy transistor is fabricated using a 2-ohm-cm n-type pellet. The emitter and collector resistivities are negligibly low. For a base width of 0.5 mil, determine whether the transistor is limited by punchthrough or avalanche breakdown at room temperature.

7-4. For a transistor whose alpha is constant for all values of current and voltage, show that the collector current in grounded-emitter configuration is $I_{CEO} = I_{CBO}/(1 - \alpha)$ for the condition that the base is open circuited.

7-5. Show that the a-c resistance of a forward-biased emitter junction is given by $r_e = kT/qI_E$, where I_E is the direct emitter current.

7-6. For the transistor of Prob. 7-3, determine the grounded-emitter reverse leakage current I_{CEO} at 70°C and $V_C = -5$ volts. Assume that the concentration of holes at the base-region side of the emitter junction is equal to 10 per cent of the equilibrium value p_{nb}, and that $L_{pb} \gg W$. The junction area is 500 mils^2. $\alpha = 0.98$ and is constant with current and temperature.

8

Junction Transistor Theory

8-1. The One-dimensional Model. An analysis of the junction transistor, exact and complete to the last detail, would be extremely complex and would actually obscure some of the more obvious mechanisms characteristic of the device. A thorough treatment would, of necessity, include motion of carriers in three dimensions, surface effects, drift currents associated with injection levels, variations of semiconductor bulk properties, frequency characteristics of distributed parameters, etc.[1,*] Such a treatment would be quite formidable, if not almost impossible, to present. Fortunately, the application of sound engineering approximations and assumptions enables us to formulate a simple and basic model of the transistor which lends itself to easy analysis. Other additional effects can be systematically considered, analyzed, and added as building blocks to the foundation afforded by the simple model.

This chapter will therefore present the one-dimensional analysis of the junction transistor, similar to that done for p-n junctions where only the effects of diffusion of carriers were considered. Although a p-n-p model is used, the results will apply equally well for an n-p-n with proper interchange of notation. The theoretical results to be obtained will be of the first order, but will significantly illustrate the theory of current gain. The theory to be presented herein is based primarily on that introduced by W. Shockley and E. L. Steele.[2–5]

The simplest structure to consider would be similar to the p-n-p model discussed in the previous introductory chapter. The one exception, however, is that in order to facilitate a one-dimensional analysis, we must formulate a semi-infinite structure as shown in Fig. 8-1, such that it can be assumed that the carriers move in the x direction only. In this semi-infinite case, the surface boundaries are sufficiently far away in the y and z

* References, indicated in the text by superscript figures, are listed at the end of the chapter.

directions that they do not contribute to the problem. In Fig. 8-1, the emitter junction at x_E is forward-biased with the potential V_E such that the total input emitter current is I_E. The voltage V_C reverse-biases the collector junction x_C; the collector current is designated I_C. Both the emitter and collector regions are of low resistivity, making them highly p type. The n region is of moderate resistivity and of thickness W, equal to $x_C - x_E$. All ohmic contacts are assumed to be infinitely far away.

8-2. p-n-p Energy-potential Diagrams. From the energy-band point of view, the Fermi levels (at thermal equilibrium) for the p regions in our transistor model would lie very close to the valence-band edge, while the

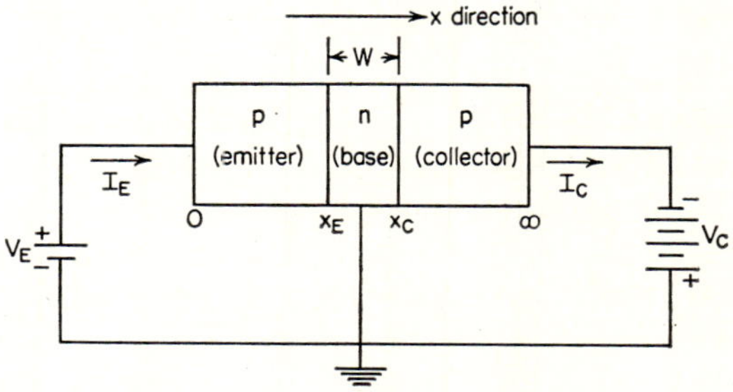

FIG. 8-1. The p-n-p junction transistor.

Fermi level for the n-type base region would be closer to the conduction band. As we have seen, the deviations of the Fermi level from the middle of the band gap (intrinsic position) are dependent on the impurity concentration and therefore on the resistivity. In order to adequately correlate the energy-potential diagram for the p-n-p transistor with the applied voltages, it becomes convenient to plot the diagram in terms of hole energies upward, rather than for electron energies, as has been done previously. This results in an inverted band picture with the valence band on top. Thus, for equilibrium, i.e., without applied voltages, the diagram for the p-n-p appears as in Fig. 8-2a. Since the Fermi level must be constant completely across, we have the contact potentials at the junctions $(\psi_n - \psi_p)$ as shown. The n-region potential is slightly more positive than that of the p regions, which we have seen is the case for equilibrium p-n junctions.

When the voltage biases shown in Fig. 8-1 are applied with the base region maintained at ground potential, equilibrium conditions no longer hold and the diagram appears as in Fig. 8-2b. The emitter region is raised by a potential V_E, thereby reducing the base-to-emitter potential, making it easier for holes to be injected into the base region. The collector region is made more negative than the base (reverse bias) by the potential V_C. Because of the concentration gradient in the base region, the injected holes diffuse across to the collector junction, whereupon they are easily swept across into the low-potential-energy region of the collector. As in Chap. 5, the departures of carrier concentration from equilibrium values are depicted by the quasi-Fermi levels, which are

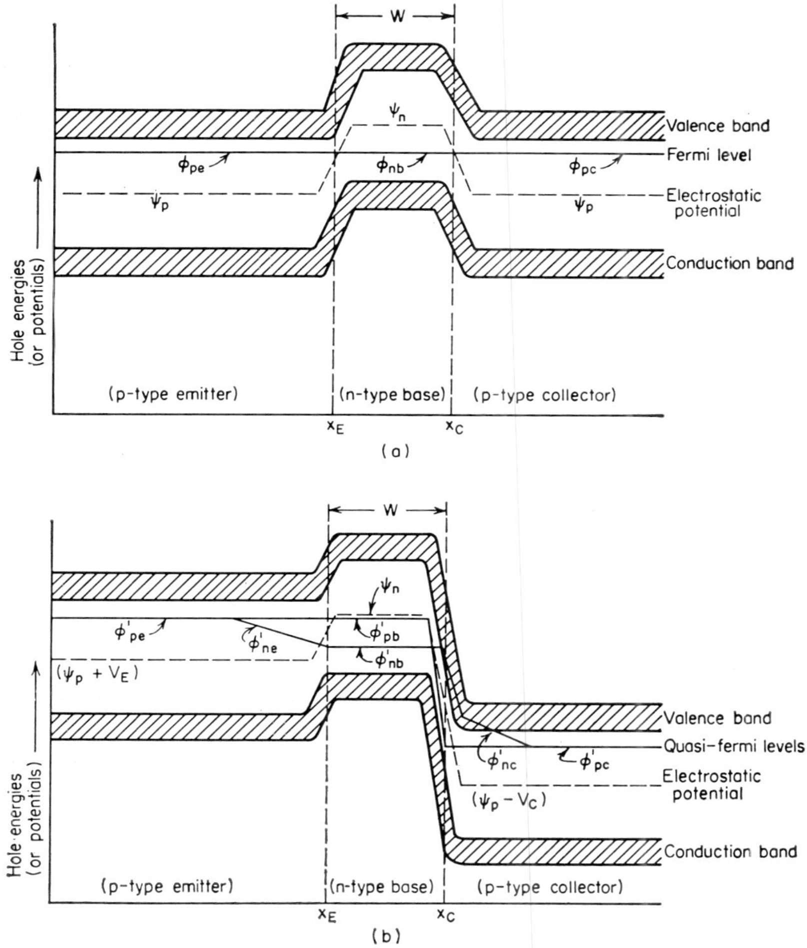

FIG. 8-2. Energy-potential diagram for the p-n-p transistor. (*a*) p-n-p transistor at equilibrium; (*b*) p-n-p transistor with applied bias. (*After W. Shockley.*)

shown in Fig. 8-2*b*. That the curves have the shapes shown will be proved in the next section on the discussion of the assumptions for the analysis.

8-3. Assumptions and Boundary Conditions. To simplify the analysis it is necessary to make important assumptions concerning the characteristics of our one-dimensional model. These assumptions are similar to those stated for the p-n junction and are as follows:

1. The conductivities of all regions are sufficiently high so that we may assume that all the applied voltage is dropped across the transition regions. This means that the carriers diffuse in regions where the electric field is zero.

2. The widths of the transition regions are very much smaller than the diffusion lengths L_p and L_n for the carriers, so that no recombination occurs within the transition regions, making the currents constant through them.

3. The lengths of the p-type emitter and collector regions are greater than the minority-carrier diffusion lengths, so that at both ends the carrier concentrations are the equilibrium values.

4. The density of holes injected into the base region is small compared to the majority-carrier density there, so that drift effects are negligible.

5. The junctions are step junctions, i.e., the transitions from p type to n type are abrupt.

Referring again to Fig. 8-2, on the basis of assumption 1 we would expect that the electrostatic potential lines (and band edges) are constant outside the transition regions as shown. Since electric field is the derivative of potential, $\mathcal{E} = 0$ outside the junctions. Assumptions 2 to 4 explain why the quasi-Fermi levels have the shapes indicated. If the hole concentrations in the emitter and collector are very much greater than the electron concentration in the base, the condition of a small injected hole density does not appreciably alter the initial concentrations from the equilibrium values. Therefore, in the emitter and collector regions, the quasi-Fermi levels for holes ϕ'_{pe} and ϕ'_{pc} are shown to be the same as the equilibrium levels ϕ_{pe} and ϕ_{pc}, respectively. To account for the injected hole density in the base region, ϕ'_{pe} is extended across to the collector, on the basis of assumptions 2 and 5. If there was appreciable recombination, ϕ'_{pb} would slope down slightly towards the collector. At the collector junction, ϕ'_{pb} drops rapidly within the transition region to the level ϕ'_{pc}. Under these applied bias conditions, electrons will flow from the collector to the base and also from the base to the emitter. Again, since we are dealing with small densities, the equilibrium concentration in the base region is hardly disturbed, and ϕ'_{nb} is shown the same as ϕ_{nb}. Since the electrons recombine with holes as they diffuse in the collector and emitter regions, the quasi-Fermi levels for the electrons ϕ'_{ne} and ϕ'_{nc} are shown decreasing to the equilibrium levels, as dictated by assumption 3.

As in Chap. 5, the quasi-Fermi-level argument enables us to postulate the important boundary conditions concerning the carrier concentrations at the junctions for nonequilibrium voltages. Although the analysis will not be repeated here, we have seen that for the holes in the base region, the concentrations are given as

$$p = p_{nb}\epsilon^{qV_E/kT} \quad \text{at emitter junction} \tag{8-1}$$
$$p = p_{nb}\epsilon^{-qV_C/kT} \quad \text{at collector junction} \tag{8-2}$$

where p_{nb} is the equilibrium hole concentration in the n-type base region and V_E and V_C are the d-c emitter and collector voltages, respectively. Similarly, the electron concentrations in the p-type emitter and collector regions will change exponentionally with voltage. For the p-n-p transistor, the boundary conditions for the minority-carrier concentrations at both sides of the junctions are summarized in Fig. 8-3. At the collector

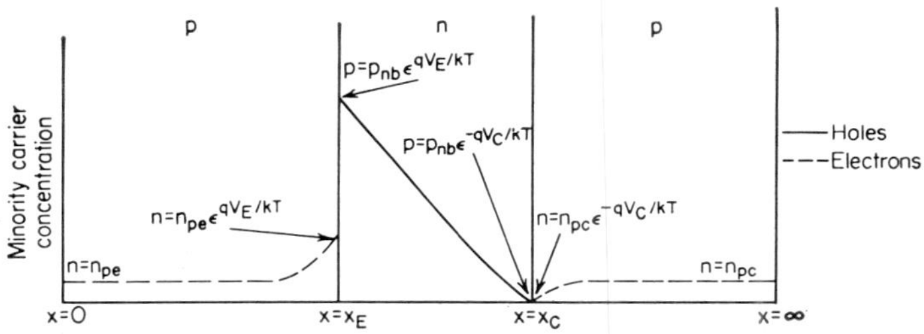

FIG. 8-3. Boundary conditions for minority-carrier concentrations for the p-n-p junction transistor.

junction, $x = x_C$, the hole and electron concentrations are almost zero because of the large reverse bias $-V_C$. It should be noted that n_{pe} and n_{pc} are the equilibrium electron concentrations (minority carriers) in the emitter and collector regions, respectively.

8-4. Emitter and Collector Currents. In order to obtain the expressions relating the input and output currents to the voltages for the p-n-p transistor, we must first solve the familiar equations for the prescribed boundary conditions. These equations, for one dimension, are:

$$-\frac{1}{q}\frac{dJ_p}{dx} - \frac{p - p_n}{\tau_p} = \frac{dp}{dt} \tag{8-3}$$

and

$$\frac{1}{q}\frac{dJ_n}{dx} - \frac{n - n_p}{\tau_n} = \frac{dn}{dt} \tag{8-4}$$

where τ_p, τ_n = lifetime for holes, electrons

J_p, J_n = hole, electron current density, amps/cm^2

From Eqs. (5-31) and (5-32), the current densities are:

$$J_p = -qD_p\frac{dp}{dx} + q\mu_p p\mathcal{E} \tag{8-5}$$

$$J_n = qD_n\frac{dn}{dx} + q\mu_n n\mathcal{E} \tag{8-6}$$

Since we are assuming that the carriers move principally by diffusion, or that the electric field $\mathcal{E}$ is zero, (8-5) and (8-6) become

$$J_p = -qD_p \frac{dp}{dx} \tag{8-7}$$

$$J_n = qD_n \frac{dn}{dx} \tag{8-8}$$

Taking the first derivatives of Eqs. (8-7) and (8-8) and substituting the results into the continuity equations (8-3) and (8-4), we obtain the simple diffusion equations for one dimension

$$D_p \frac{d^2p}{dx^2} - \frac{p - p_n}{\tau_p} = \frac{dp}{dt} \tag{8-9}$$

$$D_n \frac{d^2n}{dx^2} - \frac{n - n_p}{\tau_n} = \frac{dn}{dt} \tag{8-10}$$

The solutions of these equations yield the carrier concentrations as functions of distance x, which, when substituted into Eqs. (8-7) and (8-8), give us the desired currents.

Equation (8-9) will be solved first for the steady-state diffusion of holes in the n-type base region. Since, for the d-c case, $dp/dt = 0$, (8-9) becomes, for the base region,

$$\frac{d^2p}{dx^2} - \frac{p - p_{nb}}{L_{pb}^2} = 0 \tag{8-11}$$

where L_{pb}^2, the diffusion length, has been substituted for $D_p\tau_p$. The general solution for (8-11) is

$$p - p_{nb} = A\epsilon^{-x/L_{pb}} + B\epsilon^{x/L_{pb}} \tag{8-12}$$

To solve for the constants A and B, we make use of the boundary conditions. At $x = x_E$, the base-region side of the emitter junction,

$$x = x_E \qquad p = p_{nb}\epsilon^{qV_E/kT}$$

and (8-12) becomes

$$p_{nb}(\epsilon^{qV_E/kT} - 1) = A\epsilon^{-x_E/L_{pb}} + B\epsilon^{x_E/L_{pb}} \tag{8-13}$$

Also, at $x = x_C$, the base-region side of the collector junction,

$$x = x_C \qquad p = p_{nb}\epsilon^{-qV_C/kT}$$

and

$$p_{nb}(\epsilon^{-qV_C/kT} - 1) = A\epsilon^{-x_C/L_{pb}} + B\epsilon^{x_C/L_{pb}} \tag{8-14}$$

However, $x_C = x_E + W$, where W is the thickness of the base region, and (8-14) becomes

$$p_{nb}(\epsilon^{-qV_C/kT} - 1) = A\epsilon^{-(x_E+W)/L_{pb}} + B\epsilon^{(x_E+W)/L_{pb}} \tag{8-15}$$

Solving Eqs. (8-13) and (8-15) simultaneously for A and B, we obtain

$$A = \frac{p_{nb}(\epsilon^{qV_C/kT} - 1)\epsilon^{W/L_{pb}} - p_{nb}(\epsilon^{-qV_C/kT} - 1)}{\epsilon^{-(x_E-W)/L_{pb}} - \epsilon^{-(x_E+W)/L_{pb}}} \tag{8-16}$$

$$B = \frac{p_{nb}(\epsilon^{qV_E/kT} - 1)\epsilon^{-W/L_{pb}} - p_{nb}(\epsilon^{-qV_C/kT} - 1)}{\epsilon^{(x_E-W)/L_{pb}} - \epsilon^{(x_E+W)/L_{pb}}} \tag{8-17}$$

Factoring ϵ^{-x_E} and ϵ^{x_E} from the denominators of (8-16) and (8-17), respectively, and replacing the remainder with the hyperbolic sine, we obtain

$$A = \frac{p_{nb}(\epsilon^{qV_E/kT} - 1)\epsilon^{(x_E+W)/L_{pb}} - p_{nb}(\epsilon^{-qV_C/kT} - 1)\epsilon^{x_E/L_{pb}}}{2 \sinh (W/L_{pb})} \tag{8-18}$$

$$B = \frac{p_{nb}(\epsilon^{qV_E/kT} - 1)\epsilon^{-(x_E+W)/L_{pb}} - p_{nb}(\epsilon^{-qV_C/kT} - 1)\epsilon^{-x_E/L_{pb}}}{-2 \sinh (W/L_{pb})} \tag{8-19}$$

Substituting these terms into the original equation (8-12) and simplifying the final results, we have

$$p - p_{nb} = \frac{p_{nb}(\epsilon^{-qV_C/kT} - 1) \sinh [(x - x_E)/L_{pb}]}{\sinh (W/L_{pb})} - \frac{p_{nb}(\epsilon^{qV_E/kT} - 1) \sinh [(x - x_E - W)/L_{pb}]}{\sinh (W/L_{pb})} \tag{8-20}$$

which is the desired solution. To obtain the hole component of the emitter current, J_{pE}, we evaluate Eq. (8-7) at $x = x_E$.

$$J_{pE} = -qD_p \frac{dp}{dx}\bigg|_{x=x_E} \tag{8-21}$$

Taking the derivative of (8-20), we get

$$\frac{dp}{dx} = \frac{1}{L_{pb}} \left\{ \frac{p_{nb}(\epsilon^{-qV_C/kT} - 1) \cosh [(x - x_E)/L_{pb}]}{\sinh (W/L_{pb})} - \frac{p_{nb}(\epsilon^{qV_E/kT} - 1) \cosh [(x - x_E - W)/L_{pb}]}{\sinh (W/L_{pb})} \right\} \tag{8-22}$$

Evaluating (8-22) at $x = x_E$, substituting into (8-21), and simplifying, we obtain, finally,

$$J_{pE} = \frac{qD_{pb}p_{nb}}{L_{pb}} \left[-(\epsilon^{-qV_C/kT} - 1) \operatorname{csch} \frac{W}{L_{pb}} + (\epsilon^{qV_E/kT} - 1) \coth \frac{W}{L_{pb}} \right] \tag{8-23}$$

This equation gives the current density due to holes that diffuse into the base from the emitter junction.

The hole current at the collector junction is derived by evaluating Eq. (8-22) at $x = x_C$, noting that $x_C - x_E = W$, and substituting the result into (8-7) as before. In similar manner we obtain

$$J_{pc} = \frac{qD_{pb}p_{nb}}{L_{pb}} \left[-(\epsilon^{-qV_C/kT} - 1) \coth \frac{W}{L_{pb}} + (\epsilon^{qV_E/kT} - 1) \operatorname{csch} \frac{W}{L_{pb}} \right] \tag{8-24}$$

To complete the picture for the junction currents, it is necessary to obtain the relations for the electron components. At the emitter junction, the forward-bias current consists of electrons injected from the base into the emitter region where they recombine with holes back to the equilibrium value n_{pe}. Similarly, at the collector junction, the reverse-bias current consists of those electrons which are swept across the junction from the equilibrium level n_{pc}. These electron currents were shown in the minority-carrier diagram of Fig. 8-3. The general diffusion equation for electron flow in either the emitter or collector region is of the form

$$\frac{d^2n}{dx^2} - \frac{n - n_p}{L_n} = 0 \tag{8-25}$$

where L_n is the diffusion length for electrons in the p regions. Actually, we see that this equation is identical to that used in Chap. 5 for solving for the electron current in the p-n junction. Since our assumptions here involve similar boundary conditions as well as the fact that the thickness of the emitter and collector regions are very much greater than the diffusion lengths, we may write the solutions without further analysis. At the emitter junction, the electron current density is

$$J_{nE} = \frac{qD_{ne}n_{pe}}{L_{ne}} (\epsilon^{qV_E/kT} - 1) \tag{8-26}$$

Also, the electron current density at the collector junction becomes

$$J_{nC} = \frac{-qD_{ne}n_{pc}}{L_{nc}} (\epsilon^{-qV_C/kT} - 1) \tag{8-27}$$

Having established the relationships for the hole and electron current densities at the particular junctions, we can now write the complete expressions for the emitter and collector currents. The total current densities are given as

$$J_E = J_{pE} + J_{nE} \tag{8-28}$$
$$J_C = J_{pC} + J_{nC} \tag{8-29}$$

where J_E is the emitter current density, and J_C is the collector current density. Substituting Eqs. (8-23) and (8-26) into (8-28), and Eqs. (8-24) and (8-27) into (8-29), we get, finally,

$$J_E = \left(\frac{qD_{ne}n_{pe}}{L_{ne}} + \frac{qD_{pb}p_{nb}}{L_{pb}} \coth \frac{W}{L_{pb}}\right)(\epsilon^{qV_E/kT} - 1) - \frac{qD_{pb}p_{nb}}{L_{pb}} \operatorname{csch} \frac{W}{L_{pb}} (\epsilon^{-qV_C/kT} - 1) \qquad \text{(8-30)}$$

$$J_C = \frac{qD_{pb}p_{nb}}{L_{pb}} \operatorname{csch} \frac{W}{L_{pb}} (\epsilon^{qV_E/kT} - 1) - \left(\frac{qD_{nc}n_{pc}}{L_{nc}} + \frac{qD_{pb}p_{nb}}{L_{pb}} \coth \frac{W}{L_{pb}}\right)(\epsilon^{-qV_C/kT} - 1) \qquad \text{(8-31)}$$

These are the fundamental transistor current-density equations which characterize the operation of the p-n-p junction transistor, and from which the expression for current gain will be obtained.

8-5. Current-gain Theory for Uniform Bases. The expressions for the emitter and collector current densities obtained in the previous section, viz., Eqs. (8-30) and (8-31), are for the transistor model in which the impurity concentration is uniform throughout the base region. This is the model that one uses when analyzing nondiffused transistor structures such as alloy and double-doped transistors. In this section we shall develop the theory of transistor current gain (alpha) for the transistor with the uniform base.

In Chap. 7, the current gain α for the grounded-base configuration was defined as

$$\alpha = \left.\frac{\partial I_C}{\partial I_E}\right|_{V_C = \text{const}} \qquad \text{(8-32)}$$

which we see is simply the ratio of the collector current to the emitter current. To relate α to the physical parameters of the transistor we make use of the total current density Eqs. (8-30) and (8-31). Assuming that the magnitude of $\epsilon^{-qV_C/kT}$ is negligible under normal operation and eliminating $(\epsilon^{qV_E/kT} - 1)$ between J_E and J_C, we have

$$\frac{J_E - (qD_{pb}p_{nb}/L_{pb}) \operatorname{csch} (W/L_{pb})}{qD_{ne}n_{pe}/L_{ne} + (qD_{pb}p_{nb}/L_{pb}) \coth (W/L_{pb})} = \frac{J_C - [qD_{nc}n_{pc}/L_{nc} + (qD_{pb}p_{nb}/L_{pb}) \coth (W/L_{pb})]}{(qD_{pb}p_{nb}/L_{pb}) \operatorname{csch} (W/L_{pb})} \qquad \text{(8-33)}$$

Solving (8-33) for J_C and differentiating with respect to J_E, we obtain

$$\alpha = \frac{\partial I_C}{\partial I_E} = \frac{\partial J_C}{\partial J_E} = \frac{(qD_{pb}p_{nb}/L_{pb}) \operatorname{csch} (W/L_{pb})}{qD_{ne}n_{pe}/L_{ne} + (qD_{pb}p_{nb}/L_{pb}) \coth (W/L_{pb})} \qquad \text{(8-34)}$$

Replacing csch (W/L_{pb}) by its identity, coth (W/L_{pb}) sech (W/L_{pb}), we get, finally,

$$\alpha = \frac{\operatorname{sech} (W/L_{pb})}{1 + (D_{ne}/D_{pb})(n_{pe}/p_{nb})(L_{pb}/L_{ne}) \tanh (W/L_{pb})} \qquad \text{(8-35)}$$

For the case of $W/L_{pb} \ll 1$, Eq. (8-35) reduces to

$$\alpha = \frac{\operatorname{sech}(W/L_{pb})}{1 + (D_{ne}/D_{pb})(n_{pe}/p_{nb})(W/L_{ne})} \tag{8-36}$$

This is the general equation for alpha, the ratio of the collector current to the emitter current. Inspecting this result, it is seen that it is of the form

$$\alpha = \gamma\beta^* \tag{8-37}$$

where
$$\gamma = \frac{1}{1 + (D_{ne}/D_{pb})(n_{pe}/p_{nb})(W/L_{ne})} \tag{8-38}$$

and
$$\beta^* = \operatorname{sech}\frac{W}{L_{pb}} \tag{8-39}$$

The significance of Eqs. (8-38) and (8-39) will be studied independently in the sections to follow.

a. Emitter Efficiency. As was pointed out in Chap. 7, the emitter efficiency γ is defined as the ratio of the injected hole current to the total emitter current (for p-n-p), or in terms of densities,

$$\gamma = \frac{J_{pE}}{J_E} = \frac{J_{pE}}{J_{pE} + J_{nE}} = \frac{1}{1 + J_{nE}/J_{pE}} \tag{8-40}$$

Thus, for a p-n-p transistor, when the electron current injected into the p-type emitter approaches zero, the emitter efficiency approaches unity, indicating 100 per cent injection efficiency. If the lifetime for holes in the base region is large, so that $L_{pb} \gg W$, then the hole-current density from Eq. (8-23) is approximately

$$J_{pE} \approx \frac{qD_{pb}p_{nb}}{W}\epsilon^{qV_E/kT} \tag{8-41}$$

One recalls that this result is equivalent to that which was obtained for the p-n junction. This is understandable if we consider the collector junction to behave like an ohmic contact. The one difference, however, is that the hole concentration is considerably less than the equilibrium value because of the reverse bias. In any case, Eq. (8-41) is applicable, since the hole concentration injected into the base is much greater than either p_{nb} or $p_{nb}\epsilon^{-qV_C/kT}$.

Similarly, the emitter electron current density J_{nE} is given by Eq. (8-26) as

$$J_{nE} \approx \frac{qD_{ne}n_{pe}}{L_{ne}}\epsilon^{qV_E/kT} \tag{8-42}$$

Substituting Eqs. (8-41) and (8-42) into (8-40) and simplifying, we obtain the same result as given by Eq. (8-38). Thus, the γ term in the alpha expression corresponds to the emitter efficiency.

The emitter-efficiency expression may be more conveniently represented in terms of the emitter and base resistivities ρ_e and ρ_b, where

$$\rho_e \approx \frac{1}{q\mu_{pe}p_{pe}} = \frac{n_{pe}}{q\mu_{pe}n_i^2} \tag{8-43}$$

$$\rho_b \approx \frac{1}{q\mu_{nb}n_{nb}} = \frac{p_{nb}}{q\mu_{nb}n_i^2} \tag{8-44}$$

for the reason that $np = n_i^2$. Applying the Einstein relationship, $\mu = \frac{q}{kT} D$, to the mobilities and solving for n_{pe} and p_{nb}, we get

$$n_{pe} = \frac{q^2}{kT} n_i^2 D_{pe}\rho_e \tag{8-45}$$

$$p_{nb} = \frac{q^2}{kT} n_i^2 D_{nb}\rho_b \tag{8-46}$$

Taking the ratio n_{pe}/p_{nb} and substituting into Eq. (8-38), we get the complete expression for emitter efficiency.

$$\gamma = \frac{1}{1 + (D_{pe}D_{ne}/D_{nb}D_{pb})(\rho_e/\rho_b)(W/L_{ne})} \tag{8-47}$$

If the mobilities in the emitter and base are assumed to be equal, Eq. (8-47) reduces to the familiar form:

$$\gamma = \frac{1}{1 + \rho_e W/(\rho_b L_{ne})} \tag{8-48}$$

This is the equation used most often for calculating the emitter efficiency of a transistor. Examining this final expression, we see that for γ to approach unity, the ratio of the emitter resistivity to the base resistivity must be as small as possible. Also, the minority-carrier lifetime in the emitter region must be as large as possible. Essentially, we are minimizing the forward-biased electron current from the base region.

Although for most well-designed junction transistors γ is close to 1, its effect on alpha cannot be neglected. Particularly at higher currents, the change in γ becomes so significant that it dominates the current-gain characteristic. The variation of γ with current density will be studied in Chap. 10.

b. Base-transport Factor. Referring again to the general expression for alpha given by Eq. (8-37), we see that the remaining term is given as

$$\beta^* = \operatorname{sech} \frac{W}{L_{pb}} \tag{8-49}$$

where W is the base width and L_{pb} is the hole diffusion length in the base region. It follows from the definition of alpha that β^* must represent that fraction of the injected hole current that appears at the collector junction as current. In a sense, β^* depicts the probability that the hole carrier will reach the collector without recombining in the base region. Therefore β^* is referred to as the *base-transport factor*. It is apparent that for β^* to approach unity, W must be small compared to L_{pb}. Since $L_{pb} = \sqrt{D_{pb}\tau_{pb}}$, the minority-carrier lifetime in the base region must be as high as possible. It should be pointed out here that for a given base width W, the base-transport factor will decrease as the base impurity

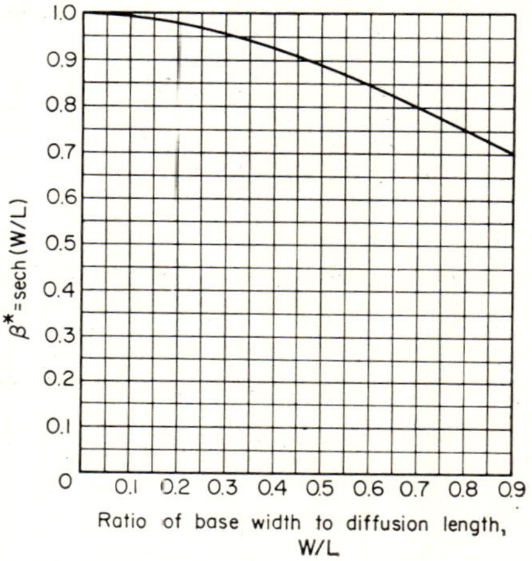

FIG. 8-4. Base-transport factor as a function of width-to-diffusion-length ratio ($\beta^* =$ sech W/L).

concentration increases, since both D_{pb} and τ_{pb} decrease with doping. For purposes of design calculations of β^*, a plot of the hyperbolic secant of W/L is given in Fig. 8-4. For β^* to exceed 0.9, it is seen that the diffusion length must be at least 2 to 3 times larger than the base width. For the small values of W/L usually encountered, it is often useful to replace sech (W/L_{pb}) by its first-order series expansion, so that

$$\beta^* \approx 1 - \frac{1}{2}\left(\frac{W}{L_{pb}}\right)^2 \tag{8-50}$$

c. Collector Multiplication. As was mentioned in the previous chapter, it is possible for the total collector current to be greater than the hole current arriving at the collector junction. This gives rise to an additional parameter, α^*, the collector multiplication ratio. This multiplication effect shows up only in junction transistors for which the collector resistivity is of the order of several ohm-centimeters or more. In such a case, the concentration of holes entering the collector region from the base may not be negligible compared to the majority hole concentration there. Consequently, charge neutrality in the collector is upset and the minority electrons must therefore establish a concentration gradient equal to that of the holes, creating an electric field. Thus we have a situation similar to that of the high-level injection condition for the p-n junction rectifier given in Chap. 6. However, because of the reverse bias at the collector junction, the electrons from the p region are swept across as current. Thus the total collector-junction current becomes greater than the hole current reaching the junction. To determine the expression for α^*, one must take into account the drift currents in the collector region. It has been shown that[6]

$$\alpha^* \approx 1 + \frac{n_{pc}\mu_{nc}}{p_{pc}\mu_{pc}} \qquad \text{p-n-p} \tag{8-51}$$

$$\alpha^* \approx 1 + \frac{p_{pc}\mu_{pc}}{n_{pc}\mu_{nc}} \qquad \text{n-p-n} \tag{8-52}$$

In other words, α^* is related to the ratio of the minority-carrier concentration to the majority-carrier concentration in the collector region. For very heavily doped collectors such as are found in alloy transistors, $\alpha^* = 1$. For epitaxial transistors, where the collector resistivity is higher than that of the base, α^* may be slightly greater than 1. Taking this into account, the over-all alpha for a junction transistor is

$$\alpha = \gamma\beta^*\alpha^* \tag{8-53}$$

One sees that if α^* is sufficiently greater than unity, the over-all alpha may exceed 1. This leads to gain instability when the transistor is operated in the grounded-emitter configuration. Nevertheless, it should be evident that α^* is closer to 1 for silicon than for germanium, since the minority-carrier concentrations are considerably smaller for a given resistivity. Further, for germanium transistors, α^* does not appreciably affect α until the collector resistivities exceed about 5 ohm-cm.

In addition to the collector multiplication just described, there exists another current-multiplication effect, due to carrier ionization in the depletion layer as the collector voltage is increased toward the avalanche-breakdown value. This was discussed in Chap. 6 under the subject of reverse-voltage breakdown. Thus, for transistor operation in the region close to breakdown, the over-all alpha must be multiplied by the factor M as given by the curves in Fig. 6-11.

$$\alpha = \gamma\beta^*\alpha^* M \tag{8-54}$$

If we assume that $\alpha^* M = 1$, we may finally write

$$\alpha \approx \left(1 + \frac{\rho_e W}{\rho_b L_{ne}}\right)^{-1} \left(1 - \frac{W^2}{2L_{pb}^2}\right) \tag{8-55}$$

which is one of the most fundamental junction-transistor design equations. In summary, it is seen from Eq. (8-55) that current gain is maximized by establishing thin base widths, high lifetimes, and low-resistivity emitter regions.

8-6. Cutoff Frequency for Uniform Bases. It is recognized that the minority carriers diffusing across the base region have a certain inertia associated with their effective masses and mobility. If the frequency of the voltage applied to the emitter junction is increased, a point is reached where the carriers do not respond instantaneously to the applied signal. This manifests itself in a physical lag or phase shift such that recombination is increased and the current gain starts to fall off. Since this phenomenon is associated with the carriers in the base region of the transistor, we must analyze the frequency response of the base transport factor β^*. Chapter 13 takes into consideration the frequency variation of γ. By definition, the frequency at which β^* falls to 0.707 of its low-frequency value β_o^* is called the base-cutoff frequency f_b.

To facilitate the frequency analysis of the one-dimensional model, we may insert, in series with the d-c voltage V_E in Fig. 8-1, a small a-c signal generator equal to $v_e\epsilon^{j\omega t}$, where v_e is the amplitude and $\omega = 2\pi f$. Thus the total voltage is of the form

$$V = V_E + v_e\epsilon^{j\omega t} \tag{8-56}$$

This equation implies that the solution to the diffusion equation will now have both d-c and a-c solutions. Repeating Eq. (8-9), we have

$$D_{pb}\frac{\partial^2 p}{\partial x^2} - \frac{p - p_n}{\tau_{pb}} = \frac{\partial p}{\partial t} \tag{8-57}$$

where p is now a function of x and t. A general solution for this differential equation for both direct current and alternating current is:

$$p = p_{\text{d-c}} + p_{\text{a-c}}\epsilon^{j\omega t} \tag{8-58}$$

If we differentiate Eq. (8-58) with respect to time, we get the general form

$$\frac{\partial p}{\partial t} = j\omega p \tag{8-59}$$

Substituting this result into Eq. (8-57) and letting $L_{pb}{}^2 = D_{pb}\tau_{pb}$, we get, finally,

$$\frac{d^2p}{dx^2} - \left(\frac{1 + j\omega\tau_{pb}}{L_{pb}{}^2}\right)p = 0 \tag{8-60}$$

Examining Eq. (8-60) and comparing it to Eq. (8-11), we see that the solutions would be identical provided that each diffusion-length term L_{pb} is multiplied by $(1 + j\omega\tau_{pb})^{-1/2}$. This is the complex function which imparts the frequency variation into the final solutions of the diffusion equation. Consequently, the exact expression for the base-transport factor becomes

$$\beta^* = \operatorname{sech}\left[\frac{W}{L_{pb}}(1 + j\omega\tau_{pb})^{1/2}\right] \tag{8-61}$$

This was obtained by substituting the complex function for L_{pb} in Eq. (8-49). At very low frequencies such that $\omega\tau_{pb} \ll 1$, Eq. (8-61) becomes approximately

$$\beta_o^* = \operatorname{sech}\frac{W}{L_{pb}} \tag{8-62}$$

where β_o^* denotes the low-frequency value. Equation (8-62) is identical to the original Eq. (8-49). If the terms of Eq. (8-61) are rearranged so that they are in the form of [sech $(a + jb)$] we may write down the hyperbolic secant of the absolute magnitude of $(a + jb)$, which becomes

$$\beta^* \approx \operatorname{sech}\left(\omega\tau_{pb}\frac{W^2}{L_{pb}^2}\right)^{1/2} = \operatorname{sech}(\omega t_b)^{1/2} \tag{8-63}$$

where ω = signal frequency
τ_{pb} = lifetime of base region
L_{pb} = diffusion length
W = thickness of base region

The term $\tau_{pb}W^2/L_{pb}^2$ may be regarded as the transit time for carriers for the base region. Thus, from Eq. (8-63), we have

$$t_b = \frac{W^2}{D_{pb}} \tag{8-64}$$

where t_b is the base-transit time. From this equation, at a frequency ω_b such that $\omega_b t_b = 2.43$, the magnitude of β^* is $1/\sqrt{2}$ and is down 0.707 or 3 db below its low-frequency value β_o^*.[7] This frequency ω_b is defined as the base-cutoff frequency for the transistor. Since $\omega_b t_b = 2.43$, we have

$$f_b = \frac{2.43 D_{pb}}{2\pi W^2} \tag{8-65}$$

Using the Einstein relation $D = \dfrac{kT}{q}\mu$, Eq. (8-65) becomes, for room temperature,

$$f_b = \frac{0.01\mu_{pb}}{W^2} \qquad \text{p-n-p} \tag{8-66}$$

$$f_b = \frac{0.01\mu_{nb}}{W^2} \qquad \text{n-p-n} \tag{8-67}$$

Thus we see that base cutoff varies inversely as the square of the base width W and is directly proportional to the minority-carrier mobility.

The important point to note, however, is that f_b is independent of the lifetime of the base region. This is not surprising, since frequency response is a measure of how fast the carriers diffuse through the base, and as long as the width is small compared to a diffusion length, f_b should depend primarily on the carrier mobility. It is apparent that for all things being equal, an n-p-n junction transistor will have a higher base-cutoff frequency than a p-n-p, since the electron mobility is higher than the hole mobility. On the same basis, a germanium transistor is superior

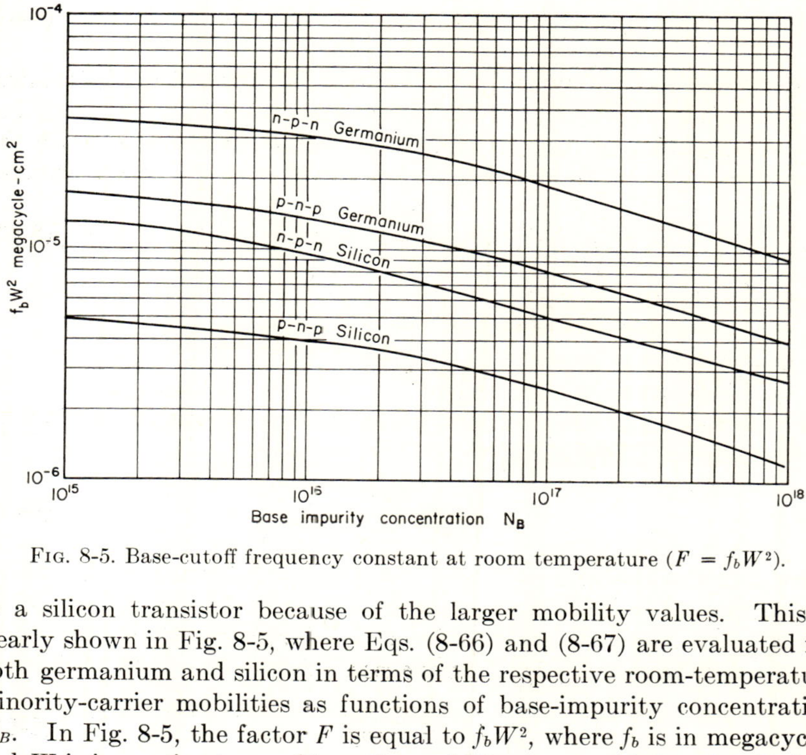

FIG. 8-5. Base-cutoff frequency constant at room temperature ($F = f_b W^2$).

to a silicon transistor because of the larger mobility values. This is clearly shown in Fig. 8-5, where Eqs. (8-66) and (8-67) are evaluated for both germanium and silicon in terms of the respective room-temperature minority-carrier mobilities as functions of base-impurity concentration N_B. In Fig. 8-5, the factor F is equal to $f_b W^2$, where f_b is in megacycles and W is in centimeters. Since the mobility drops with heavier doping, the frequency of base cutoff will decrease with resistivity, assuming that the base width is maintained constant.

8-7. Current-gain Theory for Graded Bases. In the one-dimensional analysis of current gain given in Sec. 8-5 for a p-n-p junction-transistor model, it was assumed that the impurity concentration in the base region was uniform. Further, it was assumed that the injected hole concentration was small compared to the majority-carrier concentration (this equals the base donor-impurity concentration) in the base region, so that the holes moved principally by diffusion in a region where the electric-field intensity was virtually zero. In other words, we considered a low-level injection case, where carrier drift effects were negligible. Neglecting collector multiplication, it was shown that alpha is

$$\alpha = \gamma\beta^* \tag{8-68}$$

where γ is the emitter efficiency and β^* is the base-transport factor.

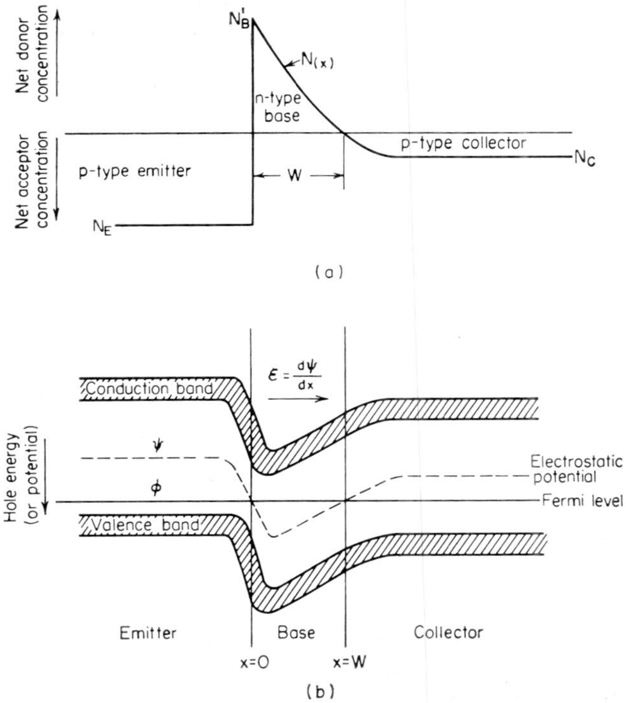

FIG. 8-6. Junction transistor having a graded base layer and a graded collector junction (p-n-p). The diagram shows the introduction of a built-in electric field. (*a*) Net impurity concentration; (*b*) energy-band diagram at equilibrium for constant Fermi level.

The situation for diffused-base transistors having a nonuniform or graded base layer is somewhat different, due to the addition of another effect. Let us consider a p-n-p junction-transistor model having a step emitter junction and a graded base layer given by the impurity distribution $N(x)$. The latter is strictly a general impurity distribution and not confined to any particular distribution such as the linear, error function, or gaussian. The net impurity concentration as a function of distance along the transistor is given in Fig. 8-6*a*, where N'_B denotes the impurity concentration just at the base side of the emitter junction. At

$x = W$, the collector junction is shown to be graded, although this is not pertinent to the problem of determining alpha.

If no bias is applied to this model, the energy-band diagram at equilibrium appears as shown in Fig. 8-6*b*. The Fermi level is drawn as a constant-energy reference across the entire length of the transistor structure. Consequently, the energy bands are shifted in energy (or potential) to account for the carrier concentrations in the various regions. In the emitter region, which is usually doped very heavily p type, the top of the valence band is drawn very close to the Fermi level. In the collector, which is not quite as p type as the emitter, the valence-band edge is further away from ϕ. Within the base region, however, the conductivity is n type and therefore the bands must shift to bring the bottom of the conduction band closer to the Fermi level. The important point to note is that because the carrier concentration in the base is nonuniform due to $N(x)$, the energy bands will bend in accordance with the grading $N(x)$. Near the emitter junction, the conduction band is closest to ϕ, since $N(x)$ is at its maximum value N'_B. As the collector is approached, the conduction band moves farther away, until at $x = W$, ϕ lies midway between the bands, corresponding to the intrinsic position. The reader will note that the base region is bounded by the points where the Fermi-level potential ϕ just equals the intrinsic electrostatic potential ψ. These are the junction crossover points.

A comparison of Fig. 8-6*b* with Fig. 8-2*a* clearly shows the effect of the base gradient. In the case of the latter, the electrostatic potential ψ is constant within the base region, so that the term $d\psi/dx$, which is the electric field, is zero in the base region. However, for the graded base, the potential ψ is higher at the emitter than at the collector. Therefore, $d\psi/dx$ has a finite value, which indicates the *existence of an electric field* within the base region. This is a built-in field which arises to prevent the majority carriers from diffusing because of their concentration gradient $dN(x)/dx$.[8,9] This field, which keeps electrons in their place, is in such a direction as to aid the transport of injected holes. Thus, for the condition of low-level injection, the holes move by both diffusion and drift. It must be emphasized that all of the above is also applicable to n-p-n structures, wherein the built-in field is in the opposite direction and aids to injected electrons.

To study the effect of the built-in electric field on α, we must first define $\mathcal{E}$ and then derive the expression for the hole current. From Eq. (5-10) we can write

$$n = n_i \epsilon^{q(\psi - \phi)/kT} \tag{8-69}$$

or, for a graded base layer,

$$n(x) = N(x) = n_i \epsilon^{q[\psi(x) - \phi]/kT} \tag{8-70}$$

If we differentiate (8-70) with respect to x, we get

$$\frac{dN(x)}{dx} = \frac{q}{kT} n_i \epsilon^{q[\psi(x)-\phi]/kT} \frac{d\psi(x)}{dx} \tag{8-71}$$

Since the built-in field $\mathcal{E}$ is

$$\mathcal{E} = -\frac{d\psi(x)}{dx} \tag{8-72}$$

we obtain, from (8-70) and (8-71),

$$\mathcal{E} = -\frac{kT}{q} \frac{1}{N(x)} \frac{dN(x)}{dx} \tag{8-73}$$

This is the general expression for electric field as a function of any impurity distribution. To account for both diffusion and drift effects, the hole current density in a nonuniform base may be given by

$$J_p = q\mu_p p\mathcal{E} - qD_p \frac{dp}{dx} \tag{8-74}$$

Substituting (8-73) into (8-74) and letting $\mu_p = (q/kT)D_p$, we have

$$J_p = -qD_p \left[\frac{dp}{dx} + \frac{p}{N(x)} \frac{dN(x)}{dx} \right] \tag{8-75}$$

Multiplying (8-75) by $N(x)$ and integrating both sides, we obtain

$$-\frac{J_p}{qD_{pb}} \int^x N(x)\, dx = \int^x \frac{d[pN(x)]}{dx}\, dx \tag{8-76}$$

If we insert the boundary condition that $p = 0$ at the collector, $x = W$, and solve for p, we get, finally,[10]

$$p = \frac{J_p}{qD_p} \frac{1}{N(x)} \int_x^W N(x)\, dx \tag{8-77}$$

This result gives us the hole concentration as a function of distance in the base layer. If we evaluate (8-77) at $x = 0$, where $N(x) = N'_B$, then we obtain the injected hole concentration.

$$p = \frac{J_p}{qD_p} \frac{1}{N'_B} \int_0^W N(x)\, dx \tag{8-78}$$

But if we assume from the quasi-Fermi-level argument that

$$p = p_{nb} \epsilon^{qV_B/kT} = \frac{n_i^2}{N'_B} \epsilon^{qV_B/kT} \tag{8-79}$$

then by setting (8-78) equal to (8-79) we find that J_p becomes

$$J_p = \frac{qD_p n_i^2 \epsilon^{qV_E/kT}}{\int_0^W N(x)\,dx} \tag{8-80}$$

We may now use these results to evaluate α for the graded base, using Eq. (8-68). The emitter efficiency γ is defined as

$$\gamma = \frac{1}{1 + J_n/J_p} \tag{8-81}$$

For a uniform emitter with impurity concentration N_E, we can write immediately the electron current density J_n.

$$J_n = \frac{qD_n n_{pe}}{L_{n\epsilon}} \epsilon^{qV_E/kT} = \frac{qD_n n_i^2}{L_{ne} N_E} \epsilon^{qV_E/kT} \tag{8-82}$$

Equation (8-82) assumes that the emitter-region thickness is much greater than the electron diffusion length L_{ne}. Inserting (8-80) and (8-82) into the expression for γ, we get

$$\gamma = \frac{1}{1 + q\mu_{nb} \int_0^W N(x)\,dx/(q\mu_{pe} N_E L_{nE})} \tag{8-83}$$

where the ratio D_n/D_p was replaced by μ_n/μ_p. Examination of this result reveals that it is of the form[10]

$$\gamma = \frac{1}{1 + R_{EE}/R_{BB}} \tag{8-84}$$

where

$$R_{EE} = \frac{1}{q\mu_{pe} N_E L_{ne}} = \frac{\rho_E}{L_{ne}} \tag{8-85}$$

$$R_{BB} = \frac{1}{q\mu_{nb} \int_0^W N(x)\,dx} = \frac{\rho_B(x)}{W} \tag{8-86}$$

The terms R_{EE} and R_{BB} are referred to as *sheet resistances* and defined as the ohmic resistances as measured from edge to edge of a square sheet of material of a certain thickness x. Thus, R_{EE} is the sheet resistance of the emitter having a uniform resistivity ρ_E and a thickness equal to L_{ne}. R_{BB} is the sheet resistance of a base having a graded resistivity $\rho_B(x)$ and a thickness W. The reader should note that for the uniform base, $R_{BB} = \rho_B/W$, and Eq. (8-83) simply reduces to the familiar form given by Eq. (8-47). (The subject of graded sheet resistance will be treated in greater detail in Sec. 9-6, where it will be shown how R_{BB} increases as a result of a graded impurity distribution.) Because of this, it is seen that the emitter efficiency is increased when the base region is graded. For linear, gaussian, or error-function base distributions, if N'_B is known, then

R_{BB} may be calculated by the relations to be given in Sec. 9-6. Regardless of any base impurity distribution, it is still required that the emitter doping be as heavy as possible for high emitter efficiency. This keeps the sheet-resistance ratio R_{EE}/R_{BB} small.

Having determined the relationship for γ, we can now turn our attention to the base-transport factor β^*. One recalls that this is a measure of the fraction of the injected current that reaches the collector junction, or which is not lost by recombination in the base. For the uniform case this was given as

$$\beta^* = \operatorname{sech} \frac{W}{L_{pb}} \approx 1 - \frac{1}{2}\left(\frac{W}{L_{pb}}\right)^2 \tag{8-87}$$

where L_{pb} is the diffusion length. For graded base layers, we would expect higher values of β^* for the same base width and lifetime as the uniform case. This is because the transit of the injected carriers is aided by the sweeping effect of the built-in field. We can estimate the increase in β^* from Eq. (8-75), which is repeated here as

$$J_p = -qD_{pb}\left[\frac{dp}{dx} + \frac{p}{N(x)}\frac{dN(x)}{dx}\right] \tag{8-88}$$

To a first approximation, the second term in brackets may be written as $p/N'_B \times N'_B/W$, which is approximately equal to dp/dx. On an average basis, therefore, Eq. (8-88) is approximated by

$$J_p \approx -qD_{pb}\frac{2dp}{dx} \tag{8-89}$$

which indicates an effective doubling of the diffusion constant. Inserting this result into (70), we obtain, for nonuniform bases only,

$$\beta^* \approx \operatorname{sech} \frac{W}{\sqrt{2}\, L_{pb}} \approx 1 - \frac{1}{4}\left(\frac{W}{L_{pb}}\right)^2 \tag{8-90}$$

In the final analysis it is seen that the introduction of a graded impurity distribution in the base layer will increase the over-all current gain α for the junction transistor.

8-8. Cutoff Frequency for Graded Bases.[10] The most important contribution that the built-in field makes to the over-all electrical characteristics of the junction transistor is a significant reduction of the transit time of the minority carriers in traversing the base region. This is a direct consequence of the superposition of the drift motion on the diffusion component, due to the additional force exerted on the carrier by the electric field. Because of this reduction in transit time, the frequency of cutoff for the base-transport factor is increased. If the emitter efficiency γ is considered to be constant for all frequencies, then for a given base width we would expect a proportional increase in the base-cutoff frequency. In this section we shall show to what extent f_b is increased for the various base-layer impurity distributions.

In Sec. 8-6 it was shown that for the uniform base wherein the carriers move by pure diffusion, the frequency of alpha cutoff is

$$f_b = \frac{2.43}{2\pi\tau_b} \tag{8-91}$$

where τ_b is the base-transit time.

$$\tau_b = \frac{W^2}{D_{pb}} \tag{8-92}$$

Equation (8-92) is applicable to p-n-p types; for n-p-n structures D_{pb} is replaced by D_{nb}. We see from (8-91) that if τ_b is halved, for instance, f_b is doubled.

From the study of drift mobility in Chap. 4, one recalls that the drift velocity for a minority carrier (e.g., holes) is

$$v_D = \mu_p \mathcal{E} \tag{8-93}$$

where $\mathcal{E}$ is the electric field. Assuming that mobility is independent of x,

$$t_b' = \frac{1}{\mu_{pb}} \int_0^W \frac{dx}{\mathcal{E}(x)} \tag{8-94}$$

On the average, however, we can write that the transit time for a carrier to cross a distance W is $t_b' = W/v_D$ or

$$t_b' = \frac{W}{\mu_{pb}\mathcal{E}} \tag{8-95}$$

In Eq. 8-73 we saw that the electric field in a base region having a graded impurity distribution $N(x)$ is given by

$$\mathcal{E} = -\frac{kT}{q}\frac{1}{N(x)}\frac{dN(x)}{dx} \tag{8-96}$$

Although the impurity gradient in a diffused-base transistor follows either an error-function or gaussian distribution, it is a good approximation to use an exponential function in order to simplify the analysis. Thus we may write

$$N(x) = N_B' \epsilon^{-x/L} - N_{BC} \tag{8-97}$$

where N_{BC} is the impurity concentration just at the collector junction such that $N(x) = 0$ at $x = W$. Substituting (8-97) into Eq. (8-96) and simplifying the result to the first order, we get

$$\mathcal{E} \approx \frac{kT}{q}\frac{1}{W}\ln\frac{N_B'}{N_{BC}} \tag{8-98}$$

Inserting this result into Eq. (8-95), we obtain, finally,

$$t_b' \approx \frac{W^2}{D_{pb} \ln (N_B'/N_{BC})} \tag{8-99}$$

wherein we see that the transit time for a graded base is reduced by the factor $\ln (N_B'/N_{BC})$. In terms of the base-cutoff frequency, as defined by Eq. (8-91),

$$f_b' \approx f_b \ln \frac{N_B'}{N_{BC}} \tag{8-100}$$

Thus we see that the drift field established by the graded impurity distribution enhances the cutoff frequency by a rather significant factor,

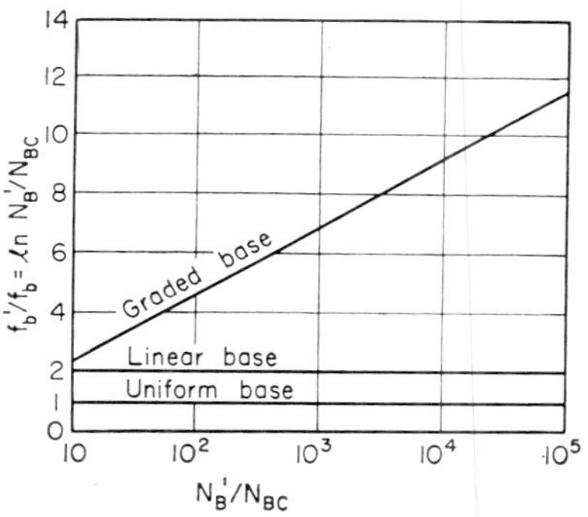

FIG. 8-7. Effect of various impurity distributions on base-cutoff frequency (low-level injection case).

particularly when the N_B'/N_{BC} ratio is high. A plot of the function is given in Fig. 8-7. Included in Fig. 8-7 is the frequency improvement obtained for the case of a linear base impurity grade.[10] For the linear case, $f_b' = 2f_b$.

It should be pointed out here that the results shown in Fig. 8-7 are fairly approximate in that it is assumed that the mobility is constant through the base layer. This is true only for the uniform case. In graded bases, where $N_B' > N_{BC}$, the impurity level at the emitter junction is greater than at the collector. Consequently, the minority-carrier mobility increases from the emitter to the collector, so that $\mu = f(x)$. This correction should be factored into the integration for transit time given by Eq. (8-94). Nevertheless, the results are reasonably accurate for most practical impurity ratios, since the mobility changes by a factor of at most 2 in this range.

Another restriction on these results is the fact that the analysis holds only for low-level conditions, that is, when the injected carrier density is small compared to N_B'. When this is not the case, the aiding effect of the built-in field is swamped out by the electric field established by the majority carriers in the base. This is similar to the problem of the p-n junction at high currents and will be discussed in Chap. 10.

Finally, it must be emphasized that it has been assumed that the emitter-to-base junction is a step junction. In certain diffused transistor structures, the emitter junction is slightly graded. Under these conditions, the maximum impurity concentration in the base N'_B does not occur at $x = 0$, but is shifted slightly towards the collector. Therefore, there exists an impurity gradient in the vicinity of the emitter junction which extends from $x = 0$ to the point where $N(x)$ equals N'_B and is

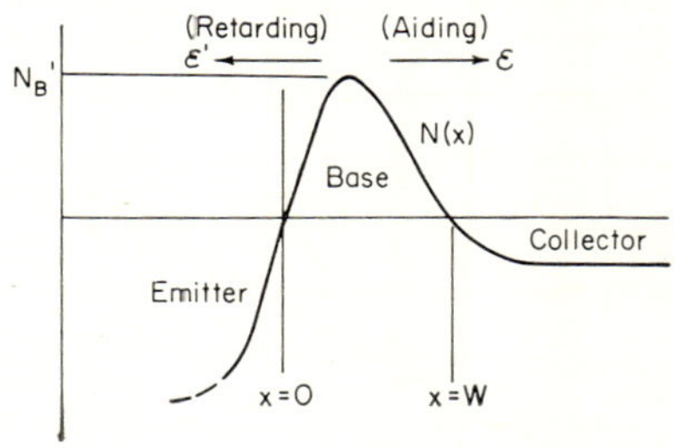

FIG. 8-8. Effects of graded emitter junctions in diffused-junction transistors, illustrating the introduction of a retarding electric field in the base.

opposite in slope to that which establishes the aiding built-in field. This opposite impurity gradient establishes, in the base region close to the emitter junction, an electric field which is retarding to the flow of injected minority carriers. Such an impurity distribution is shown in Fig. 8-8. The effect of the retarding field is to increase transit time and thereby decrease the base-cutoff frequency from the value predicted by the aiding field. Although this will not be analyzed here, it may be said that if N'_B is sufficiently close to $x = 0$ such that it is reached by the emitter depletion layer, the effect of the retarding field is negligible.[2]

8-9. Emitter and Collector Junction Capacitances. The theory for the capacitances of the junction of the transistor is identical to that developed for the p-n junction in Secs. 5-8 and 5-9. For the emitter capacitance C_{Te}, one can usually use the step-junction model for all transistor types, since the depletion-layer spreading is quite small at equilibrium or under forward-bias conditions. Thus,

$$C_{Te} \approx \sqrt{\frac{q\kappa\epsilon_o N'_B}{2V_T}} \tag{8-101}$$

where $N = N'_B$ and V_T is the emitter contact potential. For the collector, one may use Fig. 5-12 if the collector junction is a step or Figs. 5-15 to 5-30 for the graded case.

8-10. Impurity Profiles for Various Transistor Processes. To conclude this chapter on the basic theory of the junction transistor, it is appropriate to include a clarification of the nature of the base impurity distributions (i.e., uniform, graded, linear) for transistors made by the

Transistor process	Impurity profile	Base impurity distribution
Double-doped	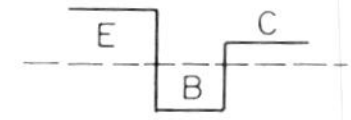	Uniform
Alloy	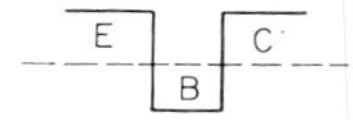	Uniform
Rate-grown	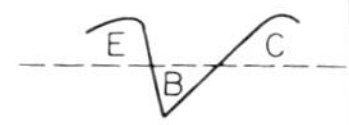	Linear
Meltback	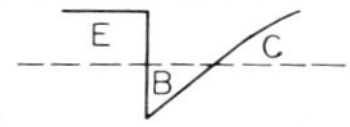	Linear
Meltback-diffused	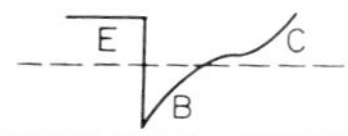	Graded
Grown-diffused	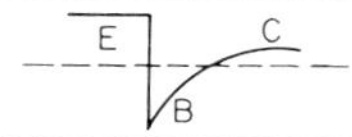	Graded
Diffused-base alloy	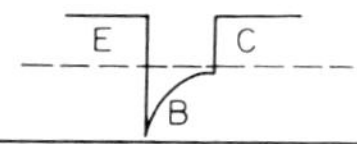	Graded
Alloy-diffused	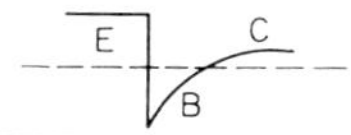	Graded
*Diffused-base mesa or planar	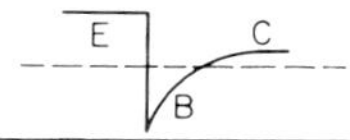	Graded
*Diffused-base epitaxial	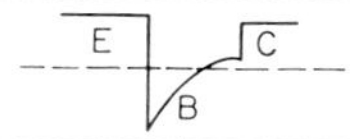	Graded

* Emitter junction may be either alloyed or diffused

FIG. 8-9. Impurity profiles for various transistor processes.

processes summarized in Table 1-2. The impurity profiles, from the emitter through the base to the collector, are summarized for each process in Fig. 8-9. No attempt was made to maintain the correct scale, particularly with regard to base width.

PROBLEMS

8-1. Determine the current gain α and base-cutoff frequency f_b for a germanium p-n-p alloy transistor having an effective base width of 0.6 mil. The base-region resistivity is 2 ohm-cm, the emitter- and collector-region resistivities are each 0.001

ohm-cm, the thickness of the emitter p-type region is 1.5 mil, and the minority-carrier lifetime in the base is 36 μsec and in the emitter is 0.01 μsec.

8-2. A high-frequency amplifier p-n-p germanium mesa transistor is fabricated by diffusing antimony from a surface concentration of 10^{18} atoms/cm^3 to a base-junction depth of 1 μ into a collector background of 1-ohm-cm p-type semiconductor. The measured base sheet resistance R_{BB} is 140 ohms. A nonpenetrating aluminum emitter stripe is formed such that the emitter resistivity is 0.0005 ohm-cm and the emitter thickness is 2,500 A. Assuming that the emitter and base lifetimes are sufficiently high so that no recombination occurs, determine the current gain and base-cutoff frequency of the structure.

8-3. An n-p-n silicon planar-epitaxial transistor is fabricated by diffusing both the base and the emitter such that an effective base width of 1 μ is obtained. If $N'_B = 10^{16}$ atoms/cm^3, what should the lifetime in the base region be to obtain a base-transport gain β^* equal to 0.975? (Assume uniform base mobility equal to that at N'_B.)

REFERENCES

1. Rittner, E. S.: Extension of the Theory of the Junction Transistor, *Phys. Rev.*, vol. 94, pp. 1161–1171, June, 1954.
2. Shockley, W., M. Sparks, and G. K. Teal: The PN Junction Transistor, *Phys. Rev.*, vol. 76, p. 459, 1949.
3. Shockley, W.: The Theory of PN Junctions in Semi-conductors and PN Junction Transistors, *Bell System Tech. J.*, vol. 28, pp. 435–489, 1949.
4. Shockley, W., M. Sparks, and G. K. Teal: P-N Junction Transistors, *Phys. Rev.*, vol. 83, pp. 151–162, 1951.
5. Steele E. L.: Theory of Alpha for PNP Diffused Junction Transistors, *Proc. IRE*, vol. 40, pp. 1424–1428, November, 1952.
6. Hunter, L. P.: "Handbook of Semiconductor Electronics," pp. 4-6, 4-7, McGraw-Hill Book Company, Inc., New York, 1956.
7. Pritchard, R. L.: Frequency Variations of Current-amplification Factor for Junction Transistors, *Proc. IRE*, vol. 40, pp. 1476–1481, November, 1952.
8. Krömer, H.: The Drift Transistor, *Naturwissenschaften*, vol. 40, pp. 578–579, 1953.
9. Krömer, H.: Theory of Diffusion and Drift Transistors, *Arch. Elect. Ubertragung*, vol. 8, pp. 363–369, 1954.
10. Moll, J. L., and I. M. Ross: The Dependence of Transistor Parameters on the Distribution of Base Layer Resistivity, *Proc. IRE*, vol. 44, pp. 72–78, January, 1956.

9

Characteristics of Junction Transistors

9-1. D-C Transistor Parameters. In this chapter, all the principles applied to the characteristics of p-n junctions will be related to the theory developed for the junction transistor in Chap. 8 in order to give a quantitative description of the key d-c transistor parameters listed in Table 7-1. This will yield design relationships for such major transistor characteristics as current gain, leakage currents, breakdown voltages, base resistance, and saturation voltages. As had been done previously, only the p-n-p model will be used as the basis of discussion; however, the reader should recognize that the results are applicable to n-p-n structures by appropriate reversal of polarities and electron and hole nomenclature.

It is apparent that one of the most important physical parameters of any transistor is the thickness of the base region. This term, denoted as the base width W, appears in many of the basic parameter equations. It is essential to recognize here that when the symbol W appears in any equation, it represents the effective base width which is usually less than the physical equilibrium base width W_o. The difference, i.e., the reduction in base width, is attributed to the spreading of the junction depletion layers which, of course, is related to the junction impurity profile and the applied voltage. Thus, the effective base width must be determined before any transistor structure is analyzed for its major electrical characteristics.

9-2. Beta, Grounded-emitter Current Gain. In circuit applications, junction transistors are commonly operated in the grounded-emitter configuration. By grounding the emitter and driving the transistor through the base connection, one gains the advantage of considerably higher current gain. In Chap. 7 we learned that if the collector-to-emitter current gain is α, the collector-to-base current gain (beta) is

$$\beta = \frac{\alpha}{1 - \alpha} \tag{9-1}$$

In this expression, as alpha approaches unity, the denominator becomes very small, and beta increases rapidly to large values. A useful plot of β as a function of α is given in Fig. 9-1. Beta will increase by an order of magnitude as α is increased from 0.91 to 0.99. In Eq. (9-1), if α is close to unity, beta is approximated by

$$\beta \approx \frac{1}{1-\alpha} \tag{9-2}$$

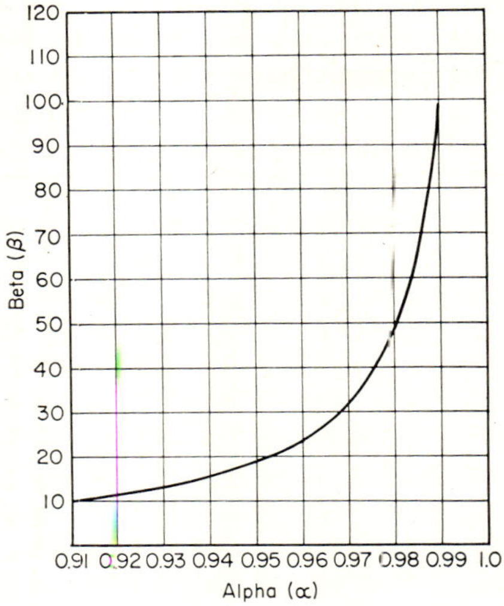

FIG. 9-1. Dependence of collector-to-base current gain on alpha. $\beta = \alpha/1 - \alpha$.

which simply means that beta is equal to the reciprocal of the base current I_B (for $I_E \approx I_C = 1$).

The final expression for α was given by Eq. (8-55) as

$$\alpha \approx \frac{1}{1 + \rho_e W/\rho_b L_{ne}} \left(1 - \frac{W^2}{2L_{pb}^2}\right) \tag{9-3}$$

If the magnitude of α in Eq. (9-3) is close to 1, we can apply Eq. (9-2) to (9-3); after simplifying, we obtain, approximately,

$$\frac{1}{\beta} \approx \frac{\rho_e W}{\rho_b L_{ne}} + \frac{W^2}{2L_{pb}^2} \tag{9-4}$$

This particular result is for a p-n-p transistor model. The first term of Eq. (9-4) results from the emitter efficiency γ and represents the ratio of the electron current I_{nE} to the injected hole current I_{pE}; the second term corresponds to the ratio of the volume recombination current in the base region I_{vB} to the hole current I_{pE}. For the condition that the magnitude of Eq. (9-4) is very small (corresponding to a small base current I_B), the total emitter current I_E is approximately equal to the injected hole current I_{pE}. Therefore, we may write (9-4) as

$$\frac{1}{\beta} \approx \frac{I_B}{I_{pE}} = \frac{I_{nE}}{I_{pE}} + \frac{I_{vB}}{I_{pE}} \tag{9-5}$$

From a design standpoint, we strive to make Eq. (9-4) or (9-5) as small as possible, making $(1 - \alpha)$ small and thereby maximizing the grounded-emitter current gain β.

One effect which has not been considered thus far in the low-level theory of current gain is that of surface recombination. In Chap. 8 we used a semi-infinite transistor model, for which surface effects were sufficiently isolated, so that only bulk recombination in the volume of the base region contributed to the total base current. A measure of this recombination was given by the base-transport factor, which we saw was related to the lifetime of the base region. In practical transistor devices, however, the semi-infinite model is far from being a truly representative model. Regardless of the device geometry, the active volume of the base region must be terminated by a semiconductor surface at some finite points. As will be discussed in greater detail in a later chapter, the effective surface area is an annular ring around the emitter junction, approximately equal in width to the transistor base width W. To remain completely general, we shall simply refer to the effective surface area as A_s for all geometries.

As was discussed in Sec. 4-10, a semiconductor surface is characterized by a surface recombination velocity s, which is a direct measure of the rate of recombination of carriers arriving at the surface. Inversely speaking, it is a measure of the effective lifetime of the surface area. A surface may have a low or high recombination velocity, depending on the nature of the surface treatment during the processing of the transistor. From a design standpoint, it is desirable to keep s as small as practicable. In spite of this, some fraction of the carriers injected into the base of a junction transistor will not reach the collector junction, but will be lost by recombination at the surface. This introduces an additional component of current to the total base current. It must be remembered that the surface recombination current adds directly to the volume recombination current as well as the current which is injected back into the emitter. Thus, for a p-n-p transistor, the total base current is

$$I_B = I_{nE} + I_{vB} + I_{sB} \tag{9-6}$$

where I_{sB} is the surface recombination current.

To correct the low-level current-gain theory for the effects of surfaces, it is necessary to modify Eq. (9-5) as follows:

$$\frac{1}{\beta} \approx \frac{I_B}{I_{pE}} = \frac{I_{nE}}{I_{pE}} + \frac{I_{vB}}{I_{pE}} + \frac{I_{sB}}{I_{pE}} \tag{9-7}$$

We see, therefore, that the effect of surface recombination is to decrease the over-all beta (current gain) of the junction transistor. In many practical devices, because of the difficulty of obtaining optimum base-region surfaces, this becomes the limiting factor for current gain. To express the third term of Eq. (9-7) in terms of the physical transistor, we may write that[1,*]

$$I_{sB} = qsA_sp \tag{9-8}$$

* References, indicated in the text by superscript figures, are listed at the end of the chapter.

where s is the surface recombination velocity. We recall from Chap. 4 that this is defined as the number of carriers recombining per second per unit surface area divided by the excess concentration over the equilibrium value at the surface. Thus, to obtain the current (charge per unit time), we multiply s by the electronic charge q, the effective area of surface recombination A_s, and the carrier concentration at the surface p, as shown by (9-8). Since most of the recombination occurs very close to the edges of the emitter junction, it is reasonable to assume that p is equal to the injected hole density p_E. From basic theory the emitter hole current is given by

$$I_{pE} \approx \frac{qAD_{pb}p_E}{W} \tag{9-9}$$

where $p_E = p_{nb}\epsilon^{qV_E/kT}$. Taking the ratio of (9-8) to (9-9), we have, finally,

$$\frac{I_{sB}}{I_{pE}} \approx \frac{sA_sW}{AD_{pb}} \tag{9-10}$$

The complete grounded-emitter beta expression becomes

$$\frac{1}{\beta} \approx 1 - \alpha \approx \frac{\rho_e W}{\rho_b L_{ne}} + \frac{W^2}{2L_{pb}{}^2} + \frac{sA_sW}{AD_{pb}} \tag{9-11}$$

In Eq. (9-11), the second and third right-hand terms represent the effects of volume and surface recombination, respectively. One interesting aspect is apparent here. If $L_{pb}{}^2 = D_{pb}\tau_{pb}$, these terms may be arranged as

$$\frac{1}{\beta} \approx \frac{\rho_e W}{\rho_b L_{ne}} + \frac{W^2}{2D_{pb}}\left(\frac{1}{\tau_{pb}} + \frac{2sA_s}{WA}\right) \tag{9-12}$$

In so doing, the second term in parenthesis appears as the reciprocal of the surface lifetime τ_{sb}. Consequently, (9-12) may be written as

$$\frac{1}{\beta} \approx \frac{\rho_e W}{\rho_b L_{ne}} + \frac{W^2}{2D_{pb}\tau_{\text{eff}}} \tag{9-13}$$

where

$$\frac{1}{\tau_{\text{eff}}} = \frac{1}{\tau_{pb}} + \frac{1}{\tau_{sb}} \tag{9-14}$$

We see that this is identical to the relation given by Shockley[2] for combining the effects of surface and volume recombination into a single effective base-region lifetime τ_{eff}. From a device-lifetime-measurement point of view this is a convenient form, since oftentimes a high base-region volume lifetime is masked by a low surface lifetime, in accordance with Eq. (9-14).

The complete expression for beta, given by Eq. (9-11), includes the effects of emitter efficiency, base recombination, and surface recombination. This result, however, is applicable only to the low-level injection case, i.e., the case in which the injected minority-carrier concentration in the base is small compared to the majority-carrier concentration. At high current densities, each term must be modified to take in the effects of high-level injection (see Chap. 10). As will be shown, beta is not constant with collector current.

9-3. Collector Reverse Current. For a junction transistor, it is important for the reverse current of the collector junction to be as small as possible. As was indicated in Chap. 7, this is the current that flows from collector to base with the emitter open-circuited; it is denoted as I_{CBO}. Since the I_{CBO} current exists as long as the collector is reverse-biased, it represents undesired power dissipation which detracts from the useful power rating of the device. Also, since it may increase by several orders of magnitude at the higher temperatures, the zero-signal collector dissipation becomes quite appreciable. It is therefore desirable to design for very low I_{CBO} at room temperature. In this section, we shall obtain the theoretical relation for I_{CBO} in terms of the p-n-p model of the previous chapter.

Let us refer now to the general current-density equations for the p-n-p transistor given by Eqs. (8-30) and (8-31). Based on typical operation for a transistor, we can make certain simplifications. Firstly, the term $\epsilon^{-qV_C/kT}$ is very small compared to 1 for reverse biases greater than 0.1 volt. Secondly, the collector region is of moderate-resistivity p type, so that the magnitude of the minority electron concentration n_{pc} cannot be neglected. Applying these considerations, we have

$$J_E = \left(\frac{qD_{ne}n_{pe}}{L_{ne}} + \frac{qD_{pb}p_{nb}}{L_{pb}} \coth \frac{W}{L_{pb}}\right)(\epsilon^{qV_E/kT} - 1) + \frac{qD_{pb}p_{nb}}{L_{pb}} \operatorname{csch} \frac{W}{L_{pb}} \tag{9-15}$$

$$J_C = \frac{qD_{pb}p_{nb}}{L_{pb}} \operatorname{csch} \frac{W}{L_{pb}} (\epsilon^{qV_E/kT} - 1) + \frac{qD_{pb}p_{nb}}{L_{pb}} \coth \frac{W}{L_{pb}} + \frac{qD_{nc}n_{pc}}{L_{nc}} \tag{9-16}$$

For the condition that the emitter is open-circuited during the I_{CBO} measurement, we set Eq. (9-15) equal to zero, i.e., $J_E = 0$. Solving for $(\epsilon^{qV_E/kT} - 1)$, we have

$$\epsilon^{qV_E/kT} - 1 = \frac{(-qD_{pb}p_{nb}/L_{pb}) \operatorname{csch} (W/L_{pb})}{qD_{ne}n_{pe}/L_{ne} + (qD_{pb}p_{nb}/L_{pb}) \coth (W/L_{pb})} \tag{9-17}$$

Examination of this result shows that it is identical to Eq. (8-34), the expression for α. Substituting this result in (9-16) and replacing csch (W/L_{pb}) by its identity, coth (W/L_{pb}) sech (W/L_{pb}), we obtain, finally,

$$J_{CBO} = \frac{qD_{pb}p_{nb}}{L_{pb}} \coth \frac{W}{L_{pb}} \left(1 - \alpha \operatorname{sech} \frac{W}{L_{pb}}\right) + \frac{qD_{nc}n_{pc}}{L_{nc}} \tag{9-18}$$

This equation, when multiplied by the junction area A, yields the current I_{CBO}. Actually, for well-designed transistors, where $W \ll L_{pb}$, we may approximate (9-18) by

$$I_{CBO} \approx \frac{qAD_{pb}p_{nb}}{W}(1-\gamma) + \frac{qAD_{nc}n_{pc}}{L_{nc}} \tag{9-19}$$

since $\alpha = \gamma \operatorname{sech}(W/L_{pb})$. Thus, for the case where the base-region lifetime is sufficiently high such that the base-transport factor is equal to unity, the I_{CBO} current is partially a function of the emitter efficiency γ. We see that, for a given base width W, to maintain I_{CBO} small it is necessary to go to lower base- and collector-region resistivities (reduce minority-carrier concentration) and smaller junction areas. If the collector region of the transistor is very heavily doped, such as in the case of alloy-type transistors, the last term of Eq. (9-19) may be neglected.

The temperature variation of Eq. (9-19) is approximately the same as that for a p-n junction as given by Fig. 6-4 for germanium, since the temperature variation of γ is not significant in this case. Finally, Eq. (9-19) is applicable to germanium transistors, but not to silicon. For the latter, the I_{CBO} current is predominantly due to depletion-layer charge generation as was explained in Chap. 6, and is given approximately by

$$I_{CBO} \approx qex_mA \tag{9-20}$$

where e = generation rate
x_m = depletion-layer width
A = collector area

Note that for both germanium and silicon, I_{CBO} will increase slightly with collector voltage, since W decreases in (9-19) and x_m increases in (9-20).

In the case of the graded-base germanium transistor, the first term in Eq. (9-19) must be modified to account for the increase in I_{CBO} due to the decreasing impurity concentration as the collector junction is approached. One will recall that for the graded base, both the mobility and minority-carrier concentration will increase with distance from the emitter junction. This will increase the magnitude of the reverse current originating from the base. An excellent approximation for the graded-base transistor may be obtained by assuming that D_{pb} and p_{nb} are each doubled as a result of the impurity gradient. This approximation is valid if one assumes a linear impurity distribution in the base. Therefore, if we denote I'_{CBO} as the reverse current for a graded-base transistor, we have

$$I'_{CBO} \approx \frac{4qAD_{pb}p_{nb}}{W}(1-\gamma) + \frac{qAD_{nc}n_{pc}}{L_{nc}} \tag{9-21}$$

In Eq. (9-21), D_{pb} and p_{nb} have values corresponding to that for the impurity concentration just at the base side of the emitter junction, i.e., N'_B. It should be pointed out here that for a graded-base silicon transistor Eq. (9-21) is not applicable since the depletion-layer generation current is still the predominant component. One may still use Eq. (9-20) for the calculation.

One important consideration, which arises when the I_{CBO} measurement on a transistor is made, concerns the potential of the emitter with respect to the base. We saw, in our examination of Eq. (9-17), that

$$\epsilon^{qV_E/kT} - 1 = -\alpha \tag{9-22}$$

where V_E is the emitter junction voltage. Because the emitter is open-circuited for determining I_{CBO}, a solution of (9-22) for V_E would yield the emitter "floating potential," viz., V_{EBO}. Thus,

$$V_{EBO} = \frac{kT}{q} \ln (1 - \alpha) \tag{9-23}$$

In other words, this is the equilibrium potential which the emitter region would assume to compensate for the nonequilibrium conditions in the base region due to the collector reverse bias. As an example, if α were 0.95 at room temperature, V_{EBO} would be -0.078 volts. This parameter yields an excellent means of ascertaining the quality of the surface treatment of practical transistors. If any channels exist across the surface of the base region of a transistor, they provide additional leakage paths which increase the magnitude of V_{EBO} above the theoretical value given by Eq. (9-23). Thus V_{EBO} can be a good indicator for product reliability.

Another point of interest is the magnitude of the collector reverse current when the emitter is shorted to the base. This is denoted as I_{CES}. For this condition, $V_E = 0$ and Eq. (9-16) simply becomes

$$J_{CES} = \frac{qD_{pb}p_{nb}}{L_{pb}} \coth \frac{W}{L_{pb}} + \frac{qD_{nc}n_{pc}}{L_{nc}} \tag{9-24}$$

Multiplying by A and letting $W \ll L_{pb}$, we get

$$I_{CES} \approx \frac{qAD_{pb}p_{nb}}{W} + \frac{qAD_{nc}n_{pc}}{L_{nc}} \tag{9-25}$$

Again, if the collector resistivity is sufficiently low such that n_{pc} is negligible, (9-25) becomes

$$I_{CES} \approx \frac{qAD_{pb}p_{nb}}{W} \tag{9-26}$$

Comparing this result with Eq. (9-19), with the second term omitted, we have

$$I_{CES} \approx \frac{I_{CBO}}{1 - \alpha} \tag{9-27}$$

for the case where $\alpha = \gamma$. In Eq. (9-27), if α is at all close to 1, then I_{CES} is considerably larger than I_{CBO}. However, it will be shown in Chap. 10 that at very low currents on the order of the reverse currents (1 μa or so), α is practically zero and $I_{CES} \approx I_{CBO}$. In certain practical transistor designs, however, there exist certain collector injection mechanisms which cause α to approach unity (collector multiplication α^*M) and thereby cause I_{CES} in Eq. (9-27) to become infinite, thereby causing breakdown of the collector junction. This phenomenon occurs when the collector voltage is increased to a value close to the true collector breakdown during the measurement of I_{CES}. This will be discussed in detail in the next section.

The measurement of both I_{CBO} and I_{CES} is best summarized by the schematics of Fig. 9-2. With the emitter open-circuited, the reverse collector current is I_{CBO}, whereas with the emitter short-circuited, the current is I_{CES}. The reason that I_{CES} is slightly larger than I_{CBO} may be explained qualitatively. When the emitter is floating, the I_{CBO} current consists mainly of minority carriers (holes) which are swept out of the base region by the collector reverse bias. However, when the emitter is shorted to the base, the junction contact potential is reduced permitting holes from the emitter region to be injected into the base. Since these additional holes are readily collected, the total current becomes slightly greater.

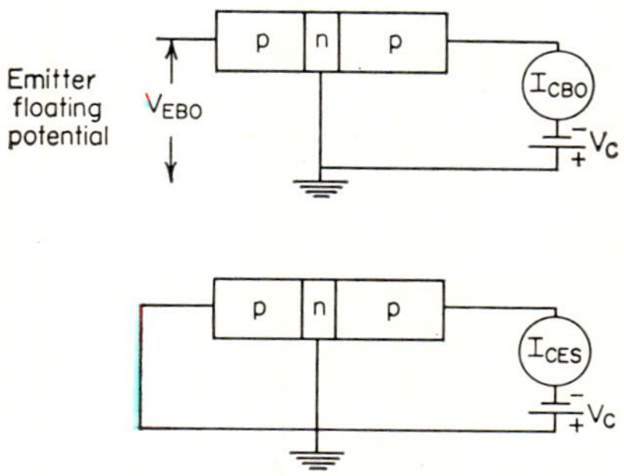

FIG. 9-2. Measurement of collector reverse currents.

9-4. Collector Breakdown Voltage. Another important d-c parameter for the junction transistor is the breakdown voltage of the collector junction. As was pointed out in Chap. 7, collector voltage breakdown is limited by either avalanche breakdown or voltage punchthrough, whichever occurs first. It is necessary to ascertain the impurity profile of the transistor design in question in order to determine which of the two is the limiting factor. Once this is established, one can apply the basic p-n junction relationships introduced in the early chapters to calculate the correct collector breakdown voltage. This voltage breakdown is denoted by BV_{CBO}, that is, the collector-to-base voltage breakdown with the emitter open-circuited.

Avalanche breakdown is a function of the high-resistivity side of the collector junction. For structures made by the double-sided impurity-contact processes, e.g., alloy or MADT, avalanche breakdown is related to the base resistivity. On the other hand, for the single-sided processes, e.g., mesa or planar, avalanche is related primarily to the collector resistivity. Thus, avalanche BV_{CBO} may be determined by referring to Fig. 6-10 for step junctions or Fig. 6-12 for graded junctions. In the latter case, for those transistor structures in which the base is diffused into a high-resistivity background, one can readily apply the step-junction approximation and use Fig. 6-10 instead of Fig. 6-12.

If the spreading of the depletion layer into the base is such that it touches the emitter junction before avalanche occurs, the transistor is limited by voltage punchthrough V_{PT}. At the condition of punchthrough, the emitter floating potential V_{EBO} will start to increase and follow the applied collector voltage. The emitter potential will continue to increase until the emitter junction breaks down due to avalanche effects. The net effect follows the relationship

$$BV_{CBO} = V_{PT} + BV_{EBO} \tag{9-28}$$

where BV_{EBO} is the avalanche-breakdown voltage of the emitter-base junction.

The magnitude of V_{PT} is a direct function of the impurity profile of the base region, wherein V_{PT} is the voltage necessary to spread the base side of the collector depletion layer a distance equal to the physical base width W_o. For a uniform base of the alloy type, the expression for V_{PT} is determined by referring to Eq. (5-77). Letting $x_m = W_o$ and solving for $V = V_{PT}$, we have

$$V_{PT} = \frac{qN_D W_o^2}{2\kappa\epsilon_o} \tag{9-29}$$

Figure 5-12, which is in the form of W_o as a function of V_{PT}/N, may be used instead of Eq. (9-29) to determine voltage punchthrough.

For a graded-base transistor fabricated by diffusion techniques, V_{PT} is determined by referring to the curves in Figs. 5-15 to 5-30, wherein $x_1 = W_o$ and $V = V_{PT}$. In using these curves, one has no choice but to use a trial-and-error procedure to obtain the solution for V_{PT}.

The determination of BV_{CBO} for a transistor made by the epitaxial process (see Sec. 1-10) presents a situation a bit different from the straightforward avalanche or punchthrough considerations of the previous paragraphs. This is attributed to the fact that the high-resistivity epitaxial layer in the collector may not be thick enough (by design) to sustain the necessary depletion-layer width corresponding to the avalanche-breakdown voltage for that resistivity. This is best explained by reference to Fig. 9-3, which shows a generalized diagram of a p-n-p epitaxial transistor. The substrate portion of the collector region is p+, which is usually a very heavily doped semiconductor. The epitaxial collector layer is generally of very high resistivity ρ_c and is denoted as having a thickness W_e.

To simplify the argument, let us assume that, whether the base is uniform or graded, all of the spreading of the collector depletion layer goes into the epitaxial region and that the latter has a uniform impurity distribution. If the epitaxial thickness W_e is equal to or greater than x_{mB}, which is the depletion-layer thickness at the voltage corresponding to the avalanche breakdown for the epitaxial-layer resistivity, the observed BV_{CBO} is limited by straightforward avalanche considerations. However, if W_e is less than x_{mB}, the actual BV_{CBO} is lower than the theoretical avalanche value because the depletion layer effectively stops spreading when it reaches the interface between the epitaxial layer and the heavily doped substrate.* This occurs at $x = W_e$ in Fig. 9-3. At this point, the electric field $\mathcal{E}$ starts to build up at a more rapid rate and causes the junction to break down prematurely.

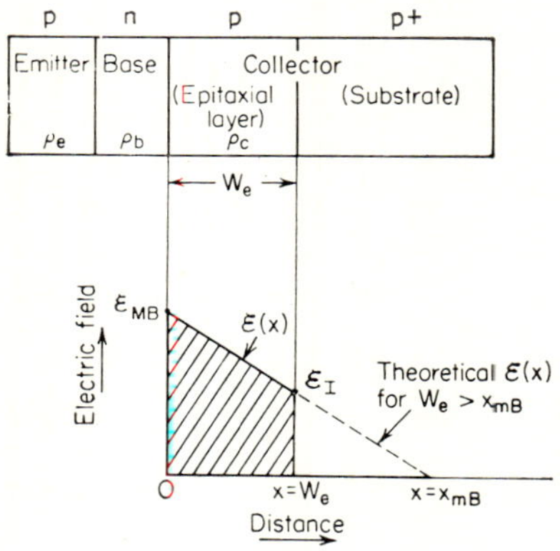

FIG. 9-3. Electric-field distribution in an epitaxial transistor at avalanche breakdown.

If it is assumed that avalanche takes place when the electric field reaches the critical value $\mathcal{E}_{MB}$† the electric-field distribution in the depletion layer at breakdown appears as shown by the solid line in the diagram of Fig. 9-3. From Poisson's equation, since $\mathcal{E}$ is the integral of the charge distribution, $\mathcal{E}(x)$ must be linear as shown because of the uniform charge density (impurity ions) in the epitaxial layer. At $x = W_e$, $\mathcal{E}(x)$ has the magnitude $\mathcal{E}_I$ and then drops quickly to zero in the substrate. If W_e is greater than x_{mB}, then $\mathcal{E}(x)$ is simply extended to zero at $x = x_{mB}$ as shown by the dashed line. The voltage being sustained at breakdown is the integral of $\mathcal{E}(x)$ and is simply equal to the area under the $\mathcal{E}(x)$ function as denoted by the shading. By applying geometrical considerations to determine the ratio of the shaded area to the total area of the right triangle $(0,\mathcal{E}_{MB},x_{mB})$, one obtains the ratio of the actual break-

* It is interesting to note that in an epitaxial transistor the collector capacitance virtually stops decreasing with collector voltage when the depletion layer reaches the interface. Normally, the capacitance variation would continue to exhibit either the square-root or cube-root dependence with voltage.

† Strictly speaking, it was shown in Chap. 6 that breakdown occurs at that value of $\mathcal{E}$ for which the ionization integral equals unity. However, the integration of $\alpha_i(\mathcal{E})$ is such that it achieves maximum value in a small range just below $\mathcal{E}_{MB}$.

down voltage to the theoretical avalanche-breakdown voltage. It is easily shown that

$$BV_{CBO} \approx \frac{W_e}{x_{mB}}\left(2 - \frac{W_e}{x_{mB}}\right) V_B \tag{9-30}$$

where V_B = avalanche-breakdown voltage (from Fig. 6-10)
x_{mB} = depletion-layer thickness at V_B (from Fig. 5-12)
W_e = thickness of epitaxial layer

A normalized plot of Eq. (9-30) is given in Fig. 9-4, e.g., for $W_e/x_{mB} = 0.5$, $BV_{CBO} = 0.75V_B$. Note that when $W_e = x_{mB}$, $BV_{CBO} = V_B$.

To complete our consideration of collector reverse voltages, we must now turn our attention to the grounded-emitter configuration to deter-

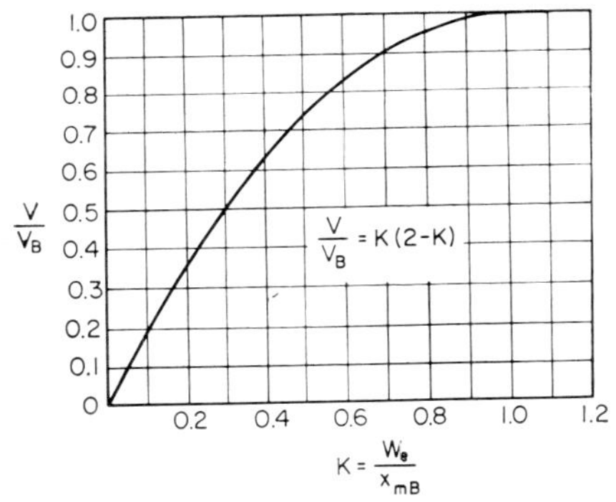

FIG. 9-4. Breakdown voltage of epitaxial layers in terms of thickness and resistivity-dependent parameters V_B and x_{mB}.

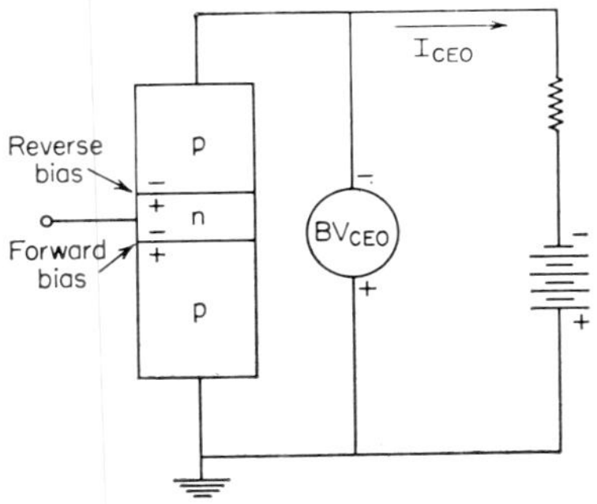

FIG. 9-5. Method of measurement of BV_{CEO} in transistors.

mine collector-to-emitter breakdown relationships, viz., BV_{CES} and BV_{CEO}. BV_{CES} is the collector-to-emitter breakdown voltage measured when the base is shorted to the emitter. If BV_{CBO} for the transistor is limited by avalanche effects, BV_{CES} is simply equal to BV_{CBO}. If BV_{CBO} for the transistor is limited by voltage punchthrough, $BV_{CES} = V_{PT}$. This is apparent from Eq. (9-28) if BV_{EBO} is equal to zero, since the emitter-base junction is shorted during the measurement of BV_{CES}. In other words, for a punchthrough-limited transistor, we have

$$BV_{CES} = BV_{CBO} - BV_{EBO} = V_{PT} \tag{9-31}$$

The remaining grounded-emitter parameter is BV_{CEO}, which is the collector-to-emitter breakdown voltage with the base open-circuited, i.e., $I_B = 0$. Under these conditions, the applied voltage across the transistor which reverse-biases the collector junction also acts to forward-bias the emitter junction, as shown in Fig. 9-5. It is interesting to note the current that flows under these conditions. In Chap. 7 we saw that

$$I_C = I_{CBO} + \alpha I_E \tag{9-32}$$

However, during the measurement of BV_{CEO}, $I_C = I_E = I_{CEO}$ and $I_{CEO} = I_{CBO} + \alpha I_{CEO}$, or

$$I_{CEO} = \frac{I_{CBO}}{1 - \alpha} \tag{9-33}$$

Equation (9-33) indicates that the grounded-emitter reverse current is larger than I_{CBO} by a factor approximately equal to the beta of the transistor at that current level. This result is understandable from a physical point of view, since the emitter is injecting additional minority carriers into the base. As the applied voltage is increased, however, I_{CEO} will begin to increase because of the increase in α due to avalanche-multiplication effects. We recall that due to multiplication effects as avalanche breakdown is approached, the current gain must be multiplied by the factor M which was given as

$$M = \frac{1}{1 - (V/V_B)^n} \tag{9-34}$$

where n has values ranging from 2 to 6 (see Eq. (6-24). If the α term in Eq. (9-33) is multiplied by Eq. (9-34), there exists some value of V for which $\alpha M = 1$ and I_{CEO} goes to infinity. This is the breakdown condition where $V = BV_{CEO}$ or

$$M = \frac{1}{\alpha} = \frac{1}{1 - (BV_{CEO}/BV_{CBO})^n} \tag{9-35}$$

Solving for BV_{CEO}, we obtain, finally,

$$BV_{CEO} = (1 - \alpha)^{1/n} BV_{CBO} \tag{9-36}$$

or, as a good approximation, for α close to 1,

$$BV_{CEO} \approx \frac{BV_{CBO}}{\sqrt[n]{\beta}} \tag{9-37}$$

Obviously this relationship holds only for the case where the transistor is not limited by V_{PT}.

9-5. Emitter Reverse Characteristics. The two significant d-c characteristics of the emitter junction of the transistor are the reverse leakage current I_{EBO} and the reverse breakdown voltage BV_{EBO}. Because of the fact that in most transistors the emitter region is very heavily doped to obtain maximum emitter efficiency, one can always use a step-junction model to determine these parameters. Thus, I_{EBO} is calculated in a manner identical to that for I_{CBO}. BV_{EBO} is simply the avalanche-breakdown voltage of the junction and is primarily a function of the impurity concentration just at the base side of the emitter junction.

9-6. Base Resistance. In the analysis presented so far, no mention has been made of the effects of the inherent resistance of the base region. In the transistor base layer, a fraction $(1 - \alpha)$ of the input emitter current constitutes a small base current I_B which flows out through the base-region contact. It should be apparent that for any practical transistor geometry wherein ohmic base contacts must be made around and/or along the periphery, the base current must flow parallel to the emitter- and collector-junction planes. Since the base layer has a specific resistivity, this current will develop a transverse voltage drop in the base region, which appears as feedback to the emitter junction. This effect may be represented by an additional term r'_B, which is called the *base spreading resistance.*[3] The term *spreading* is used here to denote the fact that both the current and voltage are distributed within the base region in some manner such that the base resistance can only be expressed by a particular average value. Thus, r'_B is determined by the ratio of the average voltage developed in the base region to the current that produced it. It should be evident that the transistor geometry is most significant in this regard; the r'_B expression is different for each structure.

Actually, the base-region resistance is common to all currents, direct or alternating, producing corresponding voltage drops. From the d-c point of view, the developed base voltage can seriously alter the action of the transistor. We recall that for typical operating current bias, the emitter forward potential is quite small, being on the order of a few tenths of a volt. If either base current or base resistivity is appreciable, it is quite possible that sufficient transverse base voltage may be developed to actually reverse-bias that part of the emitter junction farthest removed from the contact. For a constant emitter current, this mechanism increases the current density injected into the base, since part of the active emitter area is being blocked off. This effect will be considered in detail in Chap. 10. For the purposes of the analysis to follow, we assume that the entire emitter junction area is uniformly forward-biased and that the d-c voltage feedback is negligible.

To establish an analytical expression for r'_B, let us consider a transistor geometry characteristic of grown-junction types, as in Fig. 9-6. These are usually in the form of n-p-n semiconductor bars having a p-type base layer of thickness W, side dimension l, and height h, as shown in the figure. Contact to the base region is made by means of a thin-line ohmic contact along one side l, shown by the crosshatched area in Fig. 9-6. For this theoretical analysis we will assume no contact overlap and that the contact is ohmic. As we have seen, emitter current is uniformly injected into the base layer and diffuses across to the collector junction, coming out as collector current. Part of the current is lost by recombination in the base; this appears as I_B, flowing down the base region to the line contact. It is the average base voltage V_B produced by this current I_B that determines the effective base resistance r'_B.

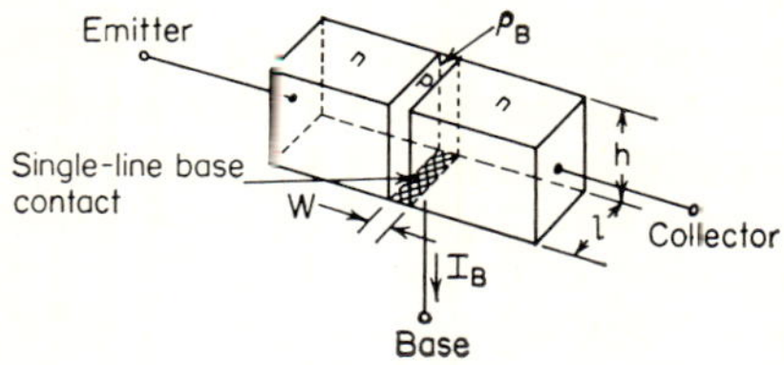

FIG. 9-6. Single-line contact-bar geometry.

In order to solve for r'_B, it is necessary first to make an assumption concerning the distribution of current I_B in the base layer, along the h dimension. In each elemental volume of the base is a uniform current moving from emitter to collector. A fraction $(1 - \alpha)$ of these currents is lost by recombination within each elemental volume. Therefore each elemental volume may be regarded as an elemental current generator, generating a current equal to

$$\Delta I_B = \Delta[(1 - \alpha)I_E] \tag{9-38}$$

and having an incremental resistance dR. If we assume that the resistivity of the base layer ρ_b is everywhere homogenous, we can depict the base-current mechanism by the simple equivalent circuit shown in Fig. 9-7b, which represents the action of the current generators within the incremental strip dW shown in Fig. 9-7a. If the strip dW is broken into

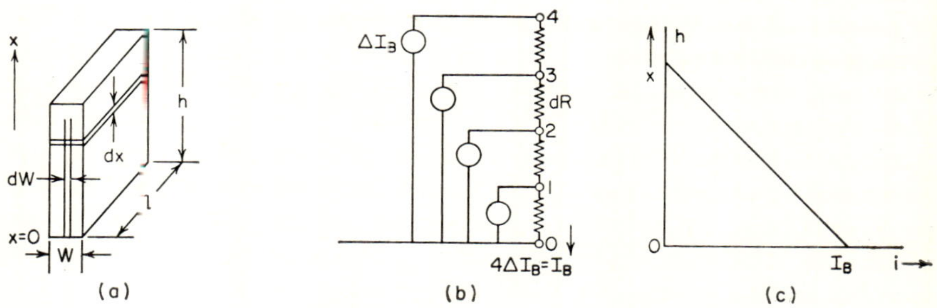

FIG. 9-7. Distribution of base current. (a) Base region; (b) equivalent circuit for dW; (c) current distribution.

resistances dR, each current generator within the respective volumes delivers a current ΔI_B to the strip. Thus the total current at the contact $x = 0$ is equal to the sum of all the current generators within the strip. As x increases toward h, the total current decreases in increments of ΔI_B toward zero. Therefore we can state that the base current has a linear distribution as shown in Fig. 9-7c. At the bar surface where $x = h$, the total current is zero. The current $I_B(x)$ is given by the linear equation

$$I_B(x) = I_B\left(1 - \frac{x}{h}\right) \tag{9-39}$$

where I_B is the total current at the line contact. Again in Fig. 9-7*a*, in the x direction the current of Eq. (9-39) flows through the incremental resistances dR, which are of thickness dx and area lW. Therefore, dR is expressed as

$$dR = \frac{\rho_b l}{A} = \frac{\rho_b\,dx}{lW} \qquad \text{ohms} \tag{9-40}$$

Multiplying (9-39) by (9-40) to obtain the incremental voltage drop dV_B, we have

$$dV_B = I_B\left(1 - \frac{x}{h}\right)\frac{\rho_b\,dx}{lW} \tag{9-41}$$

By integrating Eq. (9-41) with respect to x from 0 to x we obtain a result which gives the distribution of voltage as a function of x. That is,

$$V_B(x) = \frac{I_B\rho_b}{lW}\left(x - \frac{x^2}{2h}\right) + K_1 \tag{9-42}$$

where K_1 is the constant of integration. This is equal to zero, since at $x = 0$ we are assuming the contact to be at ground potential.

In order to obtain the spreading resistance r'_B, it is necessary first to obtain the average value of the voltage V_B across the distance h. It is to be noted that Eq. (9-42) presents a parabolic voltage distribution. The average is defined as

$$V_B(\text{average}) = \frac{1}{h}\int_0^h V_B(x)\,dx \tag{9-43}$$

or

$$V_B = \frac{I_B\rho_b h}{3Wl} \tag{9-44}$$

From Eq. (9-44), we have, finally,[3]

$$r'_B = \frac{V_B}{I_B} = \frac{\rho_b h}{3Wl} \qquad \text{ohms} \tag{9-45}$$

where ρ_b = resistivity, ohm-cm
W = base thickness
l = side dimension (contact)
h = height

For a square bar, $l = h$ and (9-45) simply becomes

$$r'_B = \frac{\rho_b}{3W} \tag{9-46}$$

Equation (9-46) may be written in the form

$$r'_B = \frac{R_{BB}}{3} \qquad \text{single-line contact} \tag{9-47}$$

where R_{BB} is the base-to-base or sheet resistance of the uniform-base region, as defined in Fig. 9-8.

We see that in order for r'_B to be as small as possible it is necessary that the base sheet resistance be as small as possible. r'_B can be reduced by utilizing a rectangular cross-sectional area such that the aspect ratio h/l is as small as practicable and by placing the contact along the l dimension.

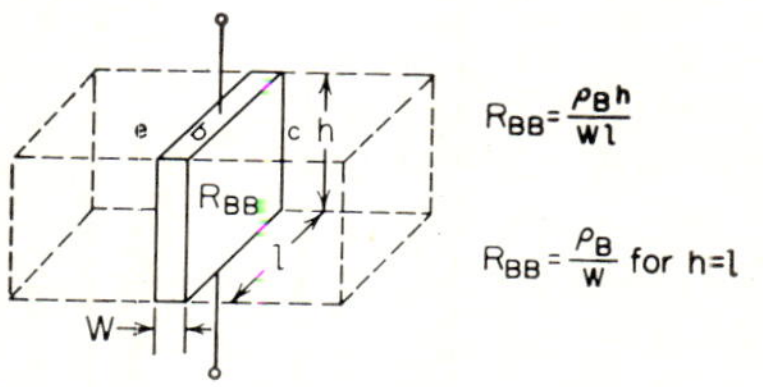

FIG. 9-8. Sheet resistance R_{BB}.

In Fig. 9-6, if an additional line contact is made on the side directly opposite the base region and both contacts are tied to the same potential point, e.g., ground, then the base spreading resistance is reduced again by a factor of four, or[3]

$$r'_B = \frac{R_{BB}}{12} \qquad \text{double-line contact} \tag{9-48}$$

In this model, the base current splits up and flows in opposite directions toward the contact, being zero at $h/2$ and increasing linearly to the contacts. From Eq. (9-41), the average voltage for each half of the base region is

$$V_B = \frac{I_B}{2} \frac{\rho_b}{3Wl} \frac{h}{2} \tag{9-49}$$

or

$$r'_B = \frac{V_B}{I_B} = \frac{\rho_b h}{12Wl} \tag{9-50}$$

It should be apparent at this point that the most ideal arrangement for minimum r'_B would be a continuous contact completely around the periphery of the base region.

Junction transistors made by such techniques as the alloy, diffused-base alloy, mesa, or planar processes may have circular or ring base contacts. A generalized model of this transistor structure is shown in Fig. 9-9. The geometry is slightly modified by showing the ring base contact around the periphery of the base rather than on one face, as is the usual practice. This simplifies the analysis for base resistance and yet does not introduce significant errors. In Fig. 9-9a, the cross-sectional area is drawn to show that the active base region consists of a uniform semiconductor base of resistivity ρ_b, thickness W, and radius r_2. The geometry is circular, having an ohmic contact all around the circumference of the base region. All pertinent dimensions are indicated in the figure. Emitter current (holes for p-n-p) is injected uniformly into the base region; it is assumed that the recombination components constituting the base current I_B or $(1 - \alpha)$ flow radially outward from $r = 0$ to the base contact. The resistance paths for I_B that make up r'_B are shown as two basic circular structures in Fig. 9-9*b*. The current originates in the disk A, which is equal in area to the emitter junction, and then flows outwardly uniformly through the annular ring B. In order to determine

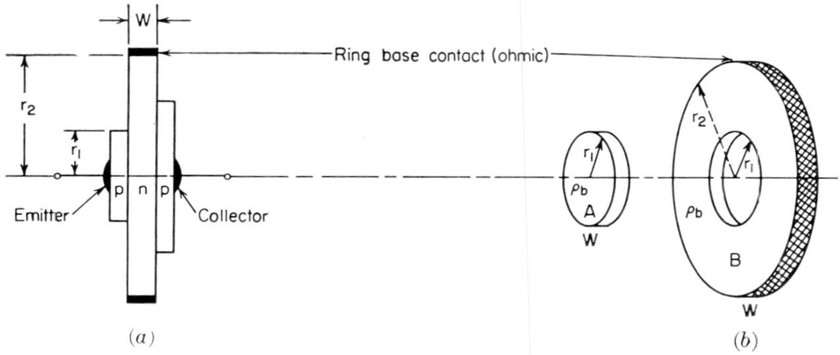

FIG. 9-9. Ring base-contact transistor geometry. (*a*) Cross-sectional geometry; (*b*) base-region components.

r'_B it is necessary merely to sum up the individual resistances of these elements. The formula for the resistance of an annular ring of resistivity ρ_b, thickness W, inner radius r_1, and outer radius r_2 is

$$R_{\text{RING}} = \frac{1}{2\pi}\frac{\rho_b}{W}\ln\frac{r_2}{r_1} \qquad \text{ohms} \tag{9-51}$$

This applies to ring B in Fig. 9-9.

For disk A, the analysis is similar to that for the single-contact bar geometry, but is complicated by the circular geometry. The formula, given by Early,[4] is simply

$$R_{\text{DISK}} = \frac{1}{8\pi}\frac{\rho_b}{W} \tag{9-52}$$

Applying Eqs. (9-51) and (9-52) to the geometry of Fig. 9-9, we obtain

$$r'_B = R_{\text{DISK}} + R_{\text{RING}} \tag{9-53}$$

or

$$r'_B = \frac{\rho_b}{W}\left(\frac{1}{8\pi} + \frac{1}{2\pi}\ln\frac{r_2}{r_1}\right) \tag{9-54}$$

One observes that for the disk term the base resistance is independent of radius. However, to minimize r'_B, the base contact must be placed as close to the active region as is practically feasible. This minimizes the ratio r_2/r_1 for the ring term.

In many practical junction transistors, the designers encounter geometries very different from the two basic types we have been concerned with thus far. Because of this, the equations for base resistance must be corrected accordingly. This is particularly true for most single-ended impurity-contact transistors.

To be inclusive of all the possible base geometries or sections thereof found in most junction transistors, a summary of the base-resistance relations introduced in this section is presented in Fig. 9-10. In the figure, the arrows denote which planes the current is flowing into and/or out of. Also, the shaded portions represent ohmic base contacts. All the equations are directly proportional to ρ/W, which was defined as the base-region sheet resistance R_{BB}. Thus, by judiciously selecting the appropriate formulas and adding them up directly, one can determine the complete expression for the base resistance of any transistor geometry. A typical illustration of the use of the formulas in Fig. 9-10 is given in Fig. 9-11a and b, wherein both a linear and a circular base geometry are presented.

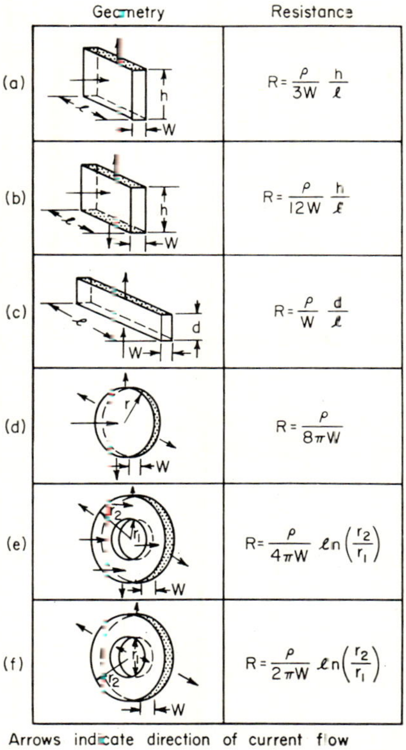

FIG. 9-10 Summary of resistance formulas for various sections of junction-transistor base-region geometries.

9-7. Sheet Resistance of Graded Bases. In the discussion of base resistance in the previous section we were concerned with the uniformly doped base region where the sheet resistance R_{BB} is simply the ratio of the base resistivity to the base width. We must now turn our attention to the case of the graded base to determine the correct designation for sheet resistance. For the graded base, sheet resistance will be denoted by R'_{BB} to distinguish it from the uniform-base case. Once R'_{BB} is determined, r'_B can be readily established simply by substituting R'_{BB} for $\rho/W = R_{BB}$ in the equations summarized in Fig. 9-10. In this section, Eq. (9-69) is the correct expression for calculating R'_{BB}.

Basically, the elemental sheet conductance $1/dR'_{BB}$ of an incremental strip dx wide is written as

$$\frac{1}{dR'_{BB}} = \frac{dx}{\rho(x)} \tag{9-55}$$

where $\rho(x)$ is the resistivity as a function of distance and may be defined as

$$\rho(x) = \frac{1}{q\mu_{nb}N(x)} \tag{9-56}$$

where $N(x)$ is the net impurity concentration in the base as a function of x. This is for a p-n-p transistor; therefore the majority-carrier mobility μ_{nb} is shown. Inserting (9-56) into (9-55) and integrating, we get

$$\frac{1}{R'_{BB}} = q\mu_{nb} \int_0^W N(x)\, dx \tag{9-57}$$

This is the general expression for the reciprocal of the transverse sheet resistance of the base region.[5] It should be evident that for a uniform base, where $N(x) = N'_B$, Eq. (9-57) reduces to $R_{BB} = \rho_b/W$. For the

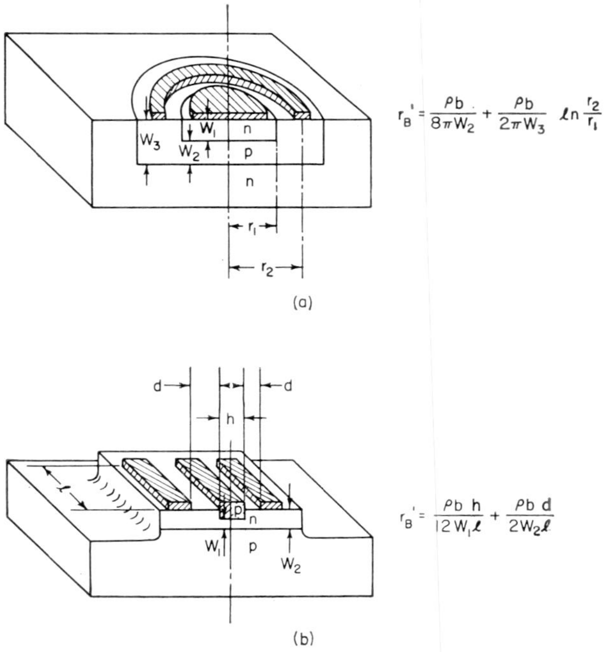

FIG. 9-11. Application of base-resistance formulas to typical transistor geometries to determine r'_B. (*a*) Circular transistor geometry; (*b*) linear transistor geometry.

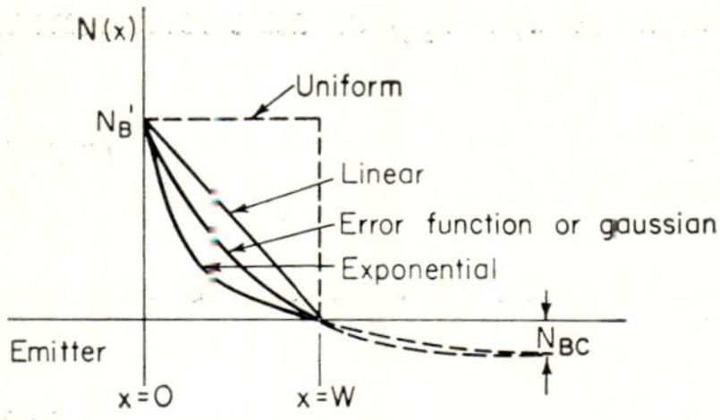

FIG. 9-12. Base impurity distribution $N(x)$ for various junction-transistor grades.

nonuniform cases, however, it is necessary to describe $N(x)$ by the proper functional relationships. The base impurity distributions corresponding to the linear, error-function or gaussian, and exponential grades are drawn in Fig. 9-12 for comparison purposes. In each case, N_B' is the impurity concentration in the base just at the emitter junction and N_{BC} is the background impurity level in the collector, compensated to establish the transition point at $x = W$. For these base layers we can establish the following quantitative expressions for $N(x)$:[5]

$$N(x) = N_B' \qquad \text{uniform} \tag{9-58}$$

$$N(x) = N_B'\left(1 - \frac{x}{W}\right) \qquad \text{linear} \tag{9-59}$$

$$N(x) = N_B'\left[1 - \operatorname{erf}\left(\frac{x}{L}\right)\right] - N_{BC} \qquad \text{error function} \tag{9-60}$$

$$N(x) = N_B'\epsilon^{-x/L} - N_{BC} \qquad \text{exponential} \tag{9-61}$$

$$N(x) = N_B'\epsilon^{-x^2/L^2} - N_{BC} \qquad \text{gaussian} \tag{9-62}$$

In the last three equations, N_{BC} is defined by letting $N(x) = 0$ at $x = W$.

To determine the sheet resistance R_{BB}' we simply insert the appropriate expression into Eq. (9-57) and carry out the integration. For example, the linear grade yields

$$\frac{1}{R_{BB}'} = q\mu_{nb}N_B'\left(x - \frac{x^2}{2W}\right)_0^W$$

$$\frac{1}{R_{BB}'} = \frac{q\mu_{nb}N_B'W}{2} \tag{9-63}$$

Note that in Eq. (9-63), we can let

$$\frac{1}{q\mu_{nb}N_B'W} = R_{BB} \tag{9-64}$$

which is the sheet resistance of a uniform base of thickness W and doping N_B'. Therefore, in our example of the linear grade, we can write

$$R_{BB}' = 2R_{BB} \qquad \text{linear grade} \tag{9-65}$$

In a similar manner we can obtain the ratio of R_{BB}' to R_{BB}, i.e., K_R, for the other graded bases given by Eqs. (9-60) through (9-62). The results of the analysis[5] are plotted in Fig. 9-13 for K_R as a function of the impurity ratio N'_B/N_{BC}. Note that the gaussian curve is the same as the curve for the error-function impurity distribution (a reasonably good approximation). From Fig. 9-13 it is evident that sheet resistance increases with base grading.

In order to calculate the absolute value of R'_{BB}, it is necessary to determine the uniform base equivalent given by Eq. (9-64). However, the question arises concerning the correct value of the mobility μ_{nb}. In a graded base, the mobility will vary with $N(x)$ such that it increases as x approaches W. In the integration of Eq. (9-57), we assumed μ_{nb} to be constant; actually, it is a function $\mu(x)$ and must be included in the integration to obtain the more accurate expression for R'_{BB}. This is rather formidable analysis, particularly for the error-function and gaussian distributions, but it has been done by Cuttriss for germanium[6] diffused bases and by Backenstoss for silicon.[7] A reasonably good approach estimating R'_{BB} accurately is to obtain an average value for the mobility and use the corrected value as the constant in the integration of Eq. (9-57). For all graded bases, let us approximate $\mu_{nb}(x)$ by a simple exponential function as follows:

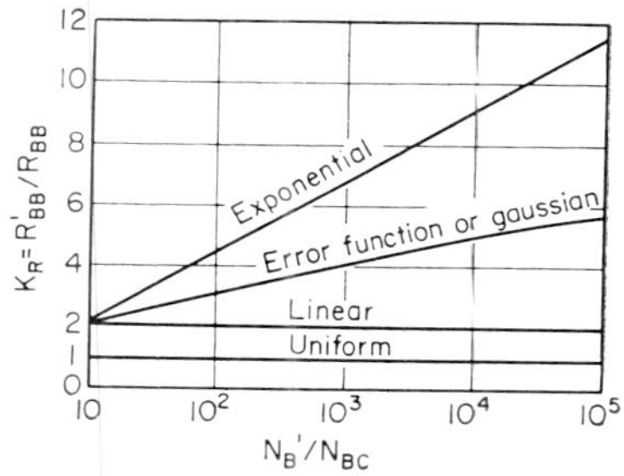

FIG. 9-13. Effect of various impurity distributions on sheet resistance normalized to equivalent uniform base. (*After J. L. Moll and I. M. Ross.*)

$$\mu_{nb}(x) = \mu_{nb}(0)\epsilon^{x/L} \tag{9-66}$$

where $\mu_{nb}(0)$ is the majority-carrier mobility corresponding to $N'_B(x = 0)$. At $x = W$, $\mu_{nb}(x) = \mu_{nb}(W)$, corresponding to the majority-carrier mobility at $N_{BC}(x = 0)$. The average mobility $\bar{\mu}_{nb}$ is obtained by integrating (9-66).

$$\bar{\mu}_{nb} = \frac{\mu_{nb}(0)}{W} \int_0^W \epsilon^{x/L}\, dx = \mu_{nb}(0)\, \frac{L}{W}\, (\epsilon^{W/L} - 1) \tag{9-67}$$

From (9-66), $\epsilon^{W/L} = \mu_{nb}(W)/\mu_{nb}(0)$ and $W/L = \ln [\mu_{nb}(W)/\mu_{nb}(0)]$. Therefore Eq. (9-67) becomes, finally,

$$\bar{\mu}_{nb} = \frac{\mu_{nb}(W) - \mu_{nb}(0)}{\ln [\mu_{nb}(W)/\mu_{nb}(0)]} \tag{9-68}$$

With this result, the complete expression for the sheet resistance of a graded base layer is

$$R'_{BB} \approx \frac{K_R}{q \dfrac{\mu_{nb}(W) - \mu_{nb}(0)}{\ln [\mu_{nb}(W)/\mu_{nb}(0)]} N'_B W} \tag{9-69}$$

where K_R is determined by using Fig. 9-13 and the impurity ratio N'_B/N_{BC}.

9-8. Base Input and Collector Saturation Voltages. Of prime importance in switching applications are the input and output voltages of the transistor when it is driven into *saturation* (or more specifically, voltage saturation) when operated in the grounded-emitter configuration. By

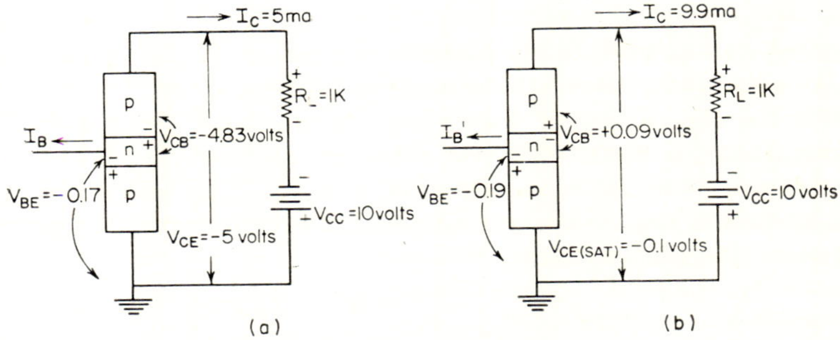

FIG. 9-14. Effect of base current drive on net junction bias potentials illustrating bias reversal at the collector junction. (*a*) Active region; (*b*) "on" region or saturation.

definition, saturation is the condition in which both the emitter and collector junctions become forward-biased. This results in the two voltages of interest in this section, viz., the base input voltage V_{BE} and the collector saturation voltage $V_{CE(\text{SAT})}$. It will now be explained on a qualitative basis how the collector-junction bias potential switches from a reverse- to a forward-bias condition in the saturation state.

Let us refer to the grounded-emitter p-n-p transistor and its V_{CE}-I_C characteristic shown in Fig. 7-10. The normal region of operation is called the active region. To illustrate operation in the active region, the junction potentials are as indicated in Fig. 9-14*a*. Let us presume an arbitrary base current such that the collector current $I_C = 5$ma, which places the operating point in the active region. For this condition, $V_{CE} = -5$ volts because of the voltage drop in the load resistor. Furthermore, as shown in Fig. 9-14*a*, $V_{CB} = -4.83$ volts and $V_{BE} = -0.17$ volts. The latter is the base input voltage. Thus we see that the emitter is forward-biased and the collector is reverse-biased, as is to be expected.

Now, if the base current I_B is increased in order to achieve a saturation or on condition, we have a situation as shown in Fig. 9-14*b*. For this case we will assume that $I_C = 9.9$ ma, which means that $V_{CE(\text{SAT})} = -0.1$

volt, or in other words, almost all the battery supply voltage is dropped across the load resistor R_L. To establish the increase in I_C from 5 ma to 9.9 ma it was necessary to increase V_{BE} from -0.17 volt to -0.19 volt. Since $V_{CE} = V_{CB} + V_{BE}$, it becomes apparent that V_{CB} must equal $+0.09$ volt, which is clearly a forward-bias condition for the collector junction. Thus we see that because of the limitations of the external load circuit the base becomes more negative than either the emitter or collector in saturation. The total voltage from collector to emitter becomes quite small, since it represents the sum of two opposing potentials. This is the collector saturation voltage $V_{CE(\text{SAT})}$; its locus of operation is shown by the $V_{CE(\text{SAT})}$ line in the collector characteristic of Fig. 7-10.

In order to quantitatively describe the behavior[8] of V_{BE} and $V_{CE(\text{SAT})}$ we can again make use of the general transistor-current Eqs. (8-30) and (8-31). In the saturation region, we permit both V_E and V_C to be forward-biased so that $\epsilon^{qV/kT} \gg 1$. We can also assume that no recombination occurs in the base region ($W \ll L_{pb}$). Finally, we can express the $qADp/W$ terms by the current designation I in order to simplify the analysis. Thus, we have

$$I_E = (I_{ne} + I_{pb})\epsilon^{qV_E/kT} - I_{pb}\epsilon^{qV_C/kT} \tag{9-70}$$

$$I_C = I_{pb}\epsilon^{qV_E/kT} - (I_{nc} + I_{pb})\epsilon^{qV_C/kT} \tag{9-71}$$

We solve Eq. (9-71) for $\epsilon^{qV_C/kT}$ and substitute the result in (9-70), which yields

$$\epsilon^{qV_E/kT} = \frac{I_E + [I_{pb}/(I_{pb} + I_{nc})]I_C}{I_{ne} + I_{pb}[1 - I_{pb}/(I_{nc} + I_{pb})]} \tag{9-72}$$

This is in the form of

$$\epsilon^{qV_E/kT} = \frac{I_E + \alpha_R I_C}{I_{ne} + I_{pb}(1 - \alpha_R)} \tag{9-73}$$

where α_R is the inverse alpha of the transistor, that is, the alpha that is measured when the collector is operated as an emitter. In this case, $\alpha_R = \gamma_R$ since $\beta^* = 1$; γ_R would be the "emitter" efficiency of the collector, or

$$\alpha_R = \gamma_R = \frac{1}{1 + R_{CC}/R_{BB}} \qquad \text{for } W \ll L_{pb} \tag{9-74}$$

where R_{CC} is the effective sheet resistance of the collector region. We also recognize that the denominator of Eq. (9-73) is the expression for the emitter reverse current I_{EBO}. Thus,

$$\epsilon^{qV_E/kT} = \frac{I_E + \alpha_R I_C}{I_{EBO}}$$

or

$$V_E = \frac{kT}{q} \ln \frac{|I_E| + |\alpha_R I_C|}{|I_{EBO}|} \tag{9-75}$$

which is the desired result for the forward-bias voltage of the base-emitter junction under saturation conditions. In a similar manner it is readily shown that the forward-biased collector voltage is given by[8]

$$V_C = \frac{kT}{q} \ln \frac{|I_C| + |\alpha I_E|}{|I_{CBO}|} \tag{9-76}$$

Now, since $V_{CE} = V_C - V_E$ in saturation, we may write

$$V_{CE} = \frac{kT}{q} \ln \frac{(I_C + \alpha I_E)I_{EBO}}{(I_E + \alpha_R I_C)I_{CBO}} \tag{9-77}$$

where the absolute magnitudes of the currents are used.

This result may be expressed in grounded-emitter terms by substituting $I_E = -(I_B + I_C)$ and the bilateral relationship $\alpha_R I_{CBO} = \alpha I_{EBO}$, which gives the final Ebers and Moll expression as[8]

$$V_{CE} = \pm \frac{kT}{q} \ln \frac{\alpha_R[1 - I_C(1 - \alpha)/(I_B\alpha)]}{1 + I_C(1 - \alpha_R)/I_B} \tag{9-78}$$

where the plus sign applies to p-n-p transistors and the minus sign to n-p-n transistors. The reader will note that in Eq. (9-78) the currents I_C and I_B are controlled by the circuitry external to the transistor.

Fig. 9-15 illustrates the manner in which Eq. (9-78) varies with base current drive for a given collector load current. The curves are for a typical p-n-p alloy junction transistor (germanium), for which $\alpha = 0.975$ and $\alpha_R = 0.9$. In the calculations it was assumed that α is constant with current and voltage. Consider, for example, the V_{CE} curve for $I_C = 10$ ma. At a base current $I_B = 0.256$ ma, we have a condition in which I_C/I_B is equal to $\alpha/(1 - \alpha)$. Consequently, V_{CE} is shown rising extremely rapidly toward infinity. This is actually the boundary between the active and saturation regions, wherein the collector junction becomes reverse-biased and V_{CE} becomes controlled by the external circuit and supply voltage. As I_B is increased, however, bringing the transistor into saturation, V_{CE} becomes smaller. This decrease is due to the fact that with increasing I_B, V_C (forward bias) increases at a faster rate than V_E, making the difference in potential smaller at the higher current. If the collector current is increased to 30 ma, we see that it requires more base current drive to saturate the device or obtain the same V_{CE}. In any case, it is seen that the resistance in saturation V_{CE}/I_C decreases with increasing circuit drive.

In our discussion thus far, we have been concerned only with the junction potentials of the transistor in the saturation state. From a practical device point of view, this is not sufficient to determine the exact saturation voltages. To be complete, it is necessary to include the ohmic body resistances of the collector and emitter regions, since these resistances are directly in series with the junction potentials. In grounded-emitter configuration, the equivalent-circuit representation would appear as shown in Fig. 9-16. The resistances r_{SC} and r_{SE} are the series resistances of the collector and emitter regions, respectively. In addition, the base resistance r'_B is included in the equivalent circuit. With reference to Fig. 9-16, we can now establish the final expressions for

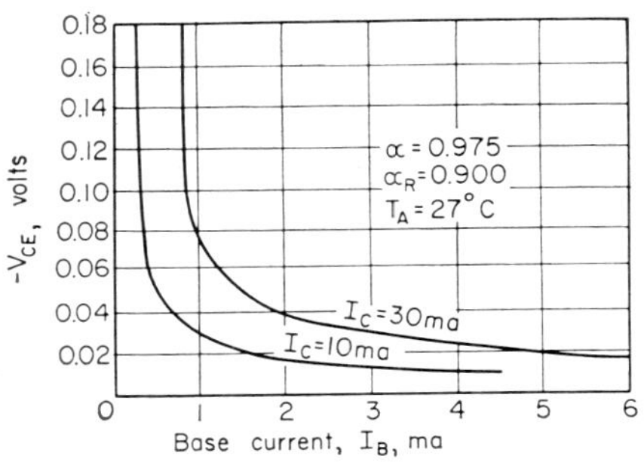

FIG. 9-15. Theoretical variation of V_{CE} with base drive for grounded-emitter p-n-p junction transistor.

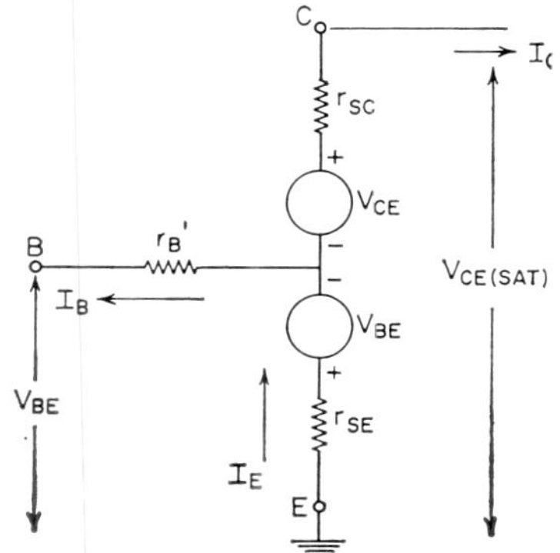

FIG. 9-16. Equivalent circuit for a grounded-emitter transistor in saturation including emitter and collector body resistance.

the saturation voltages of the transistor. The base input voltage V_{BE} becomes

$$V_{BE} = \frac{kT}{q} \ln \frac{|I_E| + |\alpha_R I_C|}{|I_{EBO}|} + I_E r_{SE} + I_B r'_B \tag{9-79}$$

and the collector saturation voltage $V_{CE(\text{SAT})}$ becomes

$$V_{CE(\text{SAT})} = \frac{kT}{q} \ln \frac{\alpha_R[1 - I_C(1 - \alpha)/(I_B\alpha)]}{1 + I_C(1 - \alpha_R)/I_B} + I_E r_{SE} + I_C r_{SC} \tag{9-80}$$

In these equations the voltage magnitudes are additive for either p-n-p or n-p-n transistor.

9-9. Parameter Temperature Dependence. Each of the major d-c transistor parameters discussed in this chapter (such as β, I_{CBO}, I_{EBO}, r'_B, V_{BE}, and $V_{CE(\text{SAT})}$) will exhibit a definite variation with or dependence on temperature. In each case, one will observe that the temperature-dependent terms in the appropriate parameter equations may be any one or more of the following: mobility, minority-carrier concentration, lifetime, and absolute temperature. The temperature behavior of these basic semiconductor characteristics has been presented in Chap. 4. In order to determine how any d-c parameter would vary with temperature, it is necessary to express the parameter equation directly in terms of

T, or[9]

$$P(T) = KT^n + \cdots \tag{9-81}$$

using a procedure identical to that which was used for the reverse current of a p-n junction in Sec. 6-3. Because the temperature variation of such parameters as mobility is a function of the carrier type, the semiconductor material, and the impurity concentration, it becomes exceedingly difficult to present any generalized relationship that satisfactorily covers all the possibilities. Each transistor structure must be treated on an individual basis.

PROBLEM

9-1. A p-n-p epitaxial germanium mesa transistor is to be analyzed for its d-c characteristics. The fabrication steps are as follows:

a. A 1-ohm-cm p-type epitaxial layer 3μ thick is grown on a 0.005-ohm-cm p-type substrate 5 mils thick. The minority-carrier lifetime in the epitaxial layer is 60 μsec.

b. An n-type base region is formed in the epitaxial layer by diffusing antimony from surface concentration of 2×10^{17} atoms/cm^3 to a junction depth of 1.5 μ. The average minority-carrier lifetime in the base layer is 4 μsec.

c. A pair of rectangular stripes is evaporated and alloyed (nonpenetrating) into the surface of the diffused base layer. The stripes are 1 mil wide and 6 mils long, separated by 0.5 mil. One stripe is an ohmic contact to the base region; the other stripe is an alloyed p-type emitter having a resistivity of 0.0002 ohm-cm, a thickness of 2,500 A, and a minority-carrier lifetime of 0.1 μsec.

d. The transistor is etched through the base and epitaxial regions, leaving a mesa with a 0.5-mil border all around the stripe pair. The final surface recombination velocity of the base is 150 cm/sec and is independent of temperature.

All bulk minority-carrier lifetimes double for every 60°C increase in temperature.

For this structure, determine the following electrical characteristics (at room temperature unless otherwise specified):

Parameter	Test conditions	Units
BV_{CBO}	$I_C = -100\ \mu a;\ I_E = 0$	volts
BV_{CES}	$I_C = -100\ \mu a;\ V_{BE} = 0$	volts
BV_{EBO}	$I_E = -100\ \mu a;\ I_C = 0$	volts
BV_{CEO}	$I_C = -1$ ma; $I_B = 0$	volts
β	$V_{CE} = -5$ volts; $I_E = *$	
β	$V_{CE} = -5$ volts; $I_E = *$; $T = 55°C$	
I_{EBO}	$V_{EB} = -1$ volt; $I_C = 0$	μa
I_{CBO}	$V_{CB} = -5$ volts; $I_C = 0$	μa
I_{CBO}	$V_{CB} = -5$ volts; $I_C = 0$; $T = 55°C$	μa
V_{BE}	$I_B = -0.5$ ma; $I_C = -10$ ma	volts
$V_{CE(SAT)}$	$I_B = -0.5$ ma; $I_C = -10$ ma	volts
r'_B	$I_E = 0;\ I_C = 0$	ohms
C_{T_e}	$V_{CB} = -5$ volts; $I_E = 0$	$\mu\mu f$
C_{T_e}	$I_E = 0;\ I_C = 0$	$\mu\mu f$

* β is based on low-level injection and is independent of current.

REFERENCES

1. Hunter, L. P.: "Handbook of Semiconductor Electronics," pp. 4-1–4-12, McGraw-Hill Book Company, Inc., New York, 1956.
2. Shockley, W.: "Electrons and Holes in Semiconductors," pp. 318–324, D. Van Nostrand Company, Inc., Princeton, N.J., 1953.
3. Pritchard, R. L., and W. N. Coffey: Small-signal Parameters of Grown-junction Transistors at High Frequencies, *IRE Convention Record*, part 3, pp. 89–98, 1954.
4. Early, J. M.: Design Theory of Junction Transistors, *Bell System Tech. J.*, vol. 32, pp. 1271–1312, November, 1953.
5. Moll, J. L., and I. M. Ross: The Dependence of Transistor Parameters on the Distribution of Base Layer Resistivity, *Proc. IRE*, vol. 44, pp. 72–78, January, 1956.
6. Cuttriss, D. B.: Relation Between Surface Concentration and Average Conductivity in Diffused Layers in Germanium, *Bell System Tech. J.*, vol. 40, pp. 509–521, March, 1961.
7. Backenstoss, G.: Evaluation of Surface Concentration of Diffused Layers in Silicon, *Bell System Tech. J.*, vol. 37, pp. 699–709, May, 1958.
8. Ebers, J. J., and J. L. Moll: Large-signal Behavior of Junction Transistors, *Proc. IRE*, vol. 42, pp. 1761–1772, December, 1954.
9. Gärtner, W. W.: Temperature Dependence of Junction Transistor Parameters, *Proc. IRE*, vol. 45, pp. 662–680, May, 1957.

10

High-current Transistor Characteristics

10-1. High-level Injection Theory. In the theory of junction transistors developed thus far, we have been concerned only with low-level injection conditions, that is, when the concentration of minority carriers injected into the base region is very small compared to the majority-carrier concentration there. Under these conditions, charge neutrality is hardly disturbed and the carriers move principally by diffusion in field-free regions.* As a result, the expression for the current gain (alpha) was shown to be independent of emitter bias current. For the grounded-emitter circuit configuration, the current gain (beta) was given as

$$\frac{1}{\beta} = \frac{\rho_e W}{\rho_b L_{ne}} + \frac{W^2}{2L_{pb}{}^2} + \frac{sA_s W}{AD_{pb}} \tag{10-1}$$

As long as low-level injection conditions are maintained, this equation is adequate. However, it is observed experimentally that the current gain of a junction transistor is not independent of emitter current I_E as is implied by Eq. (10-1), but is a function of the minority-carrier injection level in the base region. As we shall see, the current gain decreases at the higher emitter currents. Since transistors are also used in large-signal applications such as power amplifiers and high-current switches, it is important to examine the characteristics of transistors operated at high currents. Therefore, in this chapter we shall examine the variation of the grounded-emitter current gain with emitter current by correcting the first-order theory with appropriate factors that result from high-level injection conditions.

* The exception, of course, is for nonuniform base layers, wherein the injected carriers experience the built-in electric field. However, the theory was based on the fact that the injected carriers did not disturb the field distribution.

Equation (10-1) can be written in the form of current ratios as follows:

$$\frac{1}{\beta} \approx \frac{I_{sB}}{I_{pE}} + \frac{I_{vB}}{I_{pE}} + \frac{I_{nE}}{I_{pE}} \tag{10-2}$$

The order of the terms has been changed here only to suit the sections to follow. The reader will note that each term of (10-2) is divided by the emitter hole current I_{pE}. If alpha is close to unity, this current is approximately equal to the total emitter current I_E (i.e., the electron component is negligible). It must be emphasized again that in Eq. (10-2) the current gain is independent of emitter current.

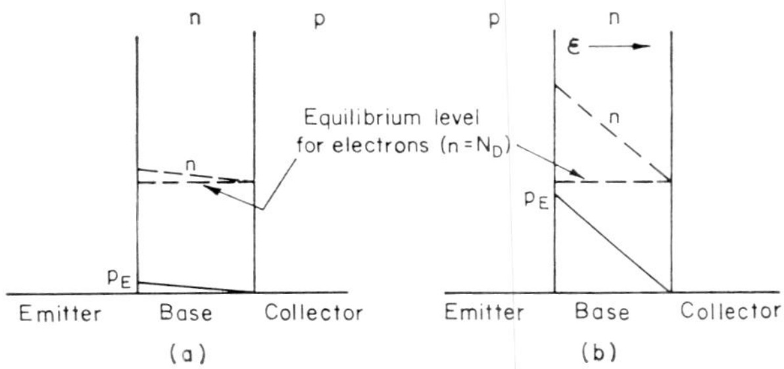

FIG. 10-1. Effect of injection level on majority-carrier distribution in the base region. (*a*) Low level; (*b*) high level.

To insert the effects of high-level injection into the base region, it is necessary to examine the carrier distributions in the base region under high-level conditions. In a p-n-p structure, for example, the majority carriers in the base are electrons and the injected minority carriers from the emitter are holes. In this chapter the concentration of the latter will be denoted as p_E. Figure 10-1 illustrates the effect of injection level on the base carrier distributions. For the low-level case shown by (*a*), p_E is small compared to n ($n \approx N_D$). Consequently, charge neutrality in the base is not disturbed appreciably by the presence of the holes, and the injected current is principally a diffusion current whose magnitude is determined by the slope of the hole-concentration gradient.

However, for the high-level case the situation is somewhat different, as shown by Fig. 10-1*b*. Here, p_E may be comparable to or greater than N_D. Under these conditions, the external circuit must supply sufficient electrons to the base layer in order to compensate for the large positive charge of the injected holes. If charge neutrality is maintained throughout the base width, the electron concentration gradient equals that of the holes. Because of the electron gradient, the electrons try to diffuse toward the collector junction. However, as soon as this occurs, charge neutrality is violated. Therefore, in order to keep the electrons in place against their gradient, an electric field must exist in the base layer.[1,*] This field is in such a direction as to aid the flow of holes to the collector junction. Thus we have a situation identical to that of the p-n junction operating at high forward currents (see Chap. 6), except that in this case the ohmic contact is replaced by the reverse-biased collector junction. Under these conditions, the emitter current cannot be determined on the basis of diffusion alone, but must also include drift effects due to the presence of the electric field.

Quantitatively, the base-region currents due to high-level injection may be expressed by

$$I_p = qA\mu_{pb}p\mathcal{E} - qAD_{pb}\frac{dp}{dx} \tag{10-3}$$

$$I_n = qA\mu_{nb}n\mathcal{E} + qAD_{nb}\frac{dn}{dx} \tag{10-4}$$

where the first right-hand term in each of the equations corresponds to the drift component due to the electric field $\mathcal{E}$. As inferred from Fig. 10-1*b*, we may write

$$\frac{dp}{dx} = \frac{dn}{dx}$$

which indicates equal concentration gradients of the carriers. Substituting this result into (10-4), letting the electron current $I_n = 0$, and solving for $\mathcal{E}$, we have

$$\mathcal{E} = \frac{-D_{nb}}{\mu_{nb}}\frac{1}{n}\frac{dp}{dx} = \frac{-kT}{q}\frac{1}{n}\frac{dp}{dx} \tag{10-5}$$

Inserting this result into Eq. (10-3) and simplifying the result, we have, finally,

$$I_p = -qAD_{pb}\frac{dp}{dx}\left(1 + \frac{p}{n}\right) \tag{10-6}$$

This is the general expression for the current due to holes (for a p-n-p transistor), where p is the injected hole concentration and n is the base majority-carrier concentration. If this expression is evaluated at the boundary ($x = 0$) corresponding to the emitter junction, we obtain the emitter current I_{pE}. For negligible recombination in the base, it is reasonable to assume that the slope of the hole concentration in the base is linear and equal to

$$\frac{dp}{dx} \approx \frac{-p_E}{W} \tag{10-7}$$

* References, indicated in the text by superscript figures, are listed at the end of the chapter.

where p_E is the injected hole density at the emitter and W is the base width. Further, in Eq. (10-6), for high-level injection conditions, we have

$$n \approx N_D + p_E \tag{10-8}$$

Combining (10-7) and (10-8) with (10-6), we have

$$I_{pE} = \frac{qAD_{pb}p_E}{W}\left(1 + \frac{p_E}{N_D + p_E}\right) \tag{10-9}$$

It should be noted that if $p_E \ll N_D$, which is the case for low-level injection, Eq. (10-9) reduces to the familiar form. However, if $p_E \gg N_D$, (10-9) becomes

$$I_{pE} = \frac{2qAD_{pb}p_E}{W} \tag{10-10}$$

indicating a twofold increase in I_{pE} in the limit. Thus we see that the term in parenthesis in Eq. (10-9) is a measure of the contribution of the electric field in the base to the total hole current as injection increases from low- to high-level conditions. A convenient way of expressing the injection level is by means of the ratio of the injected minority-carrier concentration to the majority-carrier concentration in the base. The ratio υ is defined as

$$\upsilon \equiv \frac{p_E}{N_D} \tag{10-11}$$

Thus, if Eq. (10-11) is substituted into (10-9), it becomes, after simplification,

$$I_{pE} \approx \frac{qAD_{pb}N_D}{W}\,\upsilon\,\frac{1 + 2\upsilon}{1 + \upsilon} \tag{10-12}$$

Now, if Eq. (10-12) is arranged as follows:

$$Z = \upsilon\,\frac{1 + 2\upsilon}{1 + \upsilon} = \frac{I_{pE}W}{qAD_{pb}N_D} \tag{10-13}$$

it appears in the form, suggested by Webster,[1] which relates the injection level to the actual current and characteristics of the transistor base region. By introducing the appropriate mobility factors into Eq. (10-13), we may express the Z factor directly in terms of base sheet resistance:

$$Z = \frac{qbR_{BB}W^2I_E}{kTA} \qquad \text{p-n-p} \tag{10-14}$$

$$Z = \frac{qR_{BB}W^2I_E}{kTbA} \qquad \text{n-p-n} \tag{10-15}$$

where b is the magnitude of the mobility ratio μ_n/μ_p.

In the sections to follow, the Z factors will be applied to each of the terms of the current-gain equation to illustrate the effects of injection level. However, in order to do this, it is necessary to relate v to Z, such that

$$v = f(Z) \tag{10-16}$$

This is done by solving Eq. (10-13) for v. There results an excellent approximation, which is

$$v \approx \frac{Z}{2} \tag{10-17}$$

This is a feasible result since Z is actually the ratio of the effective injected carrier density to the base impurity density that one would compute from first-order theory. However, under high-level conditions, the electric-field effect manifests itself by approximate doubling of the current. Thus it would be expected that the magnitude of Z is approximately twice that of the actual injection ratio p_E/N_D.

A convenient way of defining the emitter current at which high-level injection effects begin is to determine the magnitude of the injected minority carrier concentration and compare it to the majority-carrier concentration in the base. Let us define the beginning of high-level effects as the point when $p_E = 0.1N_D$, i.e., when $v = 0.1$. Since $Z = 2v$, high-level injection begins when $Z = 0.2$. To translate this into transistor terms, we solve Eqs. (10-14) and (10-15) for I_E and let $Z = 0.2$. We have, then,

$$I_E = \frac{0.2kTA}{qbW^2R_{BB}} \qquad \text{p-n-p} \tag{10-18}$$

$$I_E = \frac{0.2kTbA}{qW^2R_{BB}} \qquad \text{n-p-n} \tag{10-19}$$

With $b \approx 2$ for germanium and $b \approx 3$ for silicon, we get the following useful room-temperature relationships:

$$\begin{aligned}
&\text{p-n-p silicon:} && I_E \approx 1.7A/R_{BB}W^2 && \text{ma}\\
&\text{p-n-p germanium:} && I_E \approx 2.6A/R_{BB}W^2 && \text{ma}\\
&\text{n-p-n germanium:} && I_E \approx 10.4A/R_{BB}W^2 && \text{ma}\\
&\text{n-p-n silicon:} && I_E \approx 15.6A/R_{BB}W^2 && \text{ma}
\end{aligned} \tag{10-20}$$

To summarize what has been done here, Eqs. (10-20) define that emitter current I_E (in milliamperes) at which the carrier injection level becomes significant such that its effects cannot be neglected in design calculations. The relations are expressed in terms of R_{BB}, the base width, and the area A, which is the actual cross-sectional area of the transistor. One observes that an n-p-n silicon transistor will operate at currents higher than those of the other three types before high-level effects occur. Actually, what is involved here is simply a matter of a particular voltage drop in the base region. Since p-type silicon has the smallest majority-carrier mobility, it requires the largest current to produce an equivalent voltage drop.

10-2. Variation of Surface Recombination with Current. From the low-level current-gain theory we saw that the contribution of surface recombination effects to the over-all grounded-emitter current gain is given by the ratio I_{sB}/I_{pE}. This is the ratio of the current lost by surface recombination to the total emitter hole current.

$$\frac{I_{sB}}{I_{pE}} = \frac{sA_sW}{AD_{pb}} \tag{10-21}$$

It is readily seen that this result is independent of emitter current and is applicable only at very low injection levels.

To account for high-level conditions, we must now take into consideration the base-region electric-field effect at high currents. In Chap. 8 we saw that the surface recombination current may be written as

$$I_{sB} = qsA_sp_E \tag{10-22}$$

Further, in the previous section the general expression for I_{pE} was given by Eq. (10-9).

$$I_{pE} = \frac{qAD_{pb}}{W} p_E \left(1 + \frac{p_E}{N_D + p_E}\right) \tag{10-23}$$

Taking the ratio of (10-22) to (10-23), we obtain

$$\frac{I_{sB}}{I_{pE}} \approx \frac{sA_sW}{AD_{pb}} \frac{N_D + p_E}{N_D + 2p_E} \tag{10-24}$$

Examining this result, we see that the term in parenthesis may be written in terms of Z, since $p_E/N_D = v \approx Z/2$, such that (10-24) becomes

$$\frac{I_{sB}}{I_{pE}} \approx \frac{sA_sW}{AD_{pb}} \frac{1 + Z/2}{1 + Z} \tag{10-25}$$

or

$$\frac{I_{sB}}{I_{pE}} \approx \frac{sA_sW}{AD_{pb}} f'(Z) \tag{10-26}$$

The term $f'(Z)$ is referred to as the *field factor;* its magnitude as a function of Z is given in Fig. 10-2. For low values of Z, $f'(Z)$ is close to unity, such that (10-26) corresponds to the low-level case. As Z increases with increasing emitter current, $f'(Z)$ at first decreases rapidly and then asymptotically approaches ½ in the limit.

Physically, this is understandable if one realizes that as injection level is increased, the electric field established in the base increases the total emitter current. However, for any given high-level value of I_E, a smaller value of injected carrier density is required to maintain the current, thereby decreasing the proportion of injected carriers that are lost by surface recombination. Since the emitter current doubles in the limit, we can expect I_{sB}/I_{pE} to be one-half its initial value, as shown. Thus, on the basis of surface effects alone, the transistor current gain will increase with emitter current.

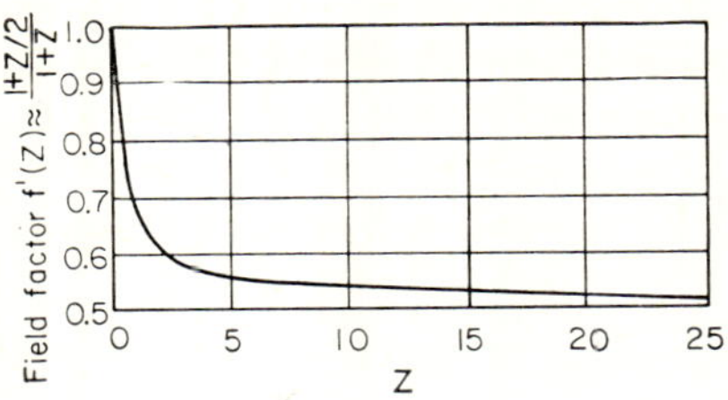

FIG. 10-2. The field factor $f'(Z)$ as a function of Z. (*After W. M. Webster.*)

10-3. Variation of Volume Recombination with Current. For the second term of Eq. (10-2), which represents the effect of volume recombination in the base region of the over-all current gain, the proportion of the emitter current lost by carrier recombination is given by

$$\frac{I_{vB}}{I_{pE}} = \frac{W^2}{2L_{pb}^2} = \frac{W^2}{2D_{pb}\tau_{pb}} \tag{10-27}$$

Since this low-level result is independent of I_E, we must now proceed to correct it to take into consideration the effects at high currents.

The volume recombination current I_{vB} may generally be expressed by the relation[2]

$$I_{vB} = qA\,\frac{p_E}{2}\,\frac{W}{\tau_{pb}} \tag{10-28}$$

where q = electron charge
AW = base-region volume
p_E/τ_{pb} = rate of volume recombination

The 2 arises from the assumption that $p_E/2$ represents the average carrier density in the base region. This is a good approximation if a linear carrier gradient is assumed.

In a manner similar to that of the previous section the ratio I_{vB}/I_{pE} is obtained by utilizing the general emitter-current expression given by Eq. (10-9). Taking the ratio of (10-28) to (10-9), we have

$$\frac{I_{vB}}{I_{pE}} = \frac{W^2}{2D_{pb}\tau_{pb}}\,\frac{N_D + p_E}{N_D + 2p_E} \tag{10-29}$$

Letting $Z/2 \approx p_E/N_D$ and simplifying, we get

$$\frac{I_{vB}}{I_{pE}} = \frac{W^2}{2D_{pb}\tau_{pb}}\,\frac{1 + Z/2}{1 + Z} \tag{10-30}$$

or

$$\frac{I_{vB}}{I_{pE}} = \frac{W^2}{2D_{pb}\tau_{pb}}\,f'(Z) \tag{10-31}$$

Thus we see that the volume-recombination term behaves in a manner identical to that of the surface-recombination term, i.e., the low-level equation is multiplied by the field factor, $f'(Z)$. As injection level increases, less carriers are needed to maintain a given current and the proportion of carriers lost by volume recombination is reduced. Here again, in the limit, the high-level result is one-half the low-level value.

In the foregoing discussion it has been implicitly assumed that the minority-carrier lifetime in the base τ_{pb} remained constant and independent of injection level. For some transistors, this may be the case. In most designs, however, the high-level lifetime will differ from the low-level lifetime because of injection effects.[3] According to the Shockley-Read-Hall recombination theory introduced in Chap. 4, the low-level-lifetime equation is modified to account for injected carriers Δc as follows:

$$\tau = \frac{\tau_{pr}(n + n_r + \Delta c) + \tau_{nr}(p + p_r + \Delta c)}{(n + p + \Delta c)} \tag{10-32}$$

For a p-n-p device having an n-type base, where $\Delta c = p_E$ and $n \approx n_D$, (10-32) becomes

$$\tau \approx \frac{\tau_{pr}(N_D + n_r + p_E) + \tau_{nr}(p_r + p_E)}{(N_D + p_E)} \tag{10-33}$$

This result can be arranged in the form

$$\tau \approx \left[\frac{\tau_{pr}(N_D + n_r) + \tau_{nr}(p_r)}{N_D} + \frac{(\tau_{pr} + \tau_{nr})p_E}{N_D}\right]\frac{1}{1 + p_E/N_D} \tag{10-34}$$

By reference to Sec. 4-8 it may be seen that the first right-hand term of (10-34) represents the low-level lifetime τ_{pb}. In the second right-hand term, $(\tau_{pr} + \tau_{nr})$ is the high-level lifetime τ_∞. We recognize also the familiar injection factor, viz., $p_E/N_D \approx Z/2$. Thus, (10-34) can be written as

$$\tau = \tau_{pb}\frac{1 + (\tau_\infty/\tau_{pb})Z/2}{1 + Z/2} \tag{10-35}$$

which is the final expression which relates the variation of volume lifetime to the injection level Z. It is interesting to note that if the high- and low-level lifetimes are equal, $\tau_\infty/\tau_{pb} = 1$ and lifetime becomes independent of injection level.

We can now apply this result to Eq. (10-31) in order to obtain the complete variation of volume recombination with current, which is

$$\frac{I_{vB}}{I_{pE}} = \frac{W^2}{2D_{pb}\tau_{pb}}f'(Z)\frac{1 + Z/2}{1 + (\tau_\infty/\tau_b)Z/2} \tag{10-36}$$

It is convenient to combine both functions of Z into a single function, viz., $f''(Z)$, which is defined as

$$f''(Z) = \frac{1 + Z/2}{1 + Z}\,\frac{1 + Z/2}{1 + (\tau_\infty/\tau_b)Z/2} \tag{10-37}$$

In Fig. 10-3 is a plot of $f''(Z)$ for different values of the lifetime ratio τ_∞/τ_b, where τ_b is the low-level lifetime applicable to either n- or p-type base regions. The curve, for which $\tau_\infty/\tau_b = 1$, is simply a plot of the function $f'(Z)$, since the lifetime-variation factor in Eq. (10-37) becomes unity. The other curves of Fig. 10-3 indicate how the variation of volume lifetime with injection level can affect the over-all transistor current gain. For example, if the base region was of sufficiently high resistivity such that the low-level lifetime was considerably higher than the high-level lifetime, we would have a situation where I_{vB}/I_{pE} might vary according to the curve where $\tau_\infty/\tau_b = 0.1$. In this case, volume recombination effects would tend to decrease the current gain. On the other hand, if $\tau_\infty/\tau_b = 2$, volume recombination effects would increase current gain, since I_{vB}/I_{pE} would decrease with emitter current. As evidenced by the $\tau_\infty/\tau_b = 0.5$ curve, it is possible to maintain a fairly constant variation of current gain due to volume effects.

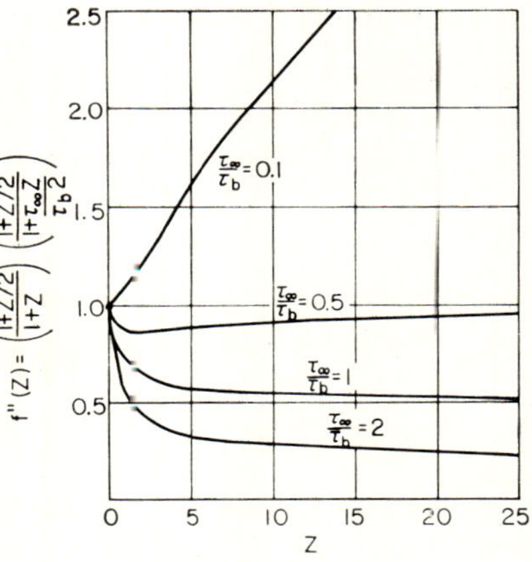

FIG. 10-3. $f''(Z)$ as a function of Z and τ_∞/τ_b.

10-4. Variation of Emitter Efficiency with Current. In a p-n-p junction transistor, the ratio of the injected hole current I_{pE} to the total emitter current $I_{pE} + I_{nE}$ is defined as the emitter efficiency. Thus,

$$\gamma = \frac{I_{pE}}{I_{pE} + I_{nE}} = \frac{1}{1 + I_{nE}/I_{pE}} \tag{10-38}$$

Solving (10-38) for I_{nE}/I_{pE}, we obtain

$$\frac{I_{nE}}{I_{pE}} \approx 1 - \gamma \qquad \text{for } \gamma \approx 1 \tag{10-39}$$

Thus we see that this result corresponds to the third term of (10-2) and represents the effect of emitter efficiency on the grounded-emitter current gain. We have, then,

$$\frac{I_{nE}}{I_{pE}} = \frac{\rho_e W}{\rho_b L_{ne}} \tag{10-40}$$

which is a result that is independent of emitter current. In order that Eq. (10-40) be small for maximum emitter efficiency, we see that the emitter resistivity ρ_e should be made small compared to the base resistivity ρ_b. We recall that

$$\rho_b \approx \frac{1}{q\mu_{nb}N_D} \tag{10-41}$$

where N_D is the donor impurity density in the base region. For low-level injection conditions, the majority electron concentration would be approximately equal to N_D.

As injection level is increased, however, the majority-carrier concentration in the base will increase to maintain charge neutrality with the large amount of injected holes. Under these conditions, the electron concentration becomes

$$n \approx N_D + p_E \tag{10-42}$$

We now have a situation where the base resistivity is decreased by the presence of injected carriers. Correcting (10-41) with (10-42) as follows,

$$\rho_b' \approx \frac{1}{q\mu_{nb}N_D}\frac{N_D}{N_D + p_E} = \rho_b \frac{1}{1 + p_E/N_D} \tag{10-43}$$

we obtain the expression for the effective base resistivity ρ_b' in which the injection factor p_E/N_D appears. Thus, wherever ρ_b appears in the emitter efficiency expression, it must be multiplied by $1/(1 + p_E/N_D)$, or

$$\frac{I_{nE}}{I_{pE}} = \frac{\rho_e W}{\rho_b L_{ne}}\left(1 + \frac{p_E}{N_D}\right) \tag{10-44}$$

Substituting $Z/2$ for p_E/N_D, we obtain, finally,[1]

$$\frac{I_{nE}}{I_{pE}} = \frac{\rho_e W}{\rho_b L_{ne}}\left(1 + \frac{Z}{2}\right) \tag{10-45}$$

In this result it is seen that, as emitter current is increased, Z becomes larger, thereby rapidly increasing the magnitude of Eq. (10-45) and causing a corresponding decrease in current gain. As will be discussed in the next section, this change in γ with I_E represents the most significant factor that effects the variation of current gain with injection level.

For the usual operating emitter-current ranges for most junction transistors, the emitter-efficiency variation described by Eq. (10-45) is quite valid. It is evident from Eq. (10-45) that as Z approaches zero, γ attains a low-level value close to unity; as Z becomes very large, γ approaches zero. Actually, at both the extreme low and high limits of emitter current, Eq. (10-45) is not applicable because of certain effects which were not considered in the development of the theory. At very high injection levels where current densities are in the order of 10^4 amp/cm^2 or more, one cannot neglect the modulation effect of the minority-carrier current on the emitter resistivity.[4] In the case of a p-n-p transistor, for example, the electron concentration injected into the emitter at very high levels may be comparable to or greater than the emitter hole concentration in spite of the heavy doping, and may contribute toward lowering the apparent emitter resistivity. Because of this effect, it has been shown that transistor emitter efficiency will tend to saturate to a finite value at high currents rather than decrease to zero.[4] This value of γ may be on the order of 0.2 to 0.4 for currents in the hundreds of amperes. In the grounded-emitter configuration, these values correspond to current gains that are less than unity.

We must now consider another effect, at the extreme opposite end of the current scale, which may alter the emitter efficiency at very low emitter currents. This is a phenomenon, associated with the presence of recombination centers in the space-charge layer of the emitter junction, which particularly occurs in silicon transistors.[5] In Sec. 6-4 it was shown that, under conditions of reverse bias, the reverse current for a silicon p-n junction is due to carrier generation from the recombination centers in the barrier. Furthermore, at ordinary temperatures, this generation current is considerably larger than the diffusion component. In like manner, under conditions of forward bias the same recombination centers act to produce a *recombination current*, due to the trapping of the injected carriers as they pass through the transition region of the junction. Thus, in the case of silicon junction transistors operated at very low emitter currents, the recombination current dominates the total current and does not contribute to transistor action. This results in a decrease in emitter efficiency, as evidenced by

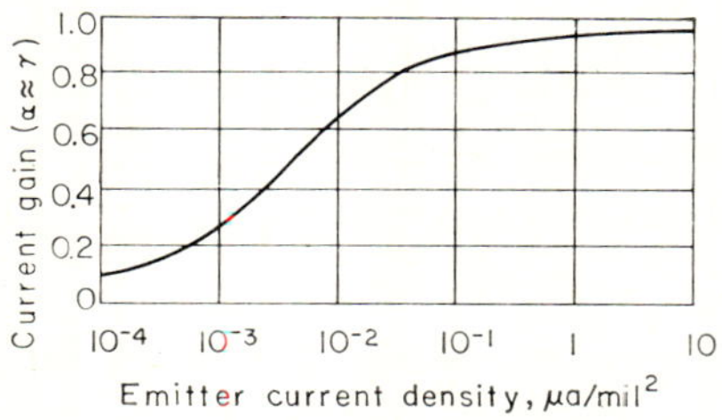

FIG. 10-4. Typical variation of current gain for silicon transistors at low emitter current densities. Falloff is attributed to impurity recombination in the emitter junction.

$$\gamma = \frac{I_{pE}}{I_{RE} + I_{pE} + I_{nE}} \tag{10-46}$$

where I_{RE} represents the emitter recombination current. At small forward biases where the normal diffusion current $I_{pE} + I_{nE}$ may be small compared to I_{RE}, it is seen that γ decreases to zero. As I_E increases with increased bias, γ increases toward unity, since the diffusion currents predominate in Eq. (10-46). A typical variation of d-c alpha, as affected by emitter efficiency at low currents, is given in Fig. 10-4. As an example for a silicon transistor with a cross-sectional area of 200 mil^2, alpha starts to maximize at approximately 1 ma emitter current. At currents less than 1 ma, alpha falls off rapidly. This falloff at low currents is much less pronounced in germanium transistors because of the fact that germanium can be processed to be relatively free of recombination centers. Consequently, in germanium devices, the barrier recombination currents are practically negligible compared to the diffusion currents.

10-5. Composite Variation of Current Gain. We will now combine the results of the preceding sections into a single relationship which will describe the variation of the grounded-emitter current gain as a function of emitter current.[1] This will combine the effects of surface recombination, volume recombination, emitter efficiency, and emitter recombination. Referring to Eqs. (10-1), (10-26), (10-36), (10-45), and (10-46), we obtain

$$\frac{1}{\beta} \approx \frac{sA_sW}{AD_{pb}} f'(Z) + \frac{W^2}{2D_{pb}\tau_{pb}} f''(Z) + \frac{\rho_e W}{\rho_b L_{ne}} \left(1 + \frac{Z}{2}\right) + \frac{I_{RE}}{I_E} \qquad (10\text{-}47)$$

where $f'(Z)$ is given by Fig. 10-2 and $f''(Z)$ by Fig. 10-3. Furthermore, Z is related to the emitter current by Eq. (10-14), which is repeated here as follows:

$$Z \approx \frac{qbR_{BB}W^2I_E}{kTA} \qquad (10\text{-}48)$$

It should be noted that these equations are for a uniform-base p-n-p transistor; for an n-p-n device it is necessary only to interchange the subscripts p and n and b appears in the denominator. For a graded transistor, it is necessary to replace the 2 in the denominator of the second term of (10-47) with a 4. Also, one uses the general form for emitter efficiency, i.e., R_{EE}/R_{BB}, instead of $\rho_e W/\rho_b L_{ne}$.

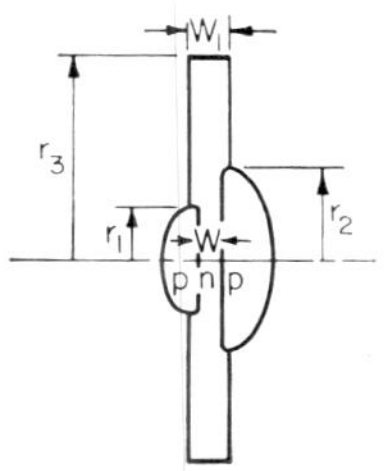

FIG. 10-5. Numerical values of physical constants for a germanium-alloy transistor to illustrate current-gain behavior.

A numerical example will best illustrate the current behavior of Eq. (10-47) and demonstrate the contributions of the separate mechanisms discussed thus far. Let us consider a p-n-p germanium-alloy transistor of the type shown in Fig. 10-5. Included in the figure are typical values for the physical parameters of interest. The value of base width indicated is the effective width at the particular collector voltage applied to the transistor. In the analysis to follow, we shall assume that the collection efficiency is 100 per cent, i.e., no carriers are lost to the collector surface. Also, we shall assume that the high-level lifetime equals the low-level value such that $\tau_\infty/\tau_b = 1$. For the surface-recombination term, the low-level value includes the term A_s, which may be determined by letting the active surface area be a ring of width W around the emitter perimeter.

Thus $A_s = 2\pi r W$. Inserting the appropriate values from Fig. 10-5, we calculate for the surface term:

$$\frac{sA_sW}{AD_{pb}} = 0.009$$

The low-level volume-recombination term is calculated to be

$$\frac{W^2}{2D_{pb}\tau_{pb}} = 0.0028$$

Lastly, using the correct values for the emitter efficiency term, we have

$$\frac{\rho_e W}{\rho_b L_{ne}} = 0.0067$$

Substituting these results into Eq. (10-47), we get

$$\frac{1}{\beta} = 0.0090\,f'(Z) + 0.0028\,f''(Z) + 0.0067\left(1 + \frac{Z}{2}\right) \qquad (10\text{-}49)$$

It is interesting to note here that surface recombination plays the most dominant role in determining current gain. At low-level conditions, the

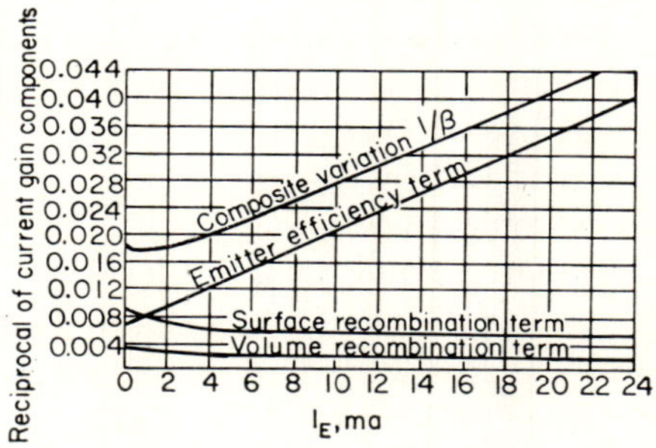

FIG. 10-6. Emitter current variation of current-gain components for illustrative example.

reciprocal of the sum of the three constants yields a current gain equal to 54. If surface recombination were negligible, the low-level gain would be equal to 105.

To extend the analysis to larger emitter currents, it is necessary to determine the magnitude of Z. In Eq. (10-48), Z becomes 412 I_E, where I_E is in amperes. For emitter currents in the range up to 25 ma, we will be concerned with values of Z up to about $Z = 10$. Thus, each term of Eq. (10-49) may be plotted as a function of Z for values of Z up to 10 by computing the magnitude of the multiplying Z functions. Furthermore, by use of Eq. (10-48) each term of (10-49) may be related directly to the emitter current I_E. This has been done; the results are shown in Fig. 10-6. Both the surface- and volume-recombination terms fall off quickly to practically constant values, whereas the emitter efficiency increases linearly at a marked slope. The composite variation curve, which is the direct sum of the three, indicates that the emitter efficiency term has the most significant effect at the higher emitter currents. In Fig. 10-7 is a plot of the grounded-emitter current gain as obtained from the reciprocal values of the composite variation curve. At very low currents, the gain is 54; it peaks to 55 at 1 ma and then falls off to 25 at 20 ma. At this point it becomes quite clear that *emitter efficiency becomes the most important factor from a transistor-design point of view if reasonable gain at high currents is desired.*

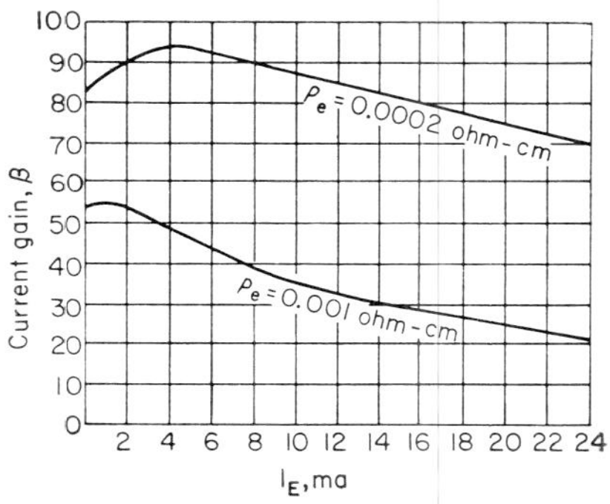

FIG. 10-7. Variation of current gain with I_E for illustrative example (p-n-p germanium-alloy transistor).

Let us critically examine, from a design viewpoint, those aspects of emitter efficiency which can be optimized in order to minimize the variation of the over-all current gain with emitter current. This problem is of particular importance in the design of power transistors, in which it is required that the current gain be as flat as possible out to currents as high as 1 amp. From Eqs. (10-47) and (10-48), the emitter-efficiency variation may be expressed as

$$\frac{R_{EE}}{R_{BB}} + \frac{qb}{kT}\frac{R_{EE}W^2I_E}{A} \tag{10-50}$$

which represents the sum of the low-level value and a term which is directly proportional to I_E. Examining the second term of (10-50), we see that emitter resistivity, base width, and junction area are the salient independent variables. We can now establish the essential criteria for maximum retention of gain with emitter current; namely, (1) heavy impurity concentration in the emitter, (2) thin bases, and (3) large junction areas. To illustrate the effect of these parameters, let us refer to the previous example and reduce the emitter resistivity from

0.001 ohm-cm to 0.0002 ohm-cm, keeping all other values constant. This decreases the low-level value of $\rho_e W/\rho_b L_{ne}$ by a factor of 5 such that the over-all current gain increases from 54 to 83. At $I_E = 20$ ma, this resistivity change increases β from 25 to 75. The complete gain characteristic for $\rho_e = 0.0002$ ohm-cm is plotted in Fig. 10-7 for purposes of comparison. The improvement derived from the emitter-resistivity change alone is quite apparent; however, both junction area and base width could also be changed to provide additional modification of the gain characteristic. An examination of Fig. 10-6 will reveal that the primary objective is to establish a composite curve which is practically independent of current. This may be accomplished by selecting suitable design values such that the absolute level and falloff of the surface and volume recombination components are somewhat compensated by the linear increase of the emitter efficiency component, wherein the latter is affected by the magnitude of Z/I_E. Thus, the design approach reduces to a graphical analysis.

One factor which has not been considered thus far is the effect of the carrier mobility in the base region. From Eqs. (10-14) and (10-15), it is seen that, for all other factors equal, the variation of emitter efficiency with I_E for a p-n-p transistor is b^2 times greater than that for an n-p-n.[1] Since $b \approx 2$ for germanium and $b \approx 3$ for silicon, we conclude that an n-p-n silicon transistor would yield the best β characteristic.

10-6. Base-region Conductivity Modulation. In Sec. 10-4 it was shown that because of operation at high current densities, the base resistivity is modified in accordance with the following relationship:

$$\rho_b' = \frac{\rho_b}{1 + Z/2} \tag{10-51}$$

This indicates an actual reduction in the effective base resistivity due to establishment of additional majority carriers under conditions of high-level injection. This effect may be expressed in terms of conductivity, i.e., the reciprocal of resistivity which results in an increase in conductivity. Thus we have the phenomenon of *conductivity modulation* of the base region.

It should therefore be apparent that the base resistance r_B' decreases as the emitter and collector currents of the transistor are increased. Since it was shown in Chap. 9 that r_B' is directly proportional to R_{BB}, we would expect the high-level base resistance to be

$$r_B'' = \frac{r_B'}{1 + Z/2} \tag{10-52}$$

Of course, it is natural to expect that Eq. (10-52) applies only to that portion of the base region that is conductivity-modulated by the injected emitter current. We shall see in later chapters that the reduction of r'_B due to high-level injection is very helpful in increasing the high-frequency power gain of transistors. Also, it reduces the magnitude of V_{BE} in saturated switching circuits.

10-7. Base-region Self-crowding. When a junction transistor is operated at high emitter currents, the base current which traverses the base region to the base contact may be very large. For example, in a power transistor operating at 0.5 amp, the base current may be as much as 50 ma. Since the base region has a finite sheet resistance, there results a finite voltage drop due to the flow of base current. This voltage appears

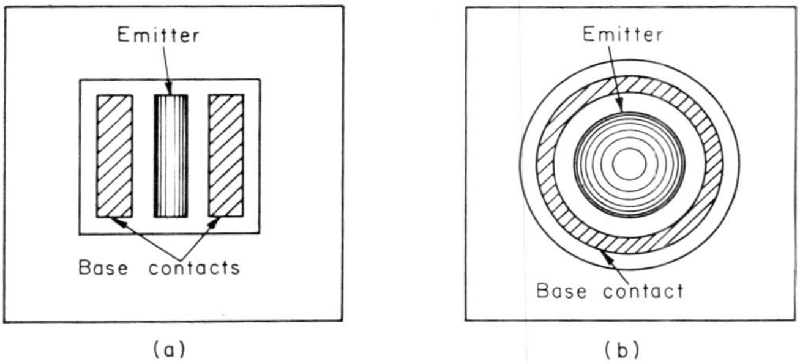

FIG. 10-8. Pictorial illustration of base-region self-crowding effect in transistors. Emitter current density is proportional to the shading intensity. (*a*) Linear transistor geometry; (*b*) circular transistor geometry.

in the form of a gradient, with the maximum potential occurring furthest away from the base contact. Since the external emitter bias voltage may be assumed to be applied uniformly over the entire emitter junction, the effect of the internal base voltage is to reduce the applied potential to those portions of the emitter junction furthest away from the base contact. The net effect is that the injected emitter current density is maximum closest to the base contact and decreases in some manner as one traverses into the transistor. Thus we have an internal self-bias or self-crowding effect in the base region of a junction transistor; this may be interpreted to be a reduction in the active cross-sectional area. This is best understood by referring to the pictorial illustration of the crowding phenomenon in Fig. 10-8. In this figure, the direction of emitter current flow is into the page and in each case the injected current density increases (darker shading) as the base contact is approached. For the geometries shown, the cutoff effect may be so severe that the injected carrier density in the center vicinity of the emitter junction may be practically zero.

Since the base-region crowding effect is primarily a voltage phenomenon, it is reasonable to expect that it can be minimized by maintaining both the sheet resistance R_{BB} and base current $(1 - \alpha)$ as small as possible. This is borne out by an analysis of the problem by Fletcher,[6] who derived the following expression:

$$J_E(x) \approx \frac{J_E(0)}{[1 + x\sqrt{(R_{BB}/2)(1 - \alpha)(qJ_E(0)/kT)}]^2} \qquad \text{for } \frac{qV(x)}{kT} \gg 1 \tag{10-53}$$

where $J_E(x)$ is the emitter current density as a function of distance x from the base contact. $J_E(0)$ is the emitter current density at $x = 0$, corresponding to the outer edge of the emitter junction. It is seen that in order to minimize the falloff of current density, the square-root term in the denominator must be made as small as possible. This involves making R_{BB} $(1 - \alpha)$ small. Actually, Eq. (10-53) is only a first-order approximation of the crowding effect, since conductivity modulation effects were not included in the analysis.

The interrelationships of conductivity modulation and base-region crowding are too complex to present quantitatively; these will be discussed here from a qualitative point of view only. Let us begin with a junction transistor operating at a moderately small I_E, with a current gain comparable to that illustrated by Fig. 10-7. For these conditions, I_B is very small and the self-crowding effect is negligible. As I_E is increased, approaching high-level conditions, several effects occur simultaneously. Firstly, I_B starts to increase, resulting in a reduction in the effective area due to base crowding. Secondly, the emitter efficiency starts to fall off due to the modulation of the base resistivity. The latter causes I_B to become even larger, which reduces A further, which in turn causes the emitter efficiency to decrease more rapidly as I_E increases further. One might expect that this interdependency would continue at an accelerated rate until the transistor would cut off completely with zero current gain. This is not the case, since a limiting factor is introduced by the effective decrease in base resistivity due to injection level, which tends to compensate for the increase in $(1 - \alpha)$ by making R_{BB} smaller. Thus, at the higher currents, all these effects come into equilibrium, yielding a current-gain characteristic somewhat comparable to that described in previous sections. Nevertheless, the base-region self-crowding effect does result in a reduction in active area, making the magnitude of Z/I_E larger, which causes β to fall off at a faster rate with emitter current.

From the point of view of practical transistor design, one can take into consideration the effects of both conductivity modulation and base-layer crowding by simply assuming that only the *periphery* of the emitter is active; this is particularly true for very-small-area high-frequency transistors. On this basis, the current-handling capacity of a transistor becomes proportional to the emitter perimeter rather than the emitter area. Furthermore, with the injected current density concentrated at the edge of the emitter, one can invariably omit the contribution of the base region directly under the emitter in the calculation of base resistance. This leaves that portion of the base between the emitter edge and the base contact as the most important segment affecting r'_B.

10-8. Effects of Current on Base-cutoff Frequency. It was shown in Secs. 8-6 and 8-8 that for conditions of low-level injection, the base-cutoff frequency for a uniform- or graded-base junction transistor is directly proportional to the minority-carrier diffusion constant D. On the basis of the theory developed in this chapter, however, it is to be expected that the base-cutoff frequency will change with injection level. This is borne out by an examination of Eq. (10-23), which is repeated here.

$$I_{pE} \approx \frac{qAD_{pb}}{W} p_E \left(1 + \frac{p_E}{N_D + p_E}\right) \tag{10-54}$$

If the term in parenthesis is simplified by the substitution

$$p_E/N_D \equiv v \approx Z/2$$

Eq. (10-54) becomes

$$I_{pE} \approx \frac{qAD_{pb}p_E}{W} \frac{1 + Z}{1 + Z/2} \tag{10-55}$$

or

$$I_{pE} \approx \frac{qAp_E}{W} \frac{D_{pb}}{f'(Z)} \tag{10-56}$$

where $f'(Z)$ is the function given previously in Fig. 10-2. In Eq. (10-56) we see that the variation of emitter current with injection level can be represented by a change in the effective diffusion constant, viz., $D_{pb}/f'(Z)$. At low levels, $f'(Z)$ is equal to unity and the familiar form for I_{pE} results. At high levels, $f'(Z)$ approaches 0.5, and we have an effective doubling of the diffusion constant[2] to $2D_{pb}$. In summary, then, the effective diffusion constant can vary between D_{pb} and $2D_{pb}$, depending on the level of injection. Therefore, because of the direct relationship, the base-cutoff frequency will approach twice the low-level value as emitter current is increased. We recall that an electric field is established in the base region in a direction to aid the flow of minority carriers; consequently, the frequency response is increased as the carrier transit time is reduced. At current levels where high-level injection is reached, this electric field is so predominant that it virtually swamps out the built-in electric field due to the impurity distribution in a graded-base transistor.

PROBLEMS

10-1. An n-p-n silicon planar transistor is fabricated with diffused base and emitter. The geometry employs the horseshoe configuration with the following significant dimensions:

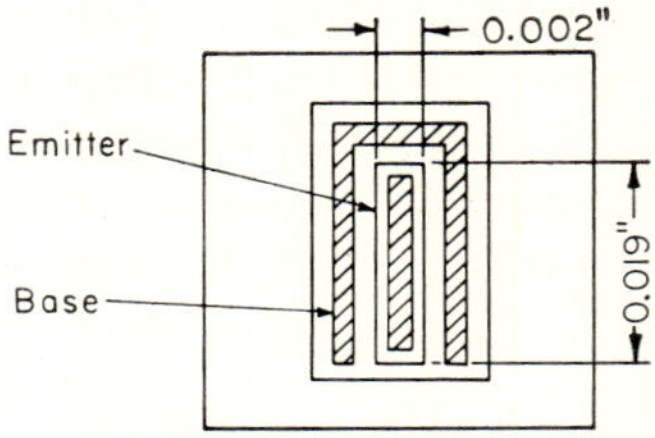

The emitter is a 2- by 19-mil rectangle having a sheet resistance of 2.8 ohms. The base region beneath the emitter has an effective width of 1 μ and a sheet resistance of 700 ohms. The average electron mobility in the base is 900 cm^2/volt-sec. The base volume lifetime is 25 mμsec and increases by a factor of 2 due to high-level injection effects. The surface-recombination velocity is 500 cm/sec. The forward-bias recombination current I_{RE}, due to impurities in the emitter junction, is equal to 1 μa.

Assuming negligible base-region crowding, calculate the room-temperature current gain β at each of the following emitter currents: 0.01 ma, 0.1 ma, 1 ma, 10 ma, 100 ma, and 1 amp.

10-2. For the p-n-p germanium-alloy transistor given in Fig. 10-5, calculate the base resistance r_B' at $I_E = 20$ ma. Assume that the self-crowding effect established an active base-region area bounded by radii of 4 and 10 mils.

REFERENCES

1. Webster, W. M.: On the Variation of Junction-transistor Current-amplification Factor with Emitter Current, *Proc. IRE*, vol. 42, pp. 914–920, June, 1954.
2. Rittner, E. S.: Extension of the Theory of the Junction Transistor, *Phys. Rev.*, vol. 94, pp. 1161–1171, June, 1954.
3. Armstrong, L. D., C. L. Carlson, and M. Bentivegna: PNP Transistors Using High-emitter-efficiency Alloy Materials, *RCA Rev.*, vol. 17, pp. 37–45, March, 1956.
4. Fletcher, N. H.: The High Current Limit for Semiconductor Junction Devices, *Proc. IRE*, vol. 45, pp. 862–872, June, 1957.
5. Sah, C., R. N. Noyce, and W. Shockley: Carrier Generation and Recombination in P-N Junctions and P-N Junction Characteristics, *Proc. IRE*, vol. 45, pp. 1228–1243, September, 1957.
6. Fletcher, N. H.: Some Aspects of the Design of Power Transistors, *Proc. IRE*, vol. 43, pp. 551–559, May, 1955.

11

Low-frequency Feedback Effects

11-1. Introduction to Frequency Analysis of Transistors. The theoretical development of the junction transistor thus far has been restricted to the d-c or steady-state analysis of a one-dimensional model on which the effects of surface boundaries and physical geometry were additionally considered. By solution of the diffusion equations for minority-carrier flow, expressions were obtained which related the direct emitter and collector currents to the respective d-c applied potential and the bulk parameters of the semiconductor regions. From these relationships, we were able to establish the appropriate relationships for the key d-c transistor parameters. Further, it was shown that various design relations for these parameters were very much a function of voltage, current, and temperature. It now remains to extend the analysis to take into consideration the effects of frequency on transistor parameters.

In this chapter, certain physical feedback effects which give rise to the low-frequency input and output impedances of the transistor will be discussed. In the chapters to follow, the frequency analysis will be extended to the h-parameter characterization of the transistor, which then will serve as the basis for the complete high-frequency analysis of the transistor impedances. Following this, a detailed study will be made of all the factors related to frequency response (transistor gain bandwidth).

11-2. Base-region Widening Effects. From a d-c point of view, we are already familiar with the effect of collector voltage on the base width of the transistor. The effective base width of the transistor is equal to the physical base width less the amount the depletion layer spreads into the base region. The distance the depletion layer spreads into the base, of course, is a function of the impurity profile of the collector junction and the resistivities on either side.

Let us now consider the effect a small-signal a-c variation of collector voltage would have on base width if it were superimposed on the d-c collector voltage. We would now expect the total depletion-layer thickness to widen and narrow slightly at a frequency equal to that of the applied a-c signal. This would impart a similar low-frequency a-c variation to the effective base thickness W. Since W appears in many design relations (particularly current gain), then the frequency variation would modify the electrical characteristics from a low-frequency point of view. In this chapter we shall see that this widening phenomenon manifests itself in the form of a-c feedback effects.

For the p-n-p transistor model having a step collector junction and a high-resistivity base compared to the collector, we can write

$$W = W_o - x_m \tag{11-1}$$

where W = effective base width
W_o = physical base width
x_m = thickness of depletion layer

From Eq. (5-77) we can write Eq. (11-1) as

$$W = W_o - \sqrt{\frac{2\kappa\epsilon_o V_C}{qN_D}} \tag{11-2}$$

where V_C is the d-c collector voltage. If a small-signal a-c variation is imparted to V_C, the effect on W is established by taking the partial derivative of Eq. (11-2) with respect to voltage.

$$\frac{\partial W}{\partial V_C} = -\frac{1}{2}\sqrt{\frac{2\kappa\epsilon_o}{qN_D V_C}} \tag{11-3}$$

If numerator and denominator are multiplied by $V_C^{1/2}$, this becomes

$$\frac{\partial W}{\partial V_C} = -\frac{x_m}{2V_C} \qquad \text{step junction} \tag{11-4}$$

Equation (11-4) gives the rate of change of base width with voltage for a step junction; x_m is the depletion-layer width at the voltage V_C. This result is generally applicable to transistors made by the double-sided impurity-contact processes, such as alloy and diffused-base alloy (MADT or drift) transistors.

For a transistor having a graded collector junction and a collector resistivity higher than that of the base, we must obtain a different relationship for the base-region widening factor $\partial W/\partial V_C$. These transistor structures generally have diffused base layers and therefore one can use the linear grade as an approximation for the impurity profile in the vicinity of the collector junction. This is especially true for the lower range of V_C, wherein the junction follows the cubic law (see Sec. 5-9). Thus we can write

$$W = W_o - \frac{x_m}{2} \tag{11-5}$$

where x_m is the total depletion-layer thickness for the linear grade region, given by Eq. (5-94) as

$$x_m = \left(\frac{12\kappa\epsilon_o V_C}{qa}\right)^{1/3} \tag{11-6}$$

Substituting this result into Eq. (11-5) and differentiating with respect to V_C, we obtain

$$\frac{\partial W}{\partial V_C} = \frac{-x_m}{6V_C} \tag{11-7}$$

It is more convenient to express Eq. (11-7) in terms of the parameter x_1, which is normally obtained from the graded junction curves of Figs. 5-15 to 5-30. Since $x_1 \approx x_m/2$ in the lower voltage cubic range, we have

$$\frac{\partial W}{\partial V_C} = \frac{-x_1}{3V_C} \qquad \begin{array}{c}\text{graded junction}\\ \text{low voltage}\end{array} \tag{11-8}$$

where x_1 is the distance the depletion layer spreads into the base at the voltage V_C. In the other ranges of voltage, Eq. (11-8) is approximately

$$\frac{\partial W}{\partial V_C} = \frac{-x_1}{2.5V_C} \qquad \begin{array}{c}\text{graded junction}\\ \text{medium voltage}\end{array} \tag{11-9}$$

$$\frac{\partial W}{\partial V_C} = \frac{-x_1}{2V_C} \qquad \begin{array}{c}\text{graded junction}\\ \text{high voltage}\end{array} \tag{11-10}$$

Equations (11-8) through (11-10) demonstrate the transition from cubic to square-root dependence with voltage for the graded junction.

The effect of these low-frequency small-signal a-c variations of the base width on the basic transistor parameters will be discussed in the forthcoming sections. The treatment is based largely on the work of J. Early, who first reported these effects in 1952.[1,*]

11-3. Collector Output Conductance. From the standpoint of maximum low-frequency amplification gain for junction transistors, it is important that the output collector resistance be as large as possible; that is, that the output conductance be small. By definition, we can write

$$g_c = \frac{1}{r_c} = \left.\frac{\partial I_C}{\partial V_C}\right|_{V_E=\text{const}} \tag{11-11}$$

If we apply this differentiation to the general current-voltage equations developed in the one-dimensional analysis of Chap. 8, viz., Eq. (8-31), we find that $g_c = 0$ or $r_c = \infty$. This comes about because, for large reverse bias on the collector, I_C is independent of collector voltage. Zero conductance is obviously not observed in transistors, for the reverse char-

* References, indicated in the text by superscript figures, are listed at the end of the chapter.

acteristics evidence a small slope (neglecting surface leakage effects) corresponding to a higher conductance.

Actual junction transistors, however, exhibit considerably higher output conductances (low frequency), corresponding to output resistances on the order of 1 to 3 megohms. These differences are first explained by Early by including the effects of the depletion-layer widening of the collector junction.[1] It should be recalled that in the one-dimensional analysis it was assumed that the base thickness W was constant and independent of voltage.

In Chap. 8, the collector current was approximately represented by

$$I_C = \alpha I_E \tag{11-12}$$

where I_E is the emitter current and α is the current gain, equal to the product of the emitter efficiency γ and the base-transport factor β^*. Since α is a function of the base thickness W, it would be expected, because of a-c widening effects, that α be a function of the collector voltage V_C. Expressed mathematically,

$$\frac{\partial \alpha}{\partial V_C} = \frac{\partial \alpha}{\partial W}\frac{\partial W}{\partial V_C} \tag{11-13}$$

From Eqs. (11-11) and (11-12), the output conductance g_c would be

$$g_c = \frac{\partial I_C}{\partial V_C} = \frac{\partial \alpha}{\partial V_C} I_E \tag{11-14}$$

Substituting (11-13) in (11-14), we obtain

$$g_c = I_E \frac{\partial \alpha}{\partial W}\frac{\partial W}{\partial V_C} \tag{11-15}$$

Here we see the introduction of the term $\partial W/\partial V_C$, for which we have already obtained expressions for step and graded junctions. These are given by Eqs. (11-4) and (11-8) to (11-10). Thus it remains to determine the expression for $\partial \alpha/\partial W$. Using the simplified expressions for α, viz.,

$$\alpha = \beta^*\gamma = \operatorname{sech}\left(\frac{W}{L_{pb}}\right)\frac{1}{1 + \rho_e W/\rho_b L_{ne}} \tag{11-16}$$

we take the partial derivative and obtain

$$\frac{\partial \alpha}{\partial W} = \frac{-\operatorname{sech}(W/L_{pb})\tanh(W/L_{pb})(1/L_{pb})}{1 + \rho_e W/\rho_b L_{ne}} - \frac{\operatorname{sech}(W/L_{pb})(\rho_e/\rho_b L_{ne})}{(1 + \rho_e W/\rho_b L_{ne})^2} \tag{11-17}$$

For the approximations that γ is close to 1 and W/L_{pb} is small, this reduces to

$$\frac{\partial \alpha}{\partial W} = -\frac{1}{W}\left(\frac{W^2}{L_{pb}^2} + \frac{\rho_e W}{\rho_b L_{ne}}\right) \tag{11-18}$$

which can also be written as

$$\frac{\partial \alpha}{\partial W} = -\frac{1}{W}[2(1 - \beta^*) + 1 - \gamma] \tag{11-19}$$

where β^* is given by $[1 - \frac{1}{2}(W/L_{pb})^2]$ as a first approximation. Inserting Eq. (11-19) into (11-15), we have, finally,

$$g_c = \frac{-I_E}{W}[2(1 - \beta^*) + (1 - \gamma)]\frac{\partial W}{\partial V_C} \tag{11-20}$$

where the appropriate equation must be used for $\partial W/\partial V_C$. This equation may be used for purposes of calculation for transistors having alphas of 0.94 or greater; otherwise the more exact expression of Eq. (11-17) should be used. Nevertheless, it is seen that the output collector conductance increases with emitter bias current, since the current change corresponding to any change in alpha (due to change in base thickness) depends on the current which is flowing. As collector bias voltage is increased, g_c becomes smaller, since the rate of change of the depletion-layer thickness decreases with increasing V_C. For all practical purposes, g_c predominates the low-frequency output impedance of the grounded-base junction transistor. For high-alpha transistors, g_c may vary from 0.1 to 1 μmho in the range of 1 to 2 ma emitter current.

11-4. Emitter Input Resistance. The a-c input resistance of the emitter junction of the transistor under conditions of forward bias is determined by the partial differentiation of Eq. (8-30), as follows:

$$\frac{1}{r_e} = \left.\frac{\partial I_E}{\partial V_E}\right|_{V_C = \text{const}} \tag{11-21}$$

Since the second term of Eq. (8-30) is a constant, we have, after differentiation,

$$\frac{\partial I_E}{\partial V_E} = \frac{1}{r_e} = \frac{qI_E}{kT} \tag{11-22}$$

or

$$r_e = \frac{kT}{qI_E} \qquad \text{ohms} \tag{11-23}$$

where kT/q equals 0.026 volt at room temperature, 300°K. We see that this result is simply the reciprocal of the small-signal slope of the V-I characteristic for a simple p-n junction. Thus the a-c input emitter resistance of the transistor will decrease as the emitter bias current I_E is increased.

In addition to its effect on the collector impedance, the small-signal widening of the collector depletion layer also affects the emitter input resistance of the transistor. For a given emitter voltage V_E, the minority-carrier current injected into the base sees an impedance the magnitude of which is determined by the thickness and impurity profile of the base region. If the collector reverse bias V_C is increased so that the effective base thickness is reduced by the widening of the depletion layer, there results a reduction in emitter resistance, permitting more emitter current to flow. Another way of looking at this change in emitter resistance with collector voltage is to consider the emitter current to be constant. The emitter current may be expressed by

$$I_E \approx -qAD_{pb}\frac{dp}{dx} \tag{11-24}$$

where dp/dx is the concentration gradient of the holes diffusing in the base region. If the injected hole concentration at the emitter is $p_{nb}\epsilon^{qV_E/kT}$ and at the collector $p = 0$ (owing to the reverse bias), then for a base width of W, Eq. (11-24) becomes

$$I_E \approx \frac{qAD_{pb}p_{nb}}{W}\epsilon^{qV_E/kT} \tag{11-25}$$

where V_E is the emitter junction voltage. If I_E is to be constant and W is reduced because of collector widening, it is apparent that V_E must decrease accordingly. In other words, the emitter potential is adjusted to maintain dp/dx constant.

For a-c operation, this effect appears in the form of a small-signal a-c feedback voltage (for constant I_E) in series with the emitter resistance r_e. Expressed mathematically, the emitter feedback voltage, as given by Early, is $\mu_{ec}v_c$, where v_c is the a-c collector voltage and μ_{ec} is the *voltage feedback factor*, defined as

$$\mu_{ec} = -\frac{\partial V_E}{\partial V_C}\bigg|_{I_E=\text{const}} \tag{11-26}$$

To obtain the expression for μ_{ec}, we make use of the derivative of Eq. (11-25). Since both W and V_E are variables, the derivative with respect to V_C is

$$\frac{\partial I_E}{\partial V_C} = \frac{qD_{pb}p_{nb}A}{W}\frac{q\epsilon^{qV_E/kT}}{kT}\frac{\partial V_E}{\partial V_C} + \epsilon^{qV_E/kT}\frac{-qD_{pb}p_{nb}A}{W^2}\frac{\partial W}{\partial V_C} \tag{11-27}$$

For constant I_E, $\partial I_E/\partial V_C$ must be zero. Therefore, μ_{ec} may be obtained by setting Eq. (11-27) equal to zero and solving for $\partial V_E/\partial V_C$. Consequently,

$$0 = \frac{q}{kT}\frac{\partial V_E}{\partial V_C} - \frac{1}{W}\frac{\partial W}{\partial V_C}$$

or

$$\mu_{ec} = -\frac{\partial V_E}{\partial V_C} = -\frac{kT}{qW}\frac{\partial W}{\partial V_C} \tag{11-28}$$

The widening-rate term appears here, too. It should be noted that the voltage feedback factor μ_{ec} is independent of emitter current because of the logarithmic relation between emitter current and emitter potential. The term W is the effective thickness of the base region at the collector bias V_C. As will be shown at the end of this chapter, emitter voltage feedback μ_{ec} appears in the equivalent circuit in the form of an a-c generator equal to $\mu_{ec}v_c$ in series with the emitter resistance r_e. The effect of this generator is to increase the low-frequency input resistance of the transistor.

11-5. Base Feedback Resistance. In the analysis of base resistance, we were concerned with resistive effects on the base-current components of the transistor. As was mentioned, there is a direct current flowing in the base region; this arises from the recombination of the emitter bias current. If the thickness of the base layer decreases and increases as the collector depletion layer widens and narrows with the a-c voltage variation, then an a-c modulation of the d-c base resistance is introduced, since the latter is inversely proportional to W. Therefore a small a-c variation is superimposed on the d-c voltage drop in the base region and appears as an additional feedback voltage generator $\mu_{bc}v_c$ in series with the base resistance r'_B. Mathematically, this is[2]

$$\mu_{bc}v_c = I_B \frac{\partial r'_B}{\partial W} \frac{\partial W}{\partial V_C} v_c \tag{11-29}$$

where I_B = d-c base current
$\partial W/\partial V_C$ = rate dependent upon nature of collector junction
v_c = a-c collector voltage

The base voltage-feedback factor is therefore

$$\mu_{bc} = I_B \frac{\partial r'_B}{\partial W} \frac{\partial W}{\partial V_C} \tag{11-30}$$

Expressions for $\partial r'_B/\partial W$ may be obtained by differentiating the appropriate r'_B equations for the various geometries. In any case, the region most affected by the modulation corresponds approximately to the area of the emitter junction. Since for any geometry r'_B is proportional to sheet resistance or $1/W$,

$$\frac{\partial r'_B}{\partial W} = -\frac{r'_B}{W} \tag{11-31}$$

Therefore, from Eqs. (11-30) and (11-31), we have

$$\mu_{bc} \approx -I_B \frac{r'_B}{W} \frac{\partial W}{\partial V_C} \tag{11-32}$$

Generally, for high-alpha transistors at typical emitter currents, I_B is very small, making μ_{bc} negligible compared to the other feedback effects at low frequencies.

11-6. Collector Diffusion Capacitance. Another effect which stems from the a-c widening of the collector space-charge barrier concerns the charge (holes) stored in the base region by the emitter bias current. As collector voltage is varied, this charge varies accordingly, giving rise to a diffusion capacitance.[2] Consider, for example, the hole distribution as a function of distance in the base region of a p-n-p junction transistor under conditions of d-c bias. For a constant emitter current the concentration gradient is approximately linear, as shown in Fig. 11-1. At $x = 0$ (the emitter junction), $p(0) = p_{nb}\epsilon^{qV_E/kT}$, which is the injected hole concentration. At $x = W$, $p(W) = 0$, since all the holes are being swept out by the reverse-biased collector junction. Since the diffusion current I_E is

$$I_E = -qAD_{pb}\frac{dp}{dx} \tag{11-33}$$

where dp/dx is the slope of $p(x)$, then, as before,

$$I_E = qAD_{pb}\frac{p_{nb}}{W}\epsilon^{qV_E/kT}$$

or

$$p_{nb}\epsilon^{qV_E/kT} = \frac{I_E W}{AqD_{pb}} \tag{11-34}$$

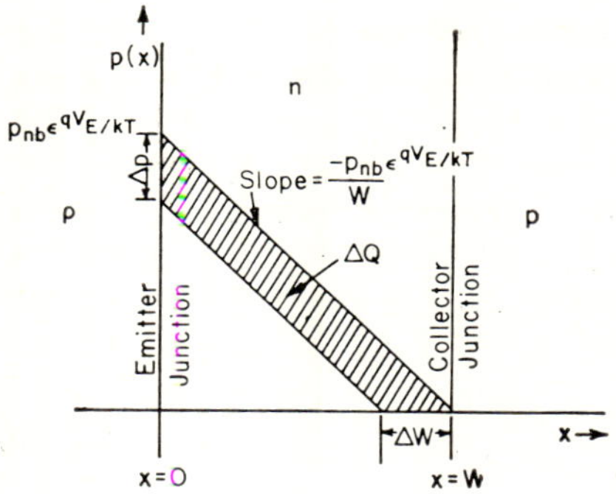

FIG. 11-1. Widening effects on base-region charge for constant emitter current. (*After R. N. Hall.*)

If the collector voltage is increased an amount ΔV_C such that the base width is reduced by ΔW as shown in Fig. 11-1, then $p(x)$ will be shifted by an amount equal to ΔW. For constant emitter current, Eq. (11-33) dictates that the slope must remain constant; therefore the emitter hole concentration must decrease by an amount equal to Δp. This accounts for the change in emitter potential, which gives rise to the μ_{ec} term described previously. As a result of this new distribution, a quantity of charge ΔQ equal to the shaded area of Fig. 11-1 must diffuse out of the base into the collector. The ratio of this charge ΔQ to the change in voltage ΔV_C is called the *collector diffusion capacitance* C_{Dc}, or[2]

$$C_{Dc} = \frac{\Delta Q}{\Delta V_C} \tag{11-35}$$

For the parallelogram (shaded area) of base ΔW and height $p_{nb}\epsilon^{qV_E/kT}$,

$$\Delta Q \approx qAp_{nb}\epsilon^{qV_E/kT}\,\Delta W$$

and

$$C_{Dc} = qAp_{nb}\epsilon^{qV_E/kT}\frac{\Delta W}{\Delta V_C} \tag{11-36}$$

Substituting Eq. (11-34) into (11-36), we get, in the limit,

$$C_{Dc} = \frac{I_E W}{D_{pb}} \frac{\partial W}{\partial V_C} \tag{11-37}$$

The diffusion capacitance adds directly to the capacitance of the collector depletion layer. It is seen that C_{Dc} is directly proportional to the emitter current I_E and inversely proportional to the collector voltage V_C. As the frequency of the a-c collector voltage is increased, the time required for the charge at the emitter end of the base to diffuse to the collector junction becomes comparable to the period of the a-c voltage and C_{Dc} effectively becomes complex.[3] The subject of diffusion admittances will be covered in greater detail by the high-frequency analysis of Chap. 13.

11-7. Low-frequency Equivalent Circuit. At this point we are in a position to write the complete a-c equivalent circuit which characterizes

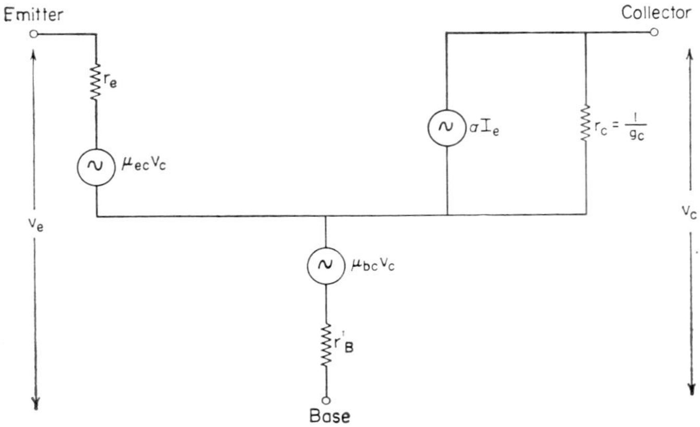

FIG. 11-2. Low-frequency junction-transistor equivalent circuit. (*After J. M. Early.*)

the operation of the junction transistor at low frequencies within the audio range.[4] By adding all the effects of collector depletion-layer widening and base resistance, we obtain the complete low-frequency equivalent circuit of Fig. 11-2. The output conductance g_c is shown in parallel with the constant-current generator αI_E. The a-c voltage-feedback generator $\mu_{ec} v_c$ is shown in series with the a-c emitter resistance r_e. Finally, the resistance r'_B and the modulation voltage generator $\mu_{bc} v_c$ are added to the base terminal. In Chap. 12, a detailed network analysis of this circuit will be performed, giving rise to the equivalent h parameters for low frequencies. Nevertheless, we can immediately make some qualitative remarks concerning the low-frequency characteristics of the transistor in the range well below the base-cutoff frequency.

At very low frequencies such that the effect of capacitance across g_c may be considered negligible, the open-circuit feedback voltage looking in at the emitter would be, by inspection, as follows:

$$v_{e(oc)} = \mu_{ec}v_c + \frac{r'_B v_c}{r'_B + r_c} + \mu_{bc}v_c \tag{11-38}$$

where $r_c = 1/g_c$. The current generator is omitted since I_E would be zero for open-circuit emitter. Since r_c is usually very much greater than r'_B, Eq. (11-38) simply becomes

$$v_{e(oc)} = v_c\left(\mu_{ec} + \mu_{bc} + \frac{r'_B}{r_c}\right)$$

$$v_{e(oc)} = [(\mu_{ec} + \mu_{bc})r_c + r'_B]\frac{v_c}{r_c} \tag{11-39}$$

Examination of this result indicates that it would hold for a circuit of the form shown in Fig. 11-3, where, for large values of r_c, the current i is

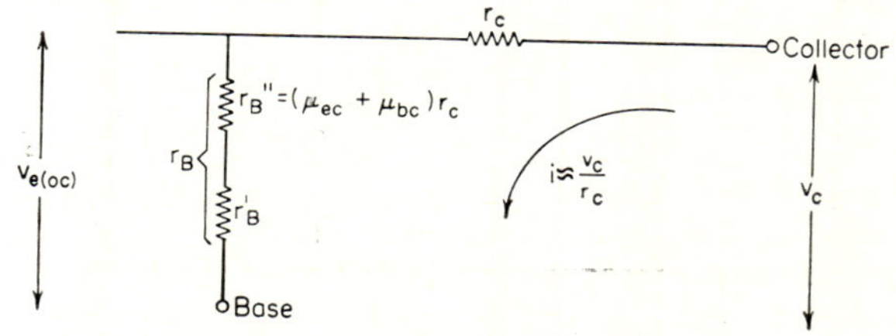

FIG. 11-3. Open-circuit equivalent for $I_E = 0$.

approximately equal to v_c/r_c. In effect, what this means is that at low frequencies the effective base resistance r_B of the transistor is not r'_B, but is

$$r_B \approx r'_B + r''_B \tag{11-40}$$

where

$$r''_B = (\mu_{ec} + \mu_{bc})r_c \tag{11-41}$$

In Eq. (11-41) $r_c = 1/g_c$; $(\mu_{ec} + \mu_{bc})$ was previously determined to be $(1/W)(kT/q + I_B r'_B)(\partial W/\partial V_C)$.

At high frequencies, the impedance of the collector capacitance is considerably smaller than r_c, such that the feedback voltage across r'_B increases appreciably, becoming much greater than $(\mu_{ec} + \mu_{bc})v_c$. This effectively swamps out the $(\mu_{ec} + \mu_{bc})r_c$ resistive term so that, at high frequencies, the base resistance becomes r'_B only.

To conclude this chapter it must be pointed out that the feedback effects due to depletion-layer widening appear primarily in junction transistors for which the base resistivity is higher than the collector resistivity. For graded transistors in which the collector resistivity is higher than the base resistivity, μ_{ec}, μ_{bc}, and g_c become considerably smaller. This is because the base width is practically constant, since the barrier spreads mostly into the collector region. Thus we see that the magnitude of the term $\partial W/\partial V_C$ becomes the most significant factor in determining the low-frequency characteristics of the transistor.

PROBLEMS

11-1. The low-frequency equivalent circuit for an alloy transistor may be approximated by the circuit in (*a*), where r_B'' is the effective base resistance introduced by widening effects. For room temperature, calculate the values of all the elements of

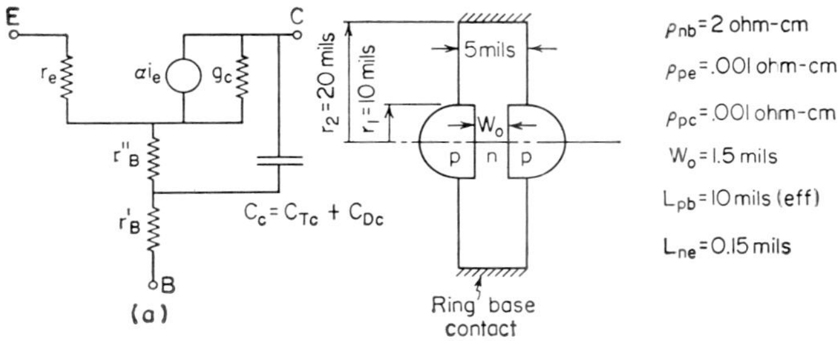

the circuit at $V_C = -5$ volts and $I_E = 1$ ma for the p-n-p germanium-alloy transistor shown. The geometry is circular and the junctions are equal in area and penetration. Assume low-level injection conditions are applicable.

11-2. Compare the magnitudes of the base-widening factor $\partial W/\partial V_C$ for a p-n-p germanium MADT transistor and a p-n-p germanium epitaxial mesa transistor for $V_C = -5$ volts. The MADT device has a base 10 μ thick, with a donor impurity concentration of 10^{15} atoms/cm^3, and an n-type layer 2 μ deep diffused into it. For the epitaxial mesa device the base is an n-type layer 2 μ deep diffused into a p-type epitaxial layer 10 μ thick and having an acceptor impurity concentration of 10^{15} atoms/cm^3. (Assume that the impurity profile of the collector junction of the epitaxial mesa device is linear, having a grade constant $a = 10^{20}$ atoms/cm^4.)

REFERENCES

1. Early, J. M.: Effects of Space-charge Layer Widening in Junction Transistors, *Proc. IRE*, vol. 40, pp. 1401–1406, November, 1952.
2. Pritchard, R. L.: Collector-base Impedance of a Junction Transistor, *Proc. IRE*, vol. 41, August, 1953.
3. Pritchard, R. L.: Frequency Variations of Junction Transistor Parameters, *Proc. IRE*, vol. 42, pp. 786–799, May, 1954.
4. Pritchard, R. L., J. B. Angell, R. B. Adler, J. M. Early, and W. M. Webster: Transistor Internal Parameters for Small-signal Representation, *Proc. IRE*, vol. 49, pp. 725–738, April, 1961.

12

Low-frequency h Parameters

12-1. Small-signal Parameters for Transistors. In developing the theory of the junction transistor, we have been able to establish an approximate equivalent circuit[1,*] which represents the electrical characteristics of the theoretical transistor. The effects of frequency on the carrier diffusion process have not as yet been introduced, so the parameters obtained thus far are assumed to be independent of frequency.† Nevertheless, for very low frequencies this equivalent circuit (Fig. 11-2) holds very well. The individual elements of this circuit, r_e, r_B', g_c, $\mu_{ec}v_c$, α, etc., were obtained from the analysis of the basic physics of the transistor and the current-voltage relationships derived from same.

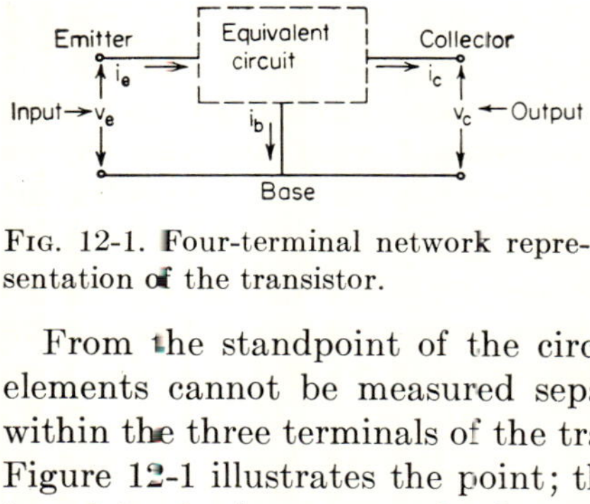

FIG. 12-1. Four-terminal network representation of the transistor.

From the standpoint of the circuit designer, these equivalent-circuit elements cannot be measured separately, since they are all contained within the three terminals of the transistor (emitter, base, and collector). Figure 12-1 illustrates the point; the equivalent circuit representing the transistor is shown completely contained within the dashed lines, and only the three outside terminals are evident. Actually, this may be considered a four-terminal network with the base terminal common to both input and output. In order to properly characterize the low-frequency performance of the junction transistor at a particular operating point (direct emitter current and collector voltage), it becomes necessary

* References, indicated in the text by superscript figures, are listed at the end of the chapter.

† The one exception is alpha, for which the frequency dependence of the base-transport factor (base-cutoff frequency) was derived in Chap. 8. The frequency variation of all parameters will be treated in Chap. 13.

to make a-c measurements at the input and output terminals and to relate the results to the equivalent circuit.

In general, if v_e and i_e are known at the input and v_c and i_c are known at the output, they may be related to each other by any one of six pairs of simultaneous circuit equations.[2] In each pair of equations there are four coefficients which are constants and are related to the elements of the network. These four coefficients are the small-signal parameters of the network and have values depending on which of the six pairs of equations is used. As far as transistors are concerned, we need be concerned with only three of the six possibilities. These three are those equations in which the coefficients are expressed in terms of either impedances Z, admittances y, or hybrid coefficients h. The latter are combinations of impedances, admittances, and constants.

Consider first the impedances Z. If the circuit equations for the transistor network of Fig. 12-1 are written, they are of the form

$$v_e = Z_{11}i_e + Z_{12}i_c \qquad (12\text{-}1a)$$
$$v_c = Z_{21}i_e + Z_{22}i_c \qquad (12\text{-}1b)$$

where the Z's represent the small-signal impedance parameters of the network. It should be noted that these equations are of a completely general form and apply to any linear four-terminal network. Suppose we are interested in finding the magnitude of Z_{11}. This may be obtained from Eq. (12-1a) by letting $i_c = 0$; $Z_{11} = v_e/i_e$. In other words, Z_{11} is the input impedance of the transistor with the collector open-circuited to alternating current. From a measurements point of view, this is rather difficult to do because of the high impedance output of the collector, which makes it difficult to apply d-c bias and still maintain true open-circuit conditions.[3] The same problem occurs also in the measurement of Z_{21} of Eq. (12-1b). Because of these measurement problems, the Z parameters are rarely used for transistors. Furthermore, it has been pointed out by Pritchard that even at low frequencies within the audio range, the small-signal Z parameters are not independent of frequency.

The second possibility might be the use of the admittance equations, or the so-called y parameters. For the junction transistor, these are of the form

$$i_e = y_{11}v_e + y_{12}v_c \qquad (12\text{-}2a)$$
$$i_c = y_{21}v_e + y_{22}v_c \qquad (12\text{-}2b)$$

where the y's are the admittance parameters. (Admittance is the reciprocal of impedance.) Here too, circuit difficulties are encountered in trying to make appropriate measurements. For example, if it is desired to measure the feedback admittance y_{12}, it is necessary to be able to measure extremely small voltages, since the y_{12} measurement requires that $v_e = 0$. It may be argued that Eqs. (12-2a) and (12-2b) are identical in form to the results obtained for the one-dimensional diffusion analysis of the transistor and would therefore truly represent the physics of the device. However, it must be realized that the analysis did not include the base resistance r'_B, which was added to the equivalent circuit as a lumped resistance.[3] Consequently, when the complete equivalent circuit including r'_B is considered from the admittance point of view, the expressions become rather cumbersome.

Finally, there remain the hybrid small-signal parameters, the so-called h parameters, which are given as

$$v_e = h_{11}i_e + h_{12}v_c \tag{12-3a}$$

$$i_c = h_{21}i_e + h_{22}v_c \tag{12-3b}$$

By inspection of these equations, it is seen that h_{12} and h_{21} are dimensionless quantities and h_{11} and h_{22} are an impedance and admittance, respectively. From the measurements standpoint, the h parameters fit in extremely well with the characteristics of the transistor, since they can be obtained by either open-circuiting the low-impedance emitter circuit or short-circuiting the high-impedance collector circuit.[3,4] It is this suitability that has led to the adoption of the h parameters for specifying the small-signal parameters of the junction transistor. As will be seen in the next section, these parameters are also consistent with the device physics, since h_{21} corresponds to alpha and h_{12} to the voltage feedback.

12-2. Grounded-base *h* Parameters. The use of h parameters is the accepted method of describing the linear small-signal parameters of the junction transistor. In order to facilitate rapid comprehension of the coefficients, the numerical subscripts have been replaced by letter designations which more clearly define both the terms and the circuit configuration. For the grounded-base circuit configuration, Eqs. (12-3*a*) and (12-3*b*) are written as

$$v_e = h_{ib}i_e + h_{rb}v_c \tag{12-4a}$$

$$i_c = h_{fb}i_e + h_{ob}v_c \tag{12-4b}$$

where the second letter in each subscript (b) denotes grounded-base parameters. The first letters read as follows: i, input; r, reverse; f, forward; o, output. If the h parameters are specified for grounded-emitter operation, the b is replaced with e, but the first letters remain the same. We can now proceed to examine each grounded-base parameter individually.

In Eq. (12-4*a*), if we let $v_c = 0$ by shorting the collector output circuit, then

$$h_{ib} = \left.\frac{v_e}{i_e}\right|_{v_c=0} \qquad \text{ohms} \tag{12-5}$$

Thus, h_{ib} represents the input impedance of the transistor, in units of ohms, with the output short-circuited.

Also from Eq. (12-4a), if we let $i_e = 0$ by open-circuiting the emitter circuit, then

$$h_{rb} = \left.\frac{v_e}{v_c}\right|_{i_e=0} \tag{12-6}$$

Since h_{rb} is a ratio of voltages, it remains dimensionless and is called the voltage feedback ratio. h_{rb} is measured by applying a voltage to the collector and measuring the open-circuit emitter voltage. From a network standpoint, it is also called the reverse-voltage transfer coefficient.

In Eq. (12-4b), if we let $v_c = 0$ by shorting the collector output circuit, then

$$h_{fb} = \left.\frac{i_c}{i_e}\right|_{v_c=0} \tag{12-7}$$

which is the ratio of the collector current to the emitter current, or more commonly, alpha (α). This, of course, is also dimensionless. The subscript fb refers to the network nomenclature, i.e., the forward-current transfer coefficient.

The fourth h parameter is determined by open-circuiting the emitter, making $i_e = 0$ in Eq. (12-4b). This gives

$$h_{ob} = \left.\frac{i_c}{v_c}\right|_{i_e=0} \quad \text{mhos} \tag{12-8}$$

which is the collector output admittance expressed in mhos.

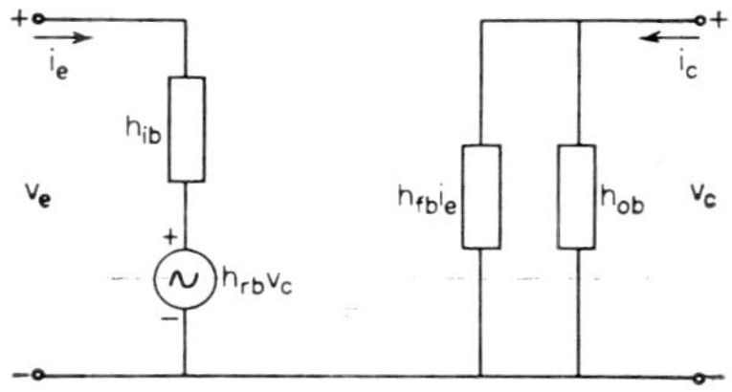

FIG. 12-2. Basic h-parameter equivalent circuit.

Thus, each of the four definitions above specifies the conditions at which any of the h parameters is measured. An examination of the circuit Eqs. (12-4a) and (12-4b) will reveal that it is possible to establish a basic equivalent circuit in which the h parameters appear as a lumped element or part. Such an arrangement is shown in Fig. 12-2, in reference to a grounded-base transistor.[4] In this circuit, h_{fb} and h_{ob} represent the input impedance and output admittance, respectively. h_{rb} appears within a feedback-voltage generator in the input circuit, and h_{fb} appears as part of the constant-current generator $h_{fb}i_e$ in the output circuit. One will note the similarity of this equivalent circuit to that of Fig. 11-3. This illustrates the close relationship of these external circuit parameters to the internal device parameters, making the h parameters convenient to the circuit and/or device designer.

12-3. Application to Equivalent Circuit. Having defined and demonstrated how the h parameters are related to the general four-terminal network, we can now apply the measurement criteria dictated by Eqs. (12-5) to (12-8) to the low-frequency equivalent circuit for the junction transistor. This circuit is repeated in Fig. 12-3 for reference, rearranged slightly to be more in the form of a four-terminal network. Both the emitter and collector capacitances are excluded, since at low frequencies their effects are negligible. Also, as we saw in Chap. 11, the parameter $\mu_{bc}v_c$, which arises from depletion-layer widening effects in the base, may be considered negligible compared to $\mu_{ec}v_c$. It must be emphasized that the values obtained for the equivalent-circuit parameters hold only for the particular d-c emitter bias, d-c collector voltage, and operating temperature of the transistor. For other operating points, a new set of values must be calculated. In the sections to follow, each of the h parameters for the circuit of Fig. 12-3 will be derived and consideration will be given to the variations with bias and temperature of each one. The treatment will be of a general nature and the results obtained will apply to both n-p-n and p-n-p types of any geometry, for grounded-base operation only.

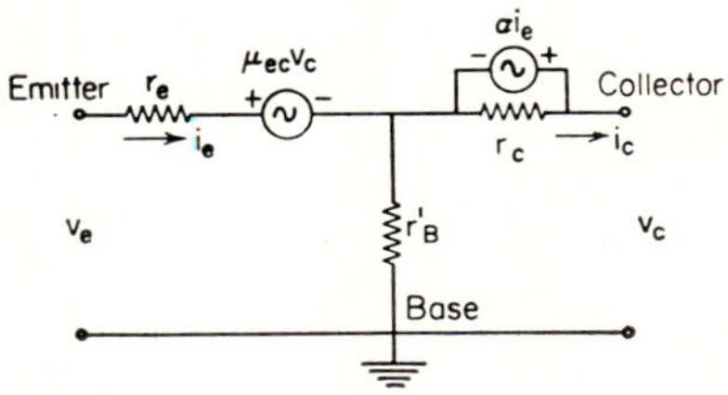

FIG. 12-3. Low-frequency equivalent circuit for the grounded-base transistor.

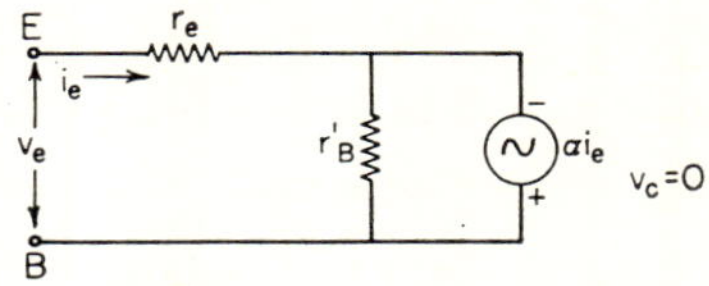

FIG. 12-4. Equivalent circuit for h_{ib}.

12-4. The Input Impedance h_{ib}. This parameter is the ratio of the a-c emitter voltage v_e to the emitter current i_e, with the output terminals of Fig. 12-3 shorted so that the collector voltage $v_c = 0$. With $v_c = 0$, the three depletion-widening terms μ_{ec}, v_c, and g_c are omitted from the circuit, which is shown in Fig. 12-4. The current generator αi_e is shunted across the base resistance r'_B. If we convert the current generator to its equivalent voltage generator,* the loop equation is simply

$$v_e = i_e(r_e + r'_B) - \alpha i_e r'_B \tag{12-9}$$

Manipulating Eq. (12-9) and dividing by i_e, we get

$$h_{ib} = \frac{v_e}{i_e} = r_e + r'_B(1 - \alpha) \qquad \text{ohms} \tag{12-10}$$

For a high-alpha transistor, the term $(1 - \alpha)$ is usually quite small, ranging from 0.05 to as little as 0.005, thereby making the second term quite small, even though r'_B may be several hundred ohms. The emitter

* A voltage generator equivalent to a current source is a voltage, equal to the product of the current αi_e and its shunt resistance r'_B, in series with the resistance r'_B.

resistance r_e is given by kT/qI_E and is seen to be inversely proportional to the direct bias current; it is about 26 ohms for $I_E = 1$ ma at room temperature. Therefore, it would be expected that h_{ib} is between 30 and 100 ohms, depending mostly on the magnitude of α. From a design point of view, h_{ib} should be as low as possible.

12-5. Reverse-voltage Feedback h_{rb}. This parameter is defined by Eq. (12-6) as the ratio of the open-circuited emitter voltage v_e to the collector voltage v_c. For all practical purposes, it is a measure of the effective base resistance of the transistor. With the emitter open-circuited, the current generator in Fig. 12-3 is set equal to zero. Thus the current i_c is equal to v_c divided by r'_B in series with the reciprocal of g_c. This current develops a voltage across r'_B equal to

$$i_c r'_B = \frac{v_c r'_B}{r'_B + 1/g_c} \tag{12-11}$$

In most junction transistors, r'_B is considerably smaller than the magnitude of the second term in the denominator of Eq. (12-11) and can be neglected. Therefore, (12-11) becomes simply

$$i_c r'_B = v_c r'_B g_c \tag{12-12}$$

At the emitter terminals, the feedback voltage is given by

$$\begin{aligned} v_{e(oc)} &= \mu_{ec} v_c + i_c r'_B \\ v_{e(oc)} &= v_c(\mu_{ec} + r'_B g_c) \end{aligned} \tag{12-13}$$

The feedback parameter h_{rb} is, finally,

$$h_{rb} = \mu_{ec} + r'_B g_c \tag{12-14}$$

where μ_{ec}, given in Chap. 11, is

$$\mu_{ec} = -\frac{kT}{qW}\frac{\partial W}{\partial V_C} \tag{12-15}$$

and

$$g_c = \frac{-I_E}{W}[2(1-\beta^*) + 1 - \gamma]\frac{\partial W}{\partial V_C} \tag{12-16}$$

Both $\partial W/\partial V_C$ and r'_B are determined by the geometry of the transistor. In general, the low-frequency value of h_{rb} might be on the order of 10^{-4}.

12-6. Forward Current Transfer h_{fb}. We have seen that this h parameter corresponds to the small-signal alpha of the transistor, defined as the ratio of the collector short-circuit current to the emitter current, or, very simply,

$$h_{fb} = \frac{i_c}{i_e} = -\alpha \tag{12-17}$$

This result may be obtained directly by inspecting the equivalent circuit in Fig. 12-3 for short-circuited output. If g_c is very small, it may be assumed that $i_c = \alpha i_e$, giving immediately the desired result for h_{fb}. Thus, h_{fb} becomes the product of the emitter efficiency γ and the base-transport factor β^*, for which the equations are given in the previous chapters.

12-7. The Output Admittance h_{ob}. From Eq. (12-8), h_{ob} is the admittance seen at the output terminals with the emitter open-circuited. It is apparent that this is equal to

$$h_{ob} = g_c \tag{12-18}$$

where g_c is the conductance given by Eq. (12-16). Note that the term r'_B has been omitted since its small magnitude does not contribute appreciably to the total output admittance.

It is convenient at this point to summarize the expressions for the low-frequency h parameters obtained thus far.

$$\begin{aligned} h_{ib} &\approx r_e + r'_B(1 - \alpha) \\ h_{rb} &\approx \mu_{ec} + r'_B g_c \qquad \text{grounded base} \\ h_{fb} &\approx -\alpha \\ h_{ob} &\approx g_c \end{aligned} \tag{12-19}$$

The significance of these small-signal h parameters should be obvious to the designer. They not only provide an easy means of evaluating the characteristics of the transistor, but also serve to indicate how well the device satisfies the criterion for signal gain. We recall that for maximum amplification the input impedance h_{ib} should be as small as possible, while the output impedance should be as large as possible. Specifically, this means that it is preferable that h_{ob} be small (less than 1 μmho). The current gain h_{fb} should be close to unity. Lastly, h_{rb} should be quite small to minimize the feedback of the collector output to the emitter. At higher frequencies, a low h_{rb} is essential for high power gain.

12-8. Variations with Emitter Current. In the foregoing treatment of the small-signal h parameters, we have been concerned only with the results at a particular d-c bias condition and operating temperature. Because the device parameters to which the h parameters are related are dependent upon either emitter current or collector voltage, it is essential to examine the manner in which the h parameters vary with bias conditions. These variations are of utmost importance to the transistor circuit designer, particularly for amplifier applications, where the gain characteristic is a function of the individual h parameters. In Sec. 12-10, the variations with temperature will be considered.

For purposes of illustration, typical grounded-base h-parameter characteristics for the type 2N335 junction transistor are shown in Fig. 12-5. This is an n-p-n silicon transistor made by the grown-diffused process (grown junction). In the figure, the variations are normalized with respect to the measured values at $I_E = 1$ ma, $V_C = 5$ volts, and $T = 25°$C. For these conditions, the typical design center values are as follows:

$h_{ib} = 42$ ohms, $h_{fb} = -0.98$, $h_{rb} = 0.4 \times 10^{-3}$, $h_{ob} = 0.4$ μmhos. These values are representative of most junction transistors, including germanium. The variations with bias, however, are representative only of transistors made by the single-ended impurity-contact processes. As a basis of comparison with those transistors made by the double-ended impurity-contact processes, Fig. 12-6 illustrates the h-parameter variations for the 2N43 p-n-p germanium-alloy transistor. As we shall see, differences in h-parameter variations among the many junction-transistor types are traced primarily to the I_{CBO} characteristic, the way base width varies with collector voltage, and the variation of alpha with temperature and emitter current.

In Fig. 12-5*a*, the room-temperature variations with emitter current are shown. Instead of h_{fb}, which corresponds to alpha, the parameter $(1 + h_{fb})$, which corresponds to $(1 - \alpha)$ is plotted. It is seen that $(1 + h_{fb})$ reaches a minimum at about 2 ma and begins increasing again. In other words, for this silicon transistor, alpha reaches its maximum at 2 ma and decreases at the higher currents. This is typical for silicon, particularly for the bar-type geometries. For p-n-p alloy transistors such as the 2N43, the parameter $(1 + h_{fb})$ is practically constant up to emitter currents in excess of 10 ma. The theoretical justification for the behavior of alpha with current was presented in Chap. 10.

The variation of the input resistance h_{ib} with I_E, shown in Fig. 12-5, is typical for all types. This is because h_{ib} is dependent on

$$h_{ib} \approx \frac{kT}{qI_E} + r'_B(1 - \alpha) \tag{12-20}$$

where the first term is the emitter resistance r_e, which is inversely proportional to I_E. At the higher currents, h_{ib} decreases more rapidly because of the reduction in r'_B due to high-level injection effects. However, if α decreases accordingly, it may compensate for the r'_B decrease, such that h_{ib} levels off to a constant value at the higher currents.

Also shown in 12-5*a* are the variations of the output admittance h_{ob}. We have seen that this is dependent on g_c.

$$h_{ob} \approx \frac{-I_E}{W}\,[2(1 - \beta^*) + 1 - \gamma]\,\frac{\partial W}{\partial V_C} \tag{12-21}$$

For all types of transistors, it is obvious that h_{ob} increases linearly with emitter current as shown above. However, if β^* and γ (or alpha) fall off with current, h_{ob} increases more rapidly with I_E. This accounts for the variation shown in (*a*) for the 2N335. For the germanium-alloy type 2N43, where $(1 + h_{fb})$ hardly changes, the h_{ob} variation is a straight line.

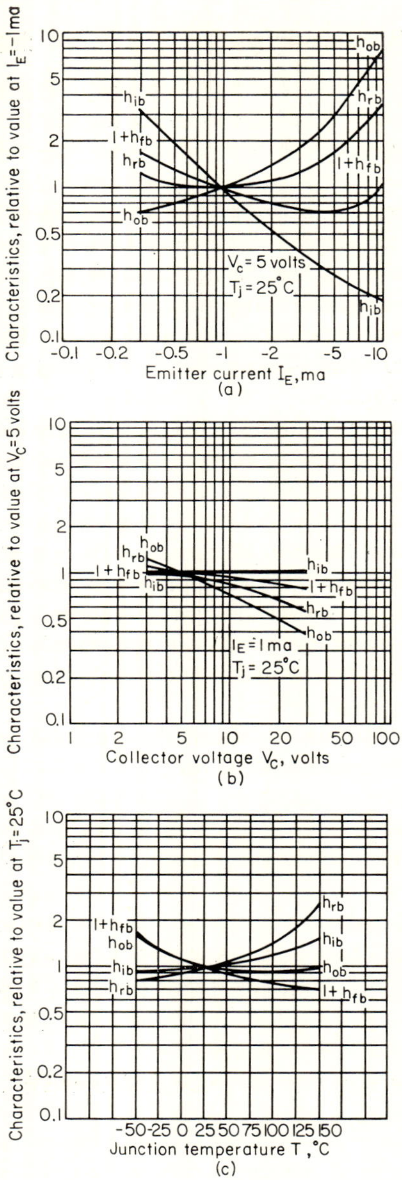

FIG. 12-5. h-Parameter variations with bias and temperature. Typical characteristics for 2N335 n-p-n grown-junction silicon transistor. (a) Common base characteristics vs. emitter current; (b) common base characteristics vs. collector voltage; (c) common base characteristics vs. junction temperature. (*Courtesy of Texas Instruments, Inc.*)

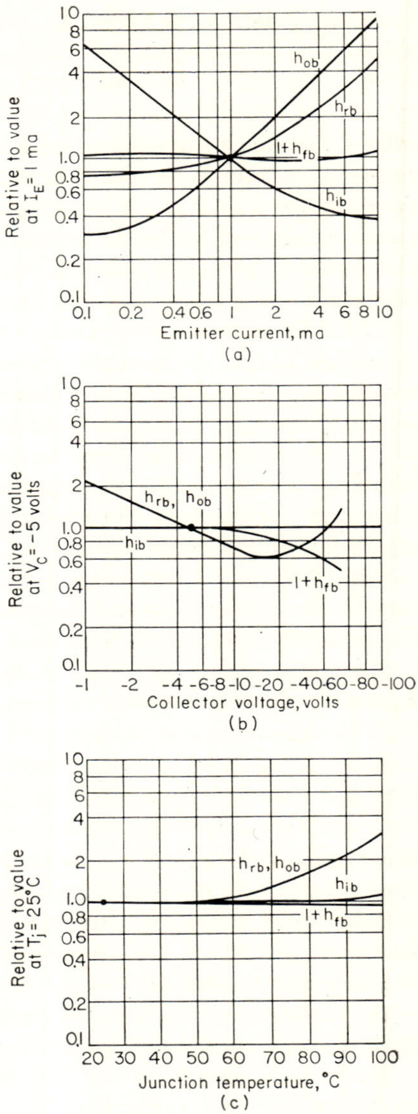

FIG. 12-6. h-Parameter variation with bias and temperature. Typical characteristics for 2N43 p-n-p germanium-alloy transistor. (*Courtesy of General Electric Co.*)

Finally, since h_{rb} is given by

$$h_{rb} \approx -\frac{kT}{qW}\frac{\partial W}{\partial V_C} + r'_B g_c \tag{12-22}$$

where the first right-hand term corresponds to the voltage-feedback factor μ_{ec}, it would be expected that h_{rb} have the same variation as g_c or, more specifically, h_{ob}. Actually, as shown in Fig. 12-5*a*, h_{rb} increases in a parallel manner but at a slower rate. This is due to the fact that r'_B decreases with emitter current, but not quite as fast as g_c increases. The variation of h_{rb} with I_E shown in (*a*) is typical of most transistor types.

12-9. Variations with Collector Voltage. The variations of the grounded-base h parameters with collector voltage for the 2N335 transistor are shown in Fig. 12-5*b*. Emitter current is kept constant at 1 ma in the figure. As is expected for almost all transistor types, the parameter $(1 + h_{fb})$ decreases with collector voltage. This is related to the increase in alpha as the effective base width decreases with barrier widening. Also, the parameter h_{ib} is shown to be practically independent of voltage; reference to Eq. (12-20) indicates why this is so. Firstly, r_e is independent of V_C; secondly, as base width decreases, the increase in r'_B due to the larger sheet resistance is compensated by the increase in α. Thus, h_{ib} remains constant as shown.

As evidenced by Eq. (12-21), the behavior of the output admittance h_{ob} with voltage is a direct function of the way base width varies with V_C. This is one parameter for which the over-all impurity distribution of the transistor becomes an important consideration. For types such as the 2N335, in which the collector resistivity is higher than the base resistivity, the collector barrier spreads further into the collector region as voltage is increased. Consequently, the slight change in W is not as effective as the decrease in $(1 + h_{fb})$ and [from Eq. (12-21)] h_{ob} decreases with collector voltage. On the other hand, for an alloy transistor such as the 2N43, the behavior of h_{ob} is slightly different. Initially, h_{ob} decreases because of the decrease in magnitude of $\partial W/\partial V_C$, since this decreases as $V_C^{-1/2}$. However, at the higher voltages, the reduction in base width W predominates, and h_{ob} starts increasing again.

Finally, as shown in Fig. 12-5*b*, the voltage-feedback parameter h_{rb} varies in the same manner as h_{ob} because of a similar dependence on g_c. This is also evident in Fig. 12-6*b*.

12-10. Variations with Temperature. The effects of temperature on the grounded-base h parameters are given by Figs. 12-5*c* and 12-6*c*. It is seen that in all cases the parameter $(1 + h_{fb})$ decreases with temperature. This is attributed to the increase in alpha due to the increase in carrier lifetime with temperature and is observed in both germanium and silicon transistors. Another parameter which behaves consistently with temperature for most junction transistor types is h_{ib}. This is dependent primarily on the magnitude of the emitter resistance r_e, which varies directly with temperature as kT/qI_E. Therefore, h_{ib} increases slightly with temperature.

The variation of h_{ob} with temperature is an excellent example of the importance of having a low I_{CBO} current rating. One recalls that g_c is defined as $\partial I_C/\partial V_C$. In the derivation, the I_{CBO} term was omitted because it was presumed to be negligible. In practical junction-transistor design, however, the effects of surface leakage cannot be completely minimized. Consequently, the output conductance is increased by the addition of a surface-leakage conductance, given as $\partial I_{CBO}/\partial V_C$, where the I_{CBO} term includes both the bulk and surface components. Although alpha tends to increase with temperature, causing g_c to decrease, it is found that the leakage conductance (corresponding to the slope of the $I_{CBO} - V_C$ characteristic) increases with temperature. This is particularly noticeable in germanium transistors. Therefore, h_{ob} increases with temperature, as noted in Fig. 12-6*c*, representing a reduction in collector resistance. Similar effects occur in silicon devices but appear at much higher temperatures.

As noted by Eq. (12-22), h_{rb} is not only dependent on the total output conductance, but on μ_{ec} as well. Because of the variation of h_{ob} with temperature just discussed and the fact that μ_{ec} varies directly with T, it follows that h_{rb} would increase with temperature faster than h_{ob}. This is demonstrated by the temperature characteristics of the 2N335 in particular.

It is appropriate at this point to emphasize what has been accomplished in these sections regarding the variations of the low-frequency h parameters with bias and temperature. We have been concerned only with two transistor types, viz., a grown-diffused n-p-n silicon and an alloy p-n-p germanium transistor. Although the results obtained are closely related, they in no way represent the characteristics for all possible transistor types. However, sufficient theory has been developed in the previous chapters to enable one to predict the behavior of the significant device parameters for almost all transistor types. If the over-all impurity distribution and structure are known, such device parameters as $\partial W/\partial V_C$, α, and g_c are readily defined. Then the h parameters are described by employing the relations and methods introduced in this chapter.

12-11. Grounded-emitter *h* Parameters. The low-frequency equivalent circuit for the grounded-base circuit configuration was shown in Fig. 12-3. This arrangement is a resistive T network having an active voltage generator $\mu_{ec}v_c$ in the emitter branch. The collector resistance r_c is the reciprocal of the conductance g_c.

When the junction transistor is operated in the grounded-emitter configuration, the low-frequency equivalent circuit appears as shown in Fig. 12-7. Here both the emitter voltage and current v_e and i_e have been replaced by the base voltage and current v_b and i_b. Since

$$i_b = (1 - \alpha)i_e \tag{12-23}$$

or

$$i_e = \frac{i_b}{(1 - \alpha)} \tag{12-24}$$

the current generator αi_e from the grounded base circuit must be replaced by $\alpha i_b/(1 - \alpha)$ for the grounded-emitter configuration. Note that $\alpha/(1 - \alpha)$ is the grounded-emitter current gain β for the grounded-emitter transistor. With the input current as a reference, regardless of

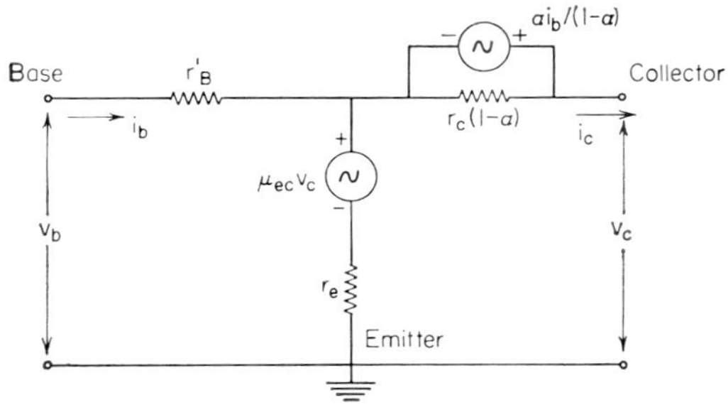

FIG. 12-7. Equivalent-circuit transformation to grounded-emitter transistor.

configuration, in order to establish the same collector voltage v_c at the output, the resistance r_c must be multiplied by $(1 - \alpha)$ for the grounded-emitter case. For high-alpha transistors the output resistance is reduced considerably, since $(1 - \alpha)$ is a small quantity. The general network equations for grounded emitter are then:

$$\begin{aligned} v_b &= h_{ie}i_b + h_{re}v_c \\ i_c &= h_{fe}i_b + h_{oe}v_c \end{aligned} \tag{12-25}$$

The same interpretations are applied to the h parameters. The only difference is in the second letter subscript, which is now e. Applying the same measurement principles to Eqs. (12-25) as for the grounded-base case, we can relate the h parameters to the equivalent-circuit elements of Fig. 12-7. For the input impedance h_{ie}, we have

$$h_{ie} = \left.\frac{v_b}{i_b}\right|_{v_c=0} \tag{12-26}$$

For the condition that $r_c \gg r_e$, it is easily shown that

$$h_{ie} \approx r'_B + \frac{r_e}{(1 - \alpha)} \tag{12-27}$$

The voltage-feedback parameter h_{re} is defined as

$$h_{re} = \left.\frac{v_b}{v_c}\right|_{i_b=0} \tag{12-28}$$

which would be

$$h_{re} = \mu_{ec} + \frac{r_e}{r_e + r_c(1-\alpha)} \tag{12-29}$$

Again, if $r_c(1-\alpha)$ is greater than r_e, we get

$$h_{re} = \mu_{ec} + \frac{r_e g_c}{(1-\alpha)} \tag{12-30}$$

The current-transfer ratio is simply

$$h_{fe} = \left.\frac{i_c}{i_b}\right|_{v_e=0} = \frac{\alpha}{1-\alpha} \tag{12-31}$$

Lastly, the grounded-emitter output admittance is

$$h_{oe} = \left.\frac{i_c}{v_c}\right|_{i_b=0} \tag{12-32}$$

or, directly from Fig. 12-7,

$$h_{oe} = \frac{1-\mu_{ec}}{r_e + r_c(1-\alpha)} \tag{12-33}$$

Since r_e is usually quite small and $\mu_{ec} \ll 1$, Eq. (12-33) becomes, approximately,

$$h_{oe} \approx \frac{g_c}{(1-\alpha)} \tag{12-34}$$

Inspecting Eqs. (12-27), (12-31), and (12-34), we see that they are related to the grounded-base parameters by the single factor[5] $(1-\alpha)$. The one exception is h_{re} [Eq. (12-30)], because of the fact that the voltage feedback μ_{ec} is independent of the current.

Summarizing the low-frequency grounded-emitter h parameters, we have the following:

$$\left.\begin{aligned} h_{ie} &\approx r'_B + \frac{r_e}{(1-\alpha)} \\ h_{re} &\approx \mu_{ec} + \frac{r_e g_c}{(1-\alpha)} \\ h_{fe} &\approx \frac{\alpha}{1-\alpha} \\ h_{oe} &\approx \frac{g_c}{(1-\alpha)} \end{aligned}\right\} \text{grounded emitter} \tag{12-35}$$

The salient feature of the grounded-emitter transistor is that only a small base current i_b is required to drive the device, giving very high current gains. Furthermore, this method of operation is characterized by higher input impedances but lower output impedances. The over-all output voltage, however, does not differ from the grounded-base case since the maximum allowable load resistor must be reduced by $(1 - \alpha)$.

The variations of the grounded-emitter h parameters may be determined by properly relating the $(1 + h_{fb})$ variations to the grounded-base h-parameter variations according to Eqs. (12-35). It should be noted that if $(1 + h_{fb})$ is somewhat constant, the normalized grounded-emitter h-parameter variations are approximately the same as the grounded-base parameter variations.

PROBLEMS

12-1. The parameters of a junction transistor are as follows: $\alpha = 0.97$, $g_c = 0.3$ μmho, $r_B' = 200$ ohms. Additional measurements indicate that the low-frequency base resistance r_B'' is 600 ohms. For this device, calculate both the grounded-base and grounded-emitter h parameters at $I_E = 1$ ma, $V_C = 5$ volts, and $T = 27°C$.

12-2. For a grounded-base transistor it is found that $h_{ib} = 55$ ohms and $h_{ob} = 0.6$ μmhos. Determine the value of h_{fb} necessary to make the input resistance equal the output resistance for grounded-emitter operation (i.e., to make $h_{ie} = 1/h_{oe}$).

12-3. A 2N335 silicon n-p-n transistor is operated as a grounded-base voltage amplifier into a 20,000-ohm load resistor R_L. At 25°C, $V_C = 5$ volts, $I_E = -1$ ma, $h_{rb} = 2 \times 10^{-4}$, $h_{fb} = -0.983$, $h_{ob} = 0.2$ μmhos, and $h_{ib} = 42$ ohms. If voltage gain is defined as

$$K = \frac{-h_{fb}}{h_{ib}/R_L + h_{ib}h_{ob} - h_{rb}h_{fb}}$$

determine the gain at $I_E = -1$ ma and at $I_E = -5$ ma (refer to Fig. 12-5).

12-4. Calculate the value of h_{rb} at low frequencies for the p-n-p germanium-alloy transistor given in Prob. 11-1.

REFERENCES

1. Early, J. M.: Effects of Space-charge Layer Widening in Junction Transistors, *Proc. IRE*, vol. 40, pp. 1401–1406, November, 1952.
2. Guillemin, E. A.: "Communication Networks," vol. II, chap. 4, John Wiley & Sons, Inc., New York, 1935.
3. Pritchard, R. L.: Small-signal Parameters for Transistors, General Electric Research Laboratory, Report RL-1086, April, 1954.
4. Pritchard, R. L.: Electric-network Representation of Transistors—A Survey, *IRE Trans.*, vol. CT-3, no. 1, March, 1956.
5. Pritchard, R. L.: Frequency Response of Grounded-base and Grounded-emitter Junction Transistors, presented at AIEE Winter Meeting, New York, January, 1954.

13

High-frequency h Parameters

13-1. Introduction. Measurements of practical junction transistors indicate that the small-signal parameters are not independent of frequency as was implied by the analyses of the previous chapters. The treatment thus far, including the effects of base-region widening, characterizes only the very-low-frequency performance of the transistor. At higher frequencies, especially in the vicinity of the base-cutoff frequency, the various relationships obtained are not completely valid, since they do not take into consideration the observed reactive effects, particularly the junction capacitances. Therefore, in order to fully describe the theoretical operation of the transistor (one-dimensional model), this chapter will present a first-order small-signal analysis which will include all the effects that were introduced heretofore. This complete treatment based on the work of Early and Pritchard[1,2,*] will yield both the direct bias and alternating signal currents including base-region widening effects, with the objective of establishing the frequency-dependent admittance parameters for the transistor. After suitable simplification of these admittance terms, a transformation to the appropriate h parameters will be made, yielding the final frequency relationships. The manner of variation of each h parameter with frequency will be discussed in turn. The effect of base resistance, however, will not be considered initially in the one-dimensional solution, but will be included at the very end.

13-2. General Solution for Minority-carrier Currents. Consider again our one-dimensional model of the p-n-p junction transistor, where we have been concerned with carrier flow in the x direction only. This is shown in Fig. 13-1, where the emitter junction is at $x = 0$ and the collector junction at $x = W$, and where W is the effective thickness of the n-type base region. In addition to the d-c bias voltages V_E and $-V_C$,

* References, indicated in the text by superscript figures, are listed at the end of the chapter.

there are impressed small-signal a-c variations of frequency ω and of amplitudes v_e and v_c at the emitter and collector, respectively. Thus, the instantaneous applied voltages are

$$V_e = V_E + v_e\epsilon^{j\omega t} \tag{13-1}$$
$$V_c = -V_C + v_c\epsilon^{j\omega t} \tag{13-2}$$

where $\epsilon^{j\omega t}$ denotes the frequency function and $v_e \ll V_E$ and $v_c \ll V_C$. The basic assumptions of the d-c analysis of Chap. 8, concerning the conductivities, dimensions, and characteristics of the various transistor regions, apply here. Particularly, it should be recalled that we are

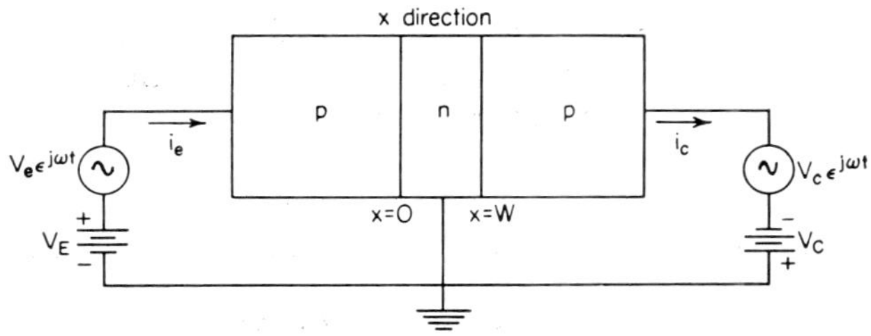

FIG. 13-1. One-dimensional p-n-p junction transistor.

assuming that the injected hole density is small compared to the majority-carrier concentration (electrons) in the base, so that only diffusion effects are considered. With these assumptions in mind, we can immediately write the steady-state boundary conditions for the carrier concentrations at the emitter and collector junctions. At the emitter junction $x = 0$,

$$p = p_{nb}\epsilon^{q(V_E+v_e\epsilon^{j\omega t})/kT} \tag{13-3}$$

and at the collector junction $x = W$,

$$p = p_{nb}\epsilon^{q(-V_C+v_c\epsilon^{j\omega t})/kT} \tag{13-4}$$

where p_{nb} is the equilibrium hole concentration in the base region. Since we are dealing with small a-c variations, Eqs. (13-3) and (13-4) may be approximated by their series expansions.

$$p\Big|_{x=0} \approx p_{nb}\epsilon^{qV_E/kT}[1 + (qv_e/kT)\epsilon^{j\omega t}] = p_E + p_e \tag{13-5}$$
$$p\Big|_{x=W} \approx p_{nb}\epsilon^{-qV_C/kT}[1 + (qv_c/kT)\epsilon^{j\omega t}] = p_C + p_c \tag{13-6}$$

Thus, the boundary concentrations have been split into d-c and a-c terms, where the capital subscripts denote the d-c and the small subscripts the a-c terms. Widening effects will not be introduced until later.

In order to solve for the hole concentration as a function of distance and time, viz., $p(x,t)$, we make use of the general continuity equation for hole diffusion, which is

$$D_{pb}\frac{\partial^2 p}{\partial x^2} - \frac{p - p_{nb}}{\tau_{pb}} = \frac{\partial p}{\partial t} \tag{13-7}$$

where D_{pb} is the diffusion constant for holes and τ_{pb} is the hole lifetime in the base. For this partial differential equation, the solution may be in the form of a Fourier series,[2] such as

$$p(x,t) = \sum_{m=-\infty}^{\infty} p_m(x)\epsilon^{j\omega t} \tag{13-8}$$

which would yield an equation for each value of m, introducing an infinite number of higher-order terms. From a practical point of view, we are interested only in the average values (d-c) and the first-order a-c terms (fundamental or first harmonic). The higher-order terms should be relatively insignificant. Thus, we shall use Eq. (13-8) as a solution to (13-7) only for the conditions that $m = 0$ and $m = 1$; this gives a solution of the form

$$p(x,t) = p_0(x) + p_1(x)\epsilon^{j\omega t} \tag{13-9}$$

From Eq. (13-8), we can write the partial derivative with respect to time as

$$\frac{\partial p}{\partial t} = j\omega p \qquad \text{for } m = 0,\ 1 \tag{13-10}$$

Substituting this result into Eq. (13-7) and putting the result into a more general form, we have

$$D_{pb}\frac{\partial^2 p}{\partial x^2} - \frac{p}{\tau_{pb}} = j\omega p$$

$$\frac{\partial^2 p}{\partial x^2} - \frac{1}{L_{pb}^2}(1 + j\omega\tau_{pb})p = 0 \tag{13-11}$$

where L_{pb}^2 is equal to $D_{pb}\tau_{pb}$, the diffusion length. The general solution of Eq. (13-11) is of the form

$$p - p_{nb} = A\epsilon^{x/L_{pb}} + B\epsilon^{-x/L_{pb}} + (C\epsilon^{Zx/L_{pb}} + D\epsilon^{-Zx/L_{pb}})\epsilon^{j\omega t} \tag{13-12}$$

where $Z = (1 + j\omega\tau_{pb})^{1/2}$ and the coefficients are to be determined from the boundary conditions. It should be noted that the first two right-hand terms correspond to the d-c solution ($m = 0$) and the last term is the a-c solution ($m = 1$). To evaluate the constants, we insert the boundary conditions of Eqs. (13-5) and (13-6). Therefore, at $x = 0$, $p = p_E + p_e$, or

$$p_E + p_e - p_{nb} = A + B + (C + D)\epsilon^{j\omega t} \tag{13-13}$$

Also, at $x = W$, $p = p_C + p_c$, or

$$p_C + p_c - p_{nb} = A\epsilon^{W/L_{pb}} + B\epsilon^{-W/L_{pb}} + (C\epsilon^{ZW/L_{pb}} + D\epsilon^{-ZW/L_{pb}})\epsilon^{j\omega t} \tag{13-14}$$

Since both the d-c and a-c solutions must hold simultaneously in steady state, these equations may be separated into two pairs of simultaneous equations in order to obtain the constants. Considering only d-c terms, we have

$$p_E - p_{nb} = A + B \tag{13-15}$$
$$p_C - p_{nb} = A\epsilon^{W/L_{pb}} + B\epsilon^{-W/L_{pb}} \tag{13-16}$$

Solving for A and B and substituting 2 sinh x for $e^x - e^{-x}$, we get, finally,

$$A = \frac{p_C - p_{nb} - (p_E - p_{nb})\epsilon^{-W/L_{pb}}}{2 \sinh (W/L_{pb})} \tag{13-17}$$
$$B = \frac{-(p_C - p_{nb}) + (p_E - p_{nb})\epsilon^{W/L_{pb}}}{2 \sinh (W/L_{pb})} \tag{13-18}$$

These results are substituted back into Eq. (13-12), thereby establishing the first two terms. It should be pointed out that this d-c solution for the hole concentration as a function of x is identical to that derived in Chap. 8.

To obtain the a-c distributions, we can equate the remaining terms of the simultaneous Eqs. (13-13) and (13-14).

$$p_e = C + D \tag{13-19}$$
$$p_c = (C\epsilon^{ZW/L_{pb}} + D\epsilon^{-ZW/L_{pb}})\epsilon^{j\omega t} \tag{13-20}$$

Solving for the constants C and D, we get

$$C = \frac{p_c - p_e\epsilon^{-ZW/L_{pb}}}{2 \sinh (ZW/L_{pb})} \tag{13-21}$$

and

$$D = \frac{-p_c + p_e\epsilon^{ZW/L_{pb}}}{2 \sinh (ZW/L_{pb})} \tag{13-22}$$

The substitution of the expressions for A, B, C, and D into Eq. (13-12) yields the complete final solution for the d-c and a-c hole concentrations. This is:

$$\begin{aligned} p - p_{nb} = {} & \frac{p_C - p_{nb} - (p_E - p_{nb})\epsilon^{-W/L_{pb}}}{2 \sinh (W/L_{pb})} \epsilon^{x/L_{pb}} \\ & + \frac{-(p_C - p_{nb}) + (p_E - p_{nb})\epsilon^{-W/L_{pb}}}{2 \sinh (W/L_{pb})} \epsilon^{-x/L_{pb}} \\ & + \frac{p_c - p_e\epsilon^{-ZW/L_{pb}}}{2 \sinh (ZW/L_{pb})} \epsilon^{Zx/L_{pb}}\epsilon^{j\omega t} + \frac{-p_c + p_e\epsilon^{ZW/L_{pb}}}{2 \sinh (ZW/L_{pb})} \epsilon^{-Zx/L_{pb}}\epsilon^{j\omega t} \end{aligned} \tag{13-23}$$

where the last two terms comprise the first-order a-c solution.

To complete the picture for this p-n-p model, it is necessary to consider the electron components. For normal operation of the junction transistor, there exist both the reverse-biased collector electron current flowing from collector to base and the forward-biased electron current flowing from base to emitter. The collector component will be neglected in this analysis. The electron emitter current is obtained by solving the diffusion equation for the electron concentration in the p-type emitter.

$$\frac{\partial^2 n}{\partial x^2} - \frac{1}{L_{ne}^2}(1 + j\omega\tau_{ne})n = 0 \qquad (13\text{-}24)$$

where L_{ne} is the diffusion length of electrons in the emitter and τ_{ne} is the electron lifetime. Again, the combined d-c and a-c solution is of the form

$$n - n_{pe} = A'\epsilon^{x/L_{ne}} + B'\epsilon^{-x/L_{ne}} + (C'\epsilon^{Z'x/L_{ne}} + D'\epsilon^{-Z'x/L_{ne}})\epsilon^{j\omega t} \qquad (13\text{-}25)$$

where $Z' = (1 + j\omega\tau_{ne})^{1/2}$. The method of evaluating the constants is identical to that of the hole flow, with the exception of the boundary conditions. At the emitter junction $x = 0$, the electron concentration is given as

$$n = n_{pe}\epsilon^{q(V_E + v_e\epsilon^{j\omega t})/kT} \qquad (13\text{-}26)$$

or, for small variations, approximately

$$n = n_{pe}\epsilon^{qV_E/kT}\left(1 + \frac{qv_e}{kT}\epsilon^{j\omega t}\right) = n_E + n_e \qquad (13\text{-}27)$$

At the emitter ohmic contact, however, the electron concentration is equal to the equilibrium value n_{pe}, so that $n = n_{pe}$ at $x = -l$. Applying these boundary conditions to Eq. (13-25), we can obtain both the d-c and a-c electron concentrations, in a manner similar to that for holes. For the a-c term,

$$n - n_{pe} = \frac{n_e \sinh\,[Z'(x + l)/L_{ne}]}{\sinh\,(Z'l/L_{ne})}\epsilon^{j\omega t} \qquad (13\text{-}28)$$

From this expression, we can proceed immediately to determine the a-c electron diffusion current by the relationship

$$i_n = AqD_{ne}\frac{dn}{dx} \qquad (13\text{-}29)$$

Differentiating (13-28) and multiplying by AqD_{ne}, we have

$$i_n = \frac{AqD_{ne}n_{pe}Z'}{L_{ne}}\,\frac{\cosh\,[Z'(x + l)/L_{ne}]}{\sinh\,(Z'l/L_{ne})}\epsilon^{j\omega t} \qquad (13\text{-}30)$$

Evaluating (13-30) at $x = 0$, we get simply

$$i_{ne} = \frac{AqD_{ne}n_{pe}Z'}{L_{ne}}\coth\left(\frac{Z'l}{L_{ne}}\right)\epsilon^{j\omega t} \qquad (13\text{-}31)$$

If $l \gg L_{ne}$, then coth $(Z'l/L_{ne})$ is close to unity, so that

$$i_{ne} = \frac{AqD_n n_{pe} Z'}{L_{ne}} \epsilon^{j\omega t} \tag{13-32}$$

Equation (13-32) represents the magnitude of the alternating electron current crossing the emitter junction, and obviously affects the emitter efficiency frequency response. The hole currents at the emitter and collector junctions may be obtained in the same way from Eq. (13-23), but this will not be done until base-region widening effects are considered.

13-3. Introduction of Base-region Widening Effects.[1] If Eq. (13-23) for the hole concentration is evaluated at the collector junction $x = W$, we observe both d-c and a-c values. These values, however, are based on a constant base thickness W and do not include the effects of the collector depletion layer which widens and narrows with the small a-c signal voltage. The variation in base thickness with frequency effectively modulates both the d-c and a-c concentrations, thereby introducing higher-order frequency terms which appear in the final solution in addition to the initial concentrations. The terms arising from the a-c modulation of the a-c hole concentration at the collector are of very small amplitude and for all practical purposes may be neglected. Thus, we are left only with the significant first-order result obtained by the a-c widening of the d-c concentration. This is similar to the widening effects discussed in Chap. 11, with the exception that the dependence on frequency will be considered now.

It will be assumed that the base-width variation is represented by the following relationship:

$$W(t) = W + W_1 \epsilon^{j\omega t} \tag{13-33}$$

where the second right-hand term represents the small-signal time variation about the average value W. The amplitude of the variation is given by W_1, which is very much smaller than W, so that it may be approximated by

$$W_1 \approx \frac{\partial W}{\partial V_C} v_c \tag{13-34}$$

where v_c is the a-c collector voltage across the junction. Since we are seeking an additional a-c solution of the form

$$p'(x,t) = (E\epsilon^{Zx/L_{pb}} + F\epsilon^{-Zx/L_{pb}})\epsilon^{j\omega t} \tag{13-35}$$

which will be added to the general solution of Eq. (13-12), the new boundary conditions for (13-35) must be determined. At the emitter junction $x = 0$, it will be assumed that widening effects are negligible. This is a good approximation since the emitter barrier is forward-biased and x_m is quite small to begin with. Therefore,

$$p'(0,t) = 0 \qquad x = 0 \tag{13-36}$$

is the first boundary condition. At the collector junction $x = W$, the boundary condition is obtained directly from the d-c part of Eq. (13-23) by letting $x = W + W_1\epsilon^{j\omega t}$. Using the first-order expansion

$$\epsilon^{(W+W_1\epsilon^{j\omega t}/L_{pb})} \approx \epsilon^{W/L_{pb}}\left(1 + \frac{W_1}{L_{pb}}\epsilon^{j\omega t}\right) \tag{13-37}$$

we obtain, from (13-23), the following:

$$\frac{p_C - p_{nb} - (p_E - p_{nb})\epsilon^{-W/L_{pb}}}{2\sinh(W/L_{pb})}\epsilon^{W/L_{pb}}\left(1 + \frac{W_1}{L_{pb}}\epsilon^{j\omega t}\right) + \frac{-(p_C - p_{nb}) + (p_E - p_{nb})\epsilon^{W/L_{pb}}}{2\sinh(W/L_{pb})}\epsilon^{-W/L_{pb}}\left(1 - \frac{W_1}{L_{pb}}\epsilon^{j\omega t}\right) \tag{13-38}$$

This result may be simplified by multiplying and regrouping terms. The modulated hole concentration at the collector is, finally,

$$p'(W + W_1\epsilon^{j\omega t}, t) = \frac{W_1}{L_{pb}}\epsilon^{j\omega t}\left[(p_E - p_{nb})\operatorname{csch}\frac{W}{L_{pb}} - (p_C - p_{nb})\coth\frac{W}{L_{pb}}\right] \tag{13-39}$$

Making the substitution for W_1 from Eq. (13-34), we get

$$p'(W + W_1\epsilon^{j\omega t}, t) = \frac{\partial W}{\partial V_C}\frac{v_c\epsilon^{j\omega t}}{L_{pb}}\left[(p_E - p_{nb})\operatorname{csch}\frac{W}{L_{pb}} - (p_C - p_{nb})\coth\frac{W}{L_{pb}}\right] \tag{13-40}$$

It should be noted that this equation is simply the d-c collector (hole) concentration multiplied by the a-c widening factor.

We can now proceed to apply these boundary values to (13-35) in order to evaluate E and F. Hence,

$$0 = E + F \tag{13-41}$$

$$p'(W + W_1\epsilon^{j\omega t}, t) = (E\epsilon^{ZW/L_{pb}} + F\epsilon^{-ZW/L_{pb}})\epsilon^{j\omega t} \tag{13-42}$$

Solving these equations simultaneously and substituting the results back into (13-35), we get the final solution for base-region widening.

$$p'(x,t) = \frac{\sinh(Zx/L_{pb})}{\sinh(ZW/L_{pb})}\left[(p_E - p_{nb})\operatorname{csch}\frac{W}{L_{pb}} - (p_C - p_{nb})\coth\frac{W}{L_{pb}}\right]\frac{\partial W}{\partial V_C}\frac{v_c\epsilon^{j\omega t}}{L_{pb}} \tag{13-43}$$

This expression, in addition to Eq. (13-23), gives us the complete solution for the hole concentration $p(x,t)$ in the base layer. By regroupment and algebraic manipulation of the terms of (13-23) and addition of Eq. (13-43), $(p - p_{nb})$ may be written as

$$p - p_{nb} = \frac{(p_C - p_{nb}) \sinh (x/L_{pb})}{\sinh (W/L_{pb})} - \frac{(p_E - p_{nb}) \sinh [(x - W)/L_{pb}]}{\sinh (W/L_{pb})} + \frac{p_e \sinh (Zx/L_{pb})}{\sinh (ZW/L_{pb})} \epsilon^{j\omega t} - \frac{p_e \sinh [Z(x - W)/L_{pb}]}{\sinh (ZW/L_{pb})} \epsilon^{j\omega t} + \frac{\sinh (Zx/L_{pb})}{\sinh (ZW/L_{pb})} \left[(p_E - p_{nb}) \operatorname{csch} \frac{W}{L_{pb}} - (p_C - p_{nb}) \coth \frac{W}{L_{pb}} \right] \frac{\partial W}{\partial V_C} \frac{v_c \epsilon^{j\omega t}}{L_{pb}} \tag{13-44}$$

We see that this expression is actually the sum of the d-c, a-c, and widening hole concentrations, in order.

The total hole current in the base region is found from the diffusion-current equation

$$I_p = -qAD_{pb} \frac{dp}{dx} \tag{13-45}$$

where D_{pb} is the hole diffusion constant and dp/dx is obtained by differentiating Eq. (13-44) with respect to x. This yields the general hole-current expression as a function of x. Performing this operation, we get

$$I_p(x) = -\frac{qAD_{pb}}{L_{pb}} \left\{ \frac{(p_C - p_{nb}) \cosh (x/L_{pb})}{\sinh (W/L_{pb})} - \frac{(p_E - p_{nb}) \cosh [(x - W)/L_{pb}]}{\sinh (W/L_{pb})} + \frac{Zp_e \cosh (Zx/L_{pb})}{\sinh (ZW/L_{pb})} \epsilon^{j\omega t} - \frac{Zp_e \cosh [Z(x - W)/L_{pb}]}{\sinh (ZW/L_{pb})} \epsilon^{j\omega t} + \frac{Z \cosh (Zx/L_{pb})}{\sinh (ZW/L_{pb})} \left[(p_E - p_{nb}) \operatorname{csch} \frac{W}{L_{pb}} - (p_C - p_{nb}) \coth \frac{W}{L_{pb}} \right] \frac{\partial W}{\partial V_C} \frac{v_c \epsilon^{j\omega t}}{L_{pb}} \right\} \tag{13-46}$$

From Eq. (13-46), the emitter and collector hole currents are determined by the substitution of $x = 0$ and $x = W$, respectively. This yields both the direct and alternating hole currents. However, because the d-c results had been determined in a previous chapter, we shall concern ourselves only with alternating currents, since the objective is to establish the small-signal frequency-dependent admittance parameters. Therefore, at $x = 0$, the emitter hole current (alternating current only) is

$$i_{pe} = -\frac{qAD_{pb}Z}{L_{pb}} \left\{ p_e \operatorname{csch} \frac{ZW}{L_{pb}} - p_e \coth \frac{ZW}{L_{pb}} + \operatorname{csch} \frac{ZW}{L_{pb}} \left[(p_E - p_{nb}) \operatorname{csch} \frac{W}{L_{pb}} - (p_C - p_{nb}) \coth \frac{W}{L_{pb}} \right] \frac{\partial W}{\partial V_C} \frac{v_c}{L_{pb}} \right\} \epsilon^{j\omega t} \tag{13-47}$$

Similarly, at the collector junction $x = W$, we have

$$i_{pc} = -\frac{qAD_{pb}Z}{L_{pb}}\left\{p_c \coth\frac{ZW}{L_{pb}} - p_e \operatorname{csch}\frac{ZW}{L_{pb}}\right.$$
$$\left. + \coth\frac{ZW}{L_{pb}}\left[(p_E - p_{nb}) \operatorname{csch}\frac{W}{L_{pb}} - (p_C - p_{nb}) \coth\frac{W}{L_{pb}}\right]\frac{\partial W}{\partial V_C}\frac{v_c}{L_{pb}}\right\}\epsilon^{j\omega t} \tag{13-48}$$

As far as the total alternating diffusion currents are concerned, we can now state that

$$i_e = i_{pe} + i_{ne} \tag{13-49}$$
$$i_c = i_{pc} \tag{13-50}$$

To obtain the total alternating emitter current i_e, we must add to Eq. (13-47) the electron component i_{ne}, which was given by Eq. (13-32) as

$$i_{ne} = \frac{qAD_{ne}Z'}{L_{ne}} n_{pe}\epsilon^{j\omega t} \tag{13-51}$$

As a good approximation, the total collector current is about equal to the hole current given by Eq. (13-48).

Lastly, there is associated with each junction of the transistor the junction or depletion-layer transition capacitance. We have seen that at the emitter, C_{Te} is quite large because of the forward-bias narrowing, while at the collector, C_{Tc} is considerably smaller because of the depletion-layer spreading with reverse bias. At high frequencies, the admittances ($j\omega C$) become appreciable enough to contribute an additional capacitive-current component to each of the total junction currents. Therefore, the complete transistor current equations must be

$$i_e = i_{pe} + i_{ne} + j\omega C_{Te}v_e \tag{13-52}$$
$$i_c = i_{pc} + j\omega C_{Tc}v_c \tag{13-53}$$

where i_e, v_e and i_c, v_c are the applied alternating currents and voltages for the emitter and collector respectively. The expressions for i_{pe}, i_{ne}, and i_{pc} are given by Eqs. (13-47), (13-51), and (13-48).

13-4. The Admittance Equations. The total alternating currents, given by Eqs. (13-52) and (13-53), are functions of the a-c voltages v_e and v_c, such that the equations may be written in the form

$$i_e = y_{11}v_e + y_{12}v_c \tag{13-54}$$
$$i_c = y_{21}v_e + y_{22}v_c \tag{13-55}$$

where the y parameters denote the various admittances. The actual expressions involved, we have seen so far, are quite lengthy. Nevertheless, by examining the current equations carefully and making appropriate substitutions, we can reduce them to very simple relationships for the admittances. To begin, we can resubstitute the proper expressions for such terms as p_E, p_C, p_e, and p_c, which were defined earlier. Note again that the capital subscripts denote d-c terms and the lower-case subscripts are for alternating components. Carrying out this procedure, we obtain

$$i_e = -\frac{qAD_{pb}Z}{L_{pb}}\left\{p_{nb}\epsilon^{-qV_C/kT}\frac{qv_c}{kT}\operatorname{csch}\frac{ZW}{L_{pb}} - p_{nb}\epsilon^{qV_E/kT}\frac{qv_e}{kT}\coth\frac{ZW}{L_{pb}} + \operatorname{csch}\frac{ZW}{L_{pb}}\left[p_{nb}(\epsilon^{qV_E/kT} - 1)\operatorname{csch}\frac{W}{L_{pb}} - p_{nb}(\epsilon^{-qV_C/kT} - 1)\coth\frac{W}{L_{pb}}\right]\frac{\partial W}{\partial V_C}\frac{v_c}{L_{pb}} - n_{pe}\epsilon^{qV_E/kT}\frac{qv_e}{kT}\frac{D_{ne}Z'L_{pb}}{D_{pb}ZL_{ne}}\right\} + j\omega C_{Te}v_e \qquad (13\text{-}56)$$

and

$$i_c = -\frac{qAD_{pb}Z}{L_{pb}}\left\{p_{nb}\epsilon^{-qV_C/kT}\frac{qv_c}{kT}\coth\frac{ZW}{L_{pb}} - p_{nb}\epsilon^{qV_E/kT}\frac{qv_e}{kT}\operatorname{csch}\frac{ZW}{L_{pb}} + \coth\frac{ZW}{L_{pb}}\left[p_{nb}(\epsilon^{qV_E/kT} - 1)\operatorname{csch}\frac{W}{L_{pb}} - p_{nb}(\epsilon^{-qV_C/kT} - 1)\coth\frac{W}{L_{pb}}\right]\frac{\partial W}{\partial V_C}\frac{v_c}{L_{pb}}\right\} + j\omega C_{Tc}v_c \qquad (13\text{-}57)$$

Recall that $Z = (1 + j\omega\tau_{pb})^{1/2}$ and that $Z' = (1 + j\omega\tau_{ne})^{1/2}$. Immediately we can make the approximation that, for reverse bias, $\epsilon^{-qV_C/kT}$ is practically zero in Eqs. (13-56) and (13-57). Grouping the terms by voltage to get the equations into the admittance form, we have, after considerable simplification,

$$i_e = \left[\frac{q^2AD_{pb}Zp_{nb}\epsilon^{qV_E/kT}}{kTL_{pb}}\left(\coth\frac{ZW}{L_{pb}} + \frac{n_{pe}D_{ne}L_{pb}}{p_{nb}D_{pb}ZL_{ne}}\right) + j\omega C_{Te}\right]v_e - \frac{qAD_{pb}Zp_{nb}}{L_{pb}^2}\frac{\partial W}{\partial V_C}\operatorname{csch}\frac{ZW}{L_{pb}}\left[(\epsilon^{qV_E/kT} - 1)\operatorname{csch}\frac{W}{L_{pb}} + \coth\frac{W}{L_{pb}}\right]v_c \qquad (13\text{-}58)$$

$$i_c = -\frac{q^2AD_{pb}Zp_{nb}\epsilon^{qV_E/kT}}{kT/L_{pb}}\operatorname{csch}\frac{ZW}{L_{pb}}v_e + \left\{\frac{qAD_{pb}Zp_{nb}}{L_{pb}^2}\frac{\partial W}{\partial V_C}\coth\frac{ZW}{L_{pb}}\left[(\epsilon^{qV_E/kT} - 1)\operatorname{csch}\frac{W}{L_{pb}} + \coth\frac{W}{L_{pb}}\right] + j\omega C_{Tc}\right\}v_c \qquad (13\text{-}59)$$

Equations (13-58) and (13-59) are now in the admittance form of Eqs. (13-54) and (13-55).

By noting certain similarities to the d-c equations derived in Chap. 8, we can make Eqs. (13-58) and (13-59) more concise.[2] Consider first that part of the widening term common to each equation, viz.,

$$\frac{qAD_{pb}p_{nb}}{L_{pb}}\left[(\epsilon^{qV_E/kT} - 1)\operatorname{csch}\frac{W}{L_{pb}} + \coth\frac{W}{L_{pb}}\right] \qquad (13\text{-}60)$$

From Eq. (8-31), as a result of the d-c analysis, we see that this is equal to the direct collector current I_C, assuming that the collector electron current is negligible. Furthermore, I_C may be expressed as $\alpha_o I_E$, where α_o is the low-frequency alpha and I_E is the direct emitter bias current.

Second, let us consider the other term common to v_e in Eqs. (13-58) and (13-59), which is

$$\frac{qAD_{pb}p_{nb}}{L_{pb}} \epsilon^{qV_E/kT} \tag{13-61}$$

From Eq. (8-30), the total direct emitter current can be arranged as follows:

$$I_E = \frac{qAD_{pb}p_{nb}}{L_{pb}} (\epsilon^{qV_E/kT} - 1) \left(\coth \frac{W}{L_{pb}} + \frac{D_{ne}n_{pe}L_{pb}}{D_{pb}p_{nb}L_{ne}} \right) + \frac{qAD_{pb}p_{nb}}{L_{pb}} \operatorname{csch} \frac{W}{L_{pb}} \tag{13-62}$$

By definition, if the emitter junction is reverse-biased so that $\epsilon^{-qV_E/kT} \approx 0$, then Eq. (13-62) becomes I_{ECS}, the emitter saturation current which flows if both junctions are reversed-biased.[2] In other words, (13-62) can be written

$$I_E = \frac{qAD_{pb}p_{nb}\epsilon^{qV_E/kT}}{L_{pb}} \left(\coth \frac{W}{L_{pb}} + \frac{D_{ne}n_{pe}L_{pb}}{D_{pb}p_{nb}L_{ne}} \right) - I_{ECS} \tag{13-63}$$

Usually, I_{ECS} is very small compared to I_E and may be omitted. Therefore, we get

$$\frac{qAD_{pb}p_{nb}\epsilon^{qV_E/kT}}{L_{pb}} = \frac{I_E}{\coth (W/L_{pb}) + D_{ne}n_{pe}L_{pb}/D_{pb}p_{nb}L_{ne}} \tag{13-64}$$

$$= \frac{I_E \tanh (W/L_{pb})}{1 + (D_{ne}n_{pe}L_{pb}/D_{pb}p_{nb}L_{ne}) \tanh (W/L_{pb})} \approx \gamma_o I_E \frac{W}{L_{pb}} \tag{13-65}$$

where it is recognized that the denominator of Eq. (13-65) is simply the low-frequency emitter efficiency denoted by γ_o.

After application of all the aforementioned modifications to the current equations, the simplified admittance equations become

$$i_e \approx \left[\frac{qI_E\gamma_o ZW}{kTL_{pb}} \left(\coth \frac{ZW}{L_{pb}} + \frac{n_{pe}D_{ne}Z'L_{pb}}{p_{nb}D_{pb}ZL_{ne}} \right) + j\omega C_{Te} \right] v_e - \left(\frac{\partial W}{\partial V_C} \frac{Z\alpha_o I_E}{L_{pb}} \operatorname{csch} \frac{ZW}{L_{pb}} \right) v_c \tag{13-66}$$

$$i_c \approx - \left(\frac{qI_E\gamma_o ZW}{kTL_{pb}} \operatorname{csch} \frac{ZW}{L_{pb}} \right) v_e + \left(\frac{\partial W}{\partial V_C} \frac{Z\gamma_o I_E}{L_{pb}} \coth \frac{ZW}{L_{pb}} + j\omega C_{Tc} \right) v_c \tag{13-67}$$

The y parameters are as follows:

$$y_{11} = \frac{\gamma_o ZW}{r_e L_{pb}} \left(\coth \frac{ZW}{L_{pb}} + \frac{n_{pe} D_{ne} Z' L_{pb}}{p_{nb} D_{pb} Z L_{ne}} \right) + j\omega C_{Te} \tag{13-68}$$

$$y_{12} = - \frac{\partial W}{\partial V_C} \alpha_o I_E \frac{Z}{L_{pb}} \operatorname{csch} \frac{ZW}{L_{pb}} \tag{13-69}$$

$$y_{21} = - \frac{\gamma_o ZW}{r_e L_{pb}} \operatorname{csch} \frac{ZW}{L_{pb}} \tag{13-70}$$

$$y_{22} = + \frac{\partial W}{\partial V_C} \alpha_o I_E \frac{Z}{L_{pb}} \coth \frac{ZW}{L_{pb}} + j\omega C_{Tc} \tag{13-71}$$

In Eqs. (13-68) and (13-70) above, the substitution $r_e = kT/qI_E$ was made. Because of the electron term appearing in Eq. (13-68), it becomes convenient to express y_{11} in terms of the frequency-dependent emitter efficiency γ, which may be extracted by rearranging (13-68) as follows:

$$y_{11} = \frac{\gamma_o ZW}{r_e L_{pb}} \left(\coth \frac{ZW}{L_{pb}} + \frac{n_{pe} D_{ne} Z' L_{pb}}{p_{nb} D_{pb} Z L_{ne}} + \frac{j\omega C_{Te} r_e L_{pb}}{\gamma_o ZW} \right)$$

Factoring out a hyperbolic tangent and rearranging, we get

$$y_{11} = \frac{\gamma_o}{r_e} \frac{ZW/L_{pb}}{\tanh (ZW/L_{pb})} \left[1 + \left(\frac{n_{pe} D_{ne} Z' W}{p_{nb} D_{pb} L_{ne}} + \frac{j\omega C_{Te} r_e}{\gamma_o} \right) \frac{\tanh (ZW/L_{pb})}{ZW/L_{pb}} \right] \tag{13-72}$$

where the bracket term is the *frequency-dependent emitter efficiency*, i.e.,

$$\gamma = \frac{1}{1 + \left(\dfrac{n_{pe} D_{ne} Z' W}{p_{nb} D_{pb} L_{ne}} + \dfrac{j\omega C_{Te} r_e}{\gamma_o} \right) \dfrac{\tanh (ZW/L_{pb})}{ZW/L_{pb}}} \tag{13-73}$$

Using the same form, e.g., $(\tanh x)/x$, for the other admittance expressions, we obtain, finally, the a-c admittance equations.

$$y_{11} = \frac{1}{r_e} \frac{ZW/L_{pb}}{\tanh (ZW/L_{pb})} \frac{\gamma_o}{\gamma} \tag{13-74}$$

$$y_{12} = - \frac{\partial W}{\partial V_C} \frac{\alpha_o I_E}{W} \frac{ZW/L_{pb}}{\sinh (ZW/L_{pb})} \tag{13-75}$$

$$y_{21} = \frac{-\gamma_o}{r_e} \frac{ZW/L_{pb}}{\sinh (ZW/L_{pb})} \tag{13-76}$$

$$y_{22} = + \frac{\partial W}{\partial V_C} \frac{\alpha_o I_E}{W} \frac{ZW/L_{pb}}{\tanh (ZW/L_{pb})} + j\omega C_{Tc} \tag{13-77}$$

13-5. Transformation to h Parameters. The admittance expressions [Eqs. (13-74) to (13-77)] could readily be expanded into first-order terms and the results analyzed to establish the frequency dependence and the high-frequency equivalent circuit. However, it would be more convenient to transform or convert the y parameters into the more familiar h parameters and study the latter. From basic network theory,[3] it is easily shown that

$$h_{11} = \frac{1}{y_{11}} \qquad h_{21} = \frac{y_{21}}{y_{11}}$$

$$h_{12} = \frac{-y_{12}}{y_{11}} \qquad h_{22} = y_{22} - \frac{y_{12}y_{21}}{y_{11}}$$

With these relationships, we can make the appropriate substitutions and arrive at each parameter in turn. The short-circuit input impedance h_{11} is simply

$$h_{11} = \frac{1}{y_{11}} = r_e \frac{\tanh\,(ZW/L_{pb})}{ZW/L_{pb}} \frac{\gamma}{\gamma_o} \tag{13-78}$$

where γ is given by Eq. (13-73). The voltage-feedback factor h_{12} becomes

$$h_{12} = \frac{-y_{12}}{y_{11}} = \frac{+\partial W}{\partial V_C} \frac{\alpha_o I_E}{W} \frac{ZW/L_{pb}}{\sinh\,(ZW/L_{pb})} r_e \frac{\tanh\,(ZW/L_{pb})}{ZW/L_{pb}} \frac{\gamma}{\gamma_o}$$

$$h_{12} = \frac{\partial W}{\partial V_C} \frac{kT}{qW} \alpha_o \operatorname{sech}\left(\frac{ZW}{L_{pb}}\right) \frac{\gamma}{\gamma_o} \tag{13-79}$$

The current-transfer ratio h_{21} is

$$h_{21} = \frac{y_{21}}{y_{11}} = \frac{-\gamma_o}{r_e} \frac{ZW/L_{pb}}{\sinh\,(ZW/L_{pb})} r_e \frac{\tanh\,(ZW/L_{pb})}{ZW/L_{pb}} \frac{\gamma}{\gamma_o}$$

$$h_{21} = -\gamma \operatorname{sech} \frac{ZW}{L_{pb}} \tag{13-80}$$

Note that this result is identical with the general expression for alpha obtained in Chap. 8, where it was assumed that the emitter efficiency was unity and independent of frequency. This complete analysis, of course, brings in the γ term.

The last h parameter, viz., the output admittance h_{22}, may be expressed by the following transformation:

$$h_{22} = y_{22} - y_{12}h_{21}$$

When the appropriate substitutions are made, we obtain

$$h_{22} = j\omega C_{Tc} + \frac{\alpha_o I_E}{W} \frac{\partial W}{\partial V_C} \left[\frac{ZW/L_{pb}}{\tanh\,(ZW/L_{pb})} - \frac{(ZW/L_{pb})\,\gamma \operatorname{sech}\,(ZW/L_{pb})}{\sinh\,(ZW/L_{pb})} \right] \tag{13-81}$$

where γ is the high-frequency emitter efficiency given by Eq. (13-73). The bracketed term may be manipulated further, such that

$$h_{22} = j\omega C_{Tc} + \frac{\alpha_o I_E}{W} \frac{\partial W}{\partial V_C} \left[\frac{ZW}{L_{pb}} \tanh \frac{ZW}{L_{pb}} + \frac{ZW}{L_{pb}} \frac{\operatorname{sech}^2\,(ZW/L_{pb})(1 - \gamma)}{\tanh\,(ZW/L_{pb})} \right] \tag{13-82}$$

Substituting Eq. (13-73) into (13-82) and simplifying the result, we obtain

$$h_{22} = j\omega C_{Tc} + \frac{\alpha_o I_E}{W}\frac{\partial W}{\partial V_C}\left[\frac{(ZW/L_{pb})^2 \tanh (ZW/L_{pb})}{ZW/L_{pb}} + \gamma \operatorname{sech}^2\left(\frac{ZW}{L_{pb}}\right)\frac{D_{ne}Z_1 n_{pe}}{D_{pb}L_{ne}p_{nb}} + j\omega\frac{C_{Te}r_\epsilon}{\gamma_o}\operatorname{sech}^2\frac{ZW}{L_{pb}}\right] \quad (13\text{-}83)$$

From Chap. 8, we recall that $\beta^* = \operatorname{sech}(ZW/L_{pb})$ and that the low-frequency emitter efficiency γ_o is approximately

$$\gamma_o = \frac{1}{1 + D_{ne}n_{pe}W/D_{pb}p_{nb}L_{ne}}$$

or

$$\frac{D_{ne}n_{pe}W}{D_{pb}p_{nb}L_{ne}} = \frac{1-\gamma_o}{\gamma_o}$$

Inserting these results into (13-83) and neglecting the last term (containing C_{Te}) which is usually quite small, we get, finally

$$h_{22} = j\omega C_{Tc} + \frac{\alpha_o I_E}{W}\frac{\partial W}{\partial V_C}\left(\frac{ZW}{L_{pb}}\right)^2\frac{\tanh (ZW/L_{pb})}{ZW/L_{pb}} + \frac{\alpha_o I_E}{W}\frac{\partial W}{\partial V_C}\frac{\gamma}{\gamma_o}(\beta^*)^2(1-\gamma_o)Z' \quad (13\text{-}84)$$

13-6. Dependence on Base-cutoff Frequency and Base Resistance. In summary, Eqs. (13-78) to (13-80) and (13-84) are the general expressions for the small-signal h parameters of the junction transistor as a function of frequency. Note that in each case the frequency dependence arises from a hyperbolic function of ZW/L_{pb} where $Z = (1 + j\omega\tau_{pb})^{1/2}$. In exactly the same manner as for the frequency variation of alpha in Chap. 8, it may be shown that[4]

$$\frac{W}{L_{pb}}(1 + j\omega\tau_{pb})^{1/2} \approx (j\omega t_b)^{1/2}$$

where t_b is the base-transit time, equal to $\tau_{pb}W^2/L_{pb}^2$. Furthermore, the base-cutoff frequency ω_b was defined as

$$\omega_b t_b = 2.43$$

at the half-power point. Therefore, we may write

$$\frac{W}{L_{pb}}(1 + j\omega\tau_{pb})^{1/2} \approx \left(j2.43\frac{\omega}{\omega_b}\right)^{1/2} \quad (13\text{-}85)$$

which is a good approximation for frequencies up to $2\omega_b$. This is a convenient form to use since it enables all the transistor parameters to be referred to the base-cutoff frequency. Applying this approximation to the h parameter equations, we obtain, finally,

$$h_{11} = r_e \frac{\tanh (j2.43\omega/\omega_b)^{1/2}}{(j2.43\omega/\omega_b)^{1/2}} \frac{\gamma}{\gamma_o} \tag{13-86}$$

$$h_{12} = \frac{kT}{qW} \frac{\partial W}{\partial V_C} \alpha_o \operatorname{sech} \left(j \frac{2.43\omega}{\omega_b} \right)^{1/2} \frac{\gamma}{\gamma_o} \tag{13-87}$$

$$h_{21} = -\gamma \operatorname{sech} \left(j \frac{2.43\omega}{\omega_b} \right)^{1/2} \tag{13-88}$$

$$h_{22} = j\omega C_{Tc} + \frac{\alpha_o I_E}{W} \frac{\partial W}{\partial V_C} \frac{W^2(1 + j\omega\tau_{pb})}{L_{pb}^2} \frac{\tanh (j2.43\omega/\omega_b)^{1/2}}{(j2.43\omega/\omega_b)^{1/2}} + \frac{\alpha_o I_E}{W} \frac{\partial W}{\partial V_C} \frac{\gamma}{\gamma_o} (\beta^*)^2(1 - \gamma_o)(1 + j\omega\tau_{ne})^{1/2} \tag{13-89}$$

where γ, the emitter efficiency, is given by

$$\gamma = \frac{1}{1 + \left[\left(\frac{D_{ne} n_{pe} W}{D_{pb} p_{nb} L_{ne}} \right) (1 + j\omega\tau_{ne})^{1/2} + \frac{j\omega\, C_{Te} r_e}{\gamma_o} \right] \frac{\tanh (j2.43\omega/\omega_b)^{1/2}}{(j2.43\omega/\omega_b)^{1/2}}} \tag{13-90}$$

It must be emphasized that, up to this point, the frequency analysis and derivation of the h parameters neglected the effects of the base resistance r'_B. It was for this reason that the numerical subscripts were used for the h parameters. From a practical point of view, the theoretical junction-transistor model must include r'_B in order to properly correlate the theory with actual devices. The manner of incorporating r'_B may best be understood by referring to Fig. 13-2. Here, Eqs. (13-86) to (13-90) for the h parameters completely describe the frequency behavior of the theoretical model. Base resistance can be factored in as shown, and its effect is determined by obtaining a new set of h parameters for the new four-terminal network of Fig. 13-2. It has been shown that[5]

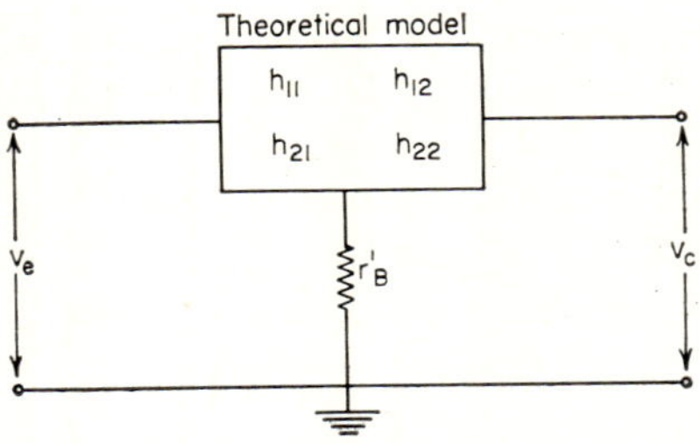

FIG. 13-2. Addition of base resistance to theoretical model. (*After R. L. Pritchard.*)

$$h_{ib} = h_{11} + \frac{r'_B(1 + h_{21})(1 - h_{12})}{1 + r'_B h_{22}} \tag{13-91}$$

$$h_{rb} = \frac{h_{12} + r'_B h_{22}}{1 + r'_B h_{22}} \tag{13-92}$$

$$h_{fb} = \frac{h_{21} - r'_B h_{22}}{1 + r'_B h_{22}} \tag{13-93}$$

$$h_{ob} = \frac{h_{22}}{1 + r'_B h_{22}} \tag{13-94}$$

In other words, the new h parameters are expressed in terms of the original parameters and r'_B. Note that, in each case, the significant term is $r'_B h_{22}$. In the sections to follow, the frequency dependence of each of the original parameters will be discussed first. Then the effect of r'_B will be included by substitution into the appropriate equation above. Finally, as far as this chapter is concerned, it will be assumed that r'_B is a constant resistance independent of frequency.

13-7. Frequency Variation of h_{fb}. Equation (13-88) is the expression for the current-amplification factor,

$$h_{21} = -\gamma \operatorname{sech} \left(\frac{j2.43\omega}{\omega_b} \right)^{1/2} \tag{13-95}$$

which is seen to be the product of the emitter efficiency γ and the base-transport factor β^*, the latter being the hyperbolic secant term. We recall from Chap. 8 that when the applied signal frequency ω is equal to the base-cutoff frequency ω_b, the magnitude of sech $(j2.43)^{1/2}$ is equal to 0.707. From a high-frequency design standpoint, the h_{21} response should be as close to unity as possible over the widest frequency range, requiring that the base-cutoff frequency be as large as possible. This, of course, is directly related to the minority-carrier mobility of the base region and the nature of the impurity profile, and is inversely related to the square of the base thickness. The response characteristic of the base-transport term of h_{21} is shown in Fig. 13-3, where it is assumed that the low-frequency $\beta_o^* = 1$. It is seen that the response is flat up to 0.1 ω_b and then decreases at a rate of 6 db per octave. Actually, the hyperbolic functions can be readily expanded into their equivalent series to make it easier to understand the high-frequency behavior of these complex relationships. For example,

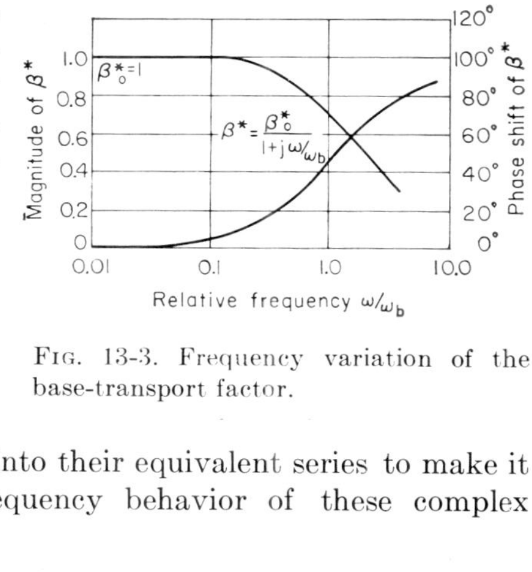

FIG. 13-3. Frequency variation of the base-transport factor.

$$\beta^* = \operatorname{sech} \left(\frac{j2.43\omega}{\omega_b} \right)^{1/2} \approx 1 - 1.2 \left(\frac{\omega}{\omega_b} \right)^2 - j1.2 \left(\frac{\omega}{\omega_b} \right) \tag{13-96}$$

If we let the coefficients in Eq. (13-96) equal approximately 1 and multiply and divide by $(1 + j\omega/\omega_b)$, we get the familiar form for the base frequency response, viz.,

$$\beta^* \approx \frac{\beta_o^*}{1 + j\omega/\omega_b} \tag{13-97}$$

Thus, when $\omega = \omega_b$, $\beta^* = \beta_o^*/\sqrt{2}$, which is the definition of base cutoff. A plot of the phase-shift angle associated with Eq. (13-97) is shown in Fig. 13-3. For this particular result, the phase shift at base cutoff is 45° and begins to level off to 90° at the higher frequencies.

In Eq. (13-95), if the emitter efficiency γ is unity for all frequencies, then the h_{21} response is primarily identical to that of Fig. 13-3. However, the frequency variation of γ is given by

$$\gamma = \frac{1}{1 + \left[\left(\dfrac{D_{ne}n_{pe}W}{D_{pb}p_{nb}L_{ne}}\right)(1 + j\omega\tau_{ne})^{1/2} + \dfrac{j\omega C_{Te}r_e}{\gamma_o}\right]\dfrac{\tanh\,(j2.43\omega/\omega_b)}{(j2.43\omega/\omega_b)^{1/2}}} \tag{13-98}$$

which is a rather complex expression to analyze directly. Nevertheless, an excellent interpretation of Eq. (13-98) can be made by considering the terms independently. Suppose, for example, that the emitter barrier capacitance C_{Te} is sufficiently small so that its effect on γ is negligible. Then (13-98) simply becomes

$$\gamma \approx \frac{1}{1 + \left[\left(\dfrac{D_{ne}n_{pe}W}{D_{pb}p_{nb}L_{ne}}\right)(1 + j\omega\tau_{ne})^{1/2}\right]\dfrac{\tanh\,(j2.43\omega/\omega_b)^{1/2}}{(j2.43\omega/\omega_b)^{1/2}}} \tag{13-99}$$

γ is limited now by the diffusion-admittance term $(1 + j\omega\tau_{ne})^{1/2}$, where τ_{ne} is the lifetime of electrons in the emitter (for p-n-p). Inspection of the frequency terms of Eq. (13-99) should indicate the manner in which it varies with frequency. Essentially, the γ response is somewhat flat, since the increase in magnitude of $(1 + j\omega\tau_{ne})^{1/2}$ with frequency is counteracted by the decrease of

$$\frac{\tanh\,(j2.43\omega/\omega_b)^{1/2}}{(j2.43\omega/\omega_b)^{1/2}}$$

Consequently, the high-frequency value of γ will remain close to unity, indicating that the magnitude of γ in (13-99) is somewhat independent of frequency. This concept may be applied to Eq. (13-98) by the fair assumption that

$$(1 + j\omega\tau_{ne})^{1/2}\frac{\tanh\,(j2.43\omega/\omega_b)^{1/2}}{(j2.43\omega/\omega_b)^{1/2}} \approx 1 \tag{13-100}$$

Since $D_{ne}n_{pe}W/D_{pb}p_{nb}L_{ne}$ equals $(1 - \gamma_o)/\gamma_o$, where γ_o is the low-frequency emitter efficiency, Eq. (13-98) simply becomes

$$\gamma \approx \frac{\gamma_o}{1 + j\omega r_e C_{Te}\dfrac{\tanh\,(j2.43\omega/\omega_b)^{1/2}}{(j2.43\omega/\omega_b)^{1/2}}} \tag{13-101}$$

If the hyperbolic tangent function is expanded into a series and the second-order terms are neglected, we have

$$\gamma \approx \frac{\gamma_o}{1 + j\omega r_e C_{Te}} \tag{13-102}$$

Thus, the frequency dependence of γ is merely a function of $j\omega r_e C_{Te}$, which is the product of the emitter transition capacitance and the a-c emitter resistance. Essentially, this term is a measure of the reactive emitter current flowing through the emitter junction capacitance. As the frequency increases, this causes an increase in the total emitter current without producing a corresponding increase in collector current, thereby decreasing $|\gamma|$, and hence h_{21}. Since the emitter junction is

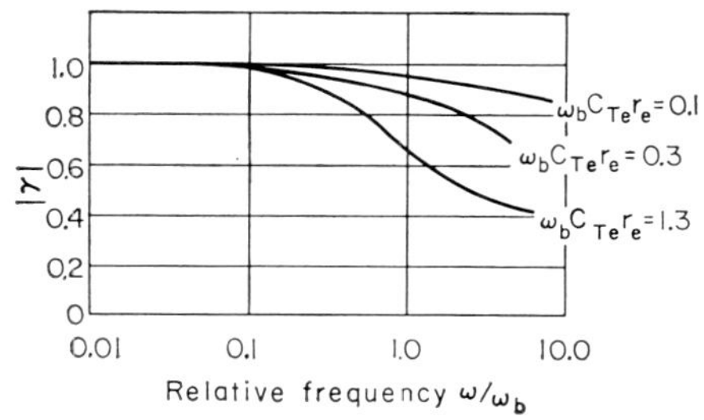

FIG. 13-4. Frequency variation of $|\gamma|$ for $\gamma_o = 1$, illustrating effect of emitter capacitance. (*After R. L. Pritchard.*)

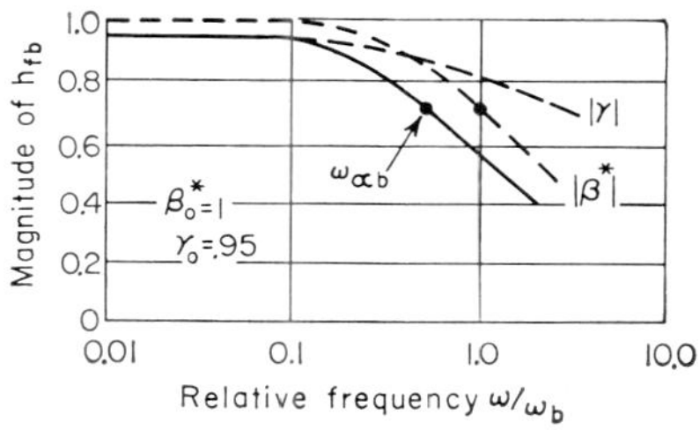

FIG. 13-5. Effect of emitter efficiency γ on frequency variation of h_{fb}.

forward-biased, C_{Te} may be large enough to reduce the over-all frequency response of the transistor. It is for this reason that very-high-frequency transistors require very small active emitter-junction areas. A plot of Eq. (13-102) as a function of relative frequency for different values of $\omega_b C_{Te} r_e$ appears in Fig. 13-4.[2] It should be noted that for a given transistor, the product can be reduced by increasing the emitter bias current I_E, since r_e is equal to kT/qI_E.

We may now combine the emitter efficiency response given by Eq. (13-102) with the base-transport response given by Eq. (13-97) and obtain the approximate over-all frequency response for alpha ($\alpha = \gamma\beta^*$).

$$h_{21} \approx \frac{-\alpha_o}{(1 + j\omega r_e C_{Te})(1 + j\omega/\omega_b)} \tag{13-103}$$

This equation is plotted graphically in Fig. 13-5, wherein an arbitrary curve for γ (limited by C_{Te}) is indicated. The resultant curve is a point-by-point product of γ and β^* as a function of frequency. It is seen that the effect of γ is to impart a gradual falloff of alpha at the lower frequencies. Furthermore, the observed cutoff frequency is reduced from the initial base-cutoff value ω_b to a lower value, $\omega_{\alpha b}$. The latter is called the *alpha-cutoff frequency.*

To consider the effect of the base resistance r'_B on the frequency response of the current-amplification factor h_{21}, we can apply Eq. (13-93), which is repeated here.

$$h_{fb} = \frac{h_{21} - r'_B h_{22}}{1 + r'_B h_{22}} \tag{13-104}$$

At the higher frequencies near alpha cutoff, h_{fb} or $-\alpha$ becomes complex, having a positive reactive component. Within this frequency range, it is most probable that h_{22} is capacitive reactive, so that $r'_B h_{22}$ becomes a positive reactive component. If r'_B is large enough so that $r'_B h_{22} > 0.1$ near $\omega_{\alpha b}$, then $r'_B h_{22}$ subtracts from h_{21}, thereby rapidly reducing the magnitude of h_{fb} (alpha) within the vicinity of alpha cutoff.[2] Therefore, the effect of the base resistance r'_B is to cause h_{fb} to fall more rapidly than predicted by the theoretical variation of $\gamma\beta^*$. In fact, if r'_B is large and the frequency is increased past alpha cutoff, then the $r'_B h_{22}$ term becomes predominant in both the numerator and denominator of Eq. (13-104), causing h_{fb} to begin increasing toward unity again. In most well-designed high-frequency transistors, the product $r'_B h_{22}$ is small compared to unity, and

$$h_{fb} \approx h_{21} \qquad r'_B h_{22} \ll 1 \tag{13-105}$$

We can then write

$$h_{fb} \approx \frac{-\alpha_o}{(1 + j\omega r_e C_{Te})(1 + j\omega/\omega_b)} \tag{13-106}$$

wherein the frequency variation of h_{fb} is limited by emitter-junction time constant and base-transit time. As an additional approximation, it is readily shown from Eq. (13-106) that the alpha-cutoff frequency is

$$\frac{1}{\omega_{\alpha b}} \approx \frac{1}{\omega_b} + r_e C_{Te} \qquad \omega_{\alpha b} = 2\pi f_{\alpha b} \tag{13-107}$$

In Chap. 14 we shall examine the behavior of $\omega_{\alpha b}$ more carefully; it will be shown that the actual $\omega_{\alpha b}$ is lower than that predicted by Eq. (13-107) because of other transistor effects that were not included in this discussion.

13-8. Frequency Variation of h_{ib}. From Eqs. (13-78) and/or (13-86), the general expression for the short-circuit input impedance of the junction transistor is seen to be

$$h_{11} = r_e \frac{\tanh (j2.43\omega/\omega_b)^{1/2}}{(j2.43\omega/\omega_b)^{1/2}} \frac{\gamma}{\gamma_o} \tag{13-108}$$

where $r_e = kT/qI_E$. If we assume that $\gamma_o = 1$ and that γ behaves according to Eq. (13-101), we have

$$h_{11} = r_e \frac{\tanh\,(j2.43\omega/\omega_b)^{1/2}}{(j2.43\omega/\omega_b)^{1/2}} \frac{1}{1 + j\omega r_e C_{Te} \tanh\,(j2.43\omega/\omega_b)^{1/2}/(j2.43\omega/\omega_b)^{1/2}} \tag{13-109}$$

If we apply the first-order series expansion to the hyperbolic functions and simplify, we have

$$h_{11} = \frac{r_e(1 - j\omega/\omega_b)}{1 + j\omega r_e C_{Te}(1 - j\omega/\omega_b)} \tag{13-110}$$

Converting this impedance expression to its equivalent admittance by taking the reciprocal and multiplying by the complex conjugate, we obtain, approximately,

$$\frac{1}{h_{11}} \approx \frac{1}{r_e} + j\left(\frac{\omega}{r_e\omega_b} + \omega C_{Te}\right) \tag{13-111}$$

Thus we see that the input admittance consists of a conductance $1/r_e$ in parallel with two susceptances which are capacitive. These are the emitter capacitance C_{Te} and the so-called *emitter diffusion capacitance*, which is equal to

$$C_{De} \approx \frac{1}{r_e\omega_b} \tag{13-112}$$

The complete equivalent circuit represented by Eq. (13-111) would appear as in Fig. 13-6.

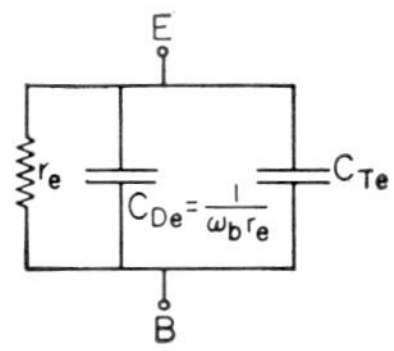

FIG. 13-6. Approximate equivalent circuit for h_{11}, illustrating emitter diffusion capacitance.

To consider the effect of base resistance on h_{11}, we must analyze Eq. (13-91), which is

$$h_{ib} = h_{11} + \frac{r'_B(1 + h_{21})(1 - h_{12})}{1 + r'_B h_{22}} \tag{13-113}$$

For frequencies of ω_b or less, we can assume that both h_{12} and $r'_B h_{22}$ are small compared to unity. Also, since h_{21} is $-\alpha$, Eq. (13-113) becomes, simply,

$$h_{ib} \approx h_{11} + r'_B(1 - \alpha) \tag{13-114}$$

This result is similar in form to the low-frequency expression obtained for h_{ib} in Chap. 12, except that (13-114) contains the reactive terms arising from the frequency variation.

If we assume that I_E is large enough to minimize the effect of $r_e C_{Te}$ on h_{11} and α, Eq. (13-114) can be written as

$$h_{ib} \approx r_e\left(1 - \frac{j\omega}{\omega_b}\right) + r'_B\left(1 - \frac{1}{1 + j\omega/\omega_b}\right) \tag{13-115}$$

or, after additional manipulation,

$$h_{ib} \approx r_e + r'_B \left(\frac{\omega}{\omega_b}\right)^2 + j\left(r'_B \frac{\omega}{\omega_b} - r_e \frac{\omega}{\omega_b}\right) \tag{13-116}$$

Since r'_B is greater than r_e, we observe from Eq. (13-116) that h_{ib} is predominantly inductive because of the effect of $r'_B(1 - \alpha)$. Also, the real resistive component will increase with frequency because of $r'_B(\omega/\omega_b)^2$. One will observe that at high frequencies it is possible for input resonances to occur. In summary, the complete expression for h_{ib} is given as

$$h_{ib} \approx \frac{1}{1/r_e + j(\omega/r_e\omega_b + \omega C_{Te})} + r'_B(1 + h_{fb}) \tag{13-117}$$

where $h_{fb} = -\alpha$ as given by Eq. (13-106). Thus we see that h_{ib} increases with frequency primarily because of the decrease in α and its effect on r'_B.

13-9. Frequency Variation of h_{rb}. Equation (13-87) gives the open-circuit voltage-feedback factor for the theoretical model as

$$h_{12} = \left(\frac{kT}{qW}\frac{\partial W}{\partial V_C}\right) \alpha_o \operatorname{sech}\left(\frac{j2.43\omega}{\omega_b}\right)^{1/2} \frac{\gamma}{\gamma_o} \tag{13-118}$$

In this result, we recognize that the first term in parenthesis is the low-frequency feedback factor μ_{ec} and that the remaining multiplier is simply the frequency variation of alpha. In other words, (13-118) becomes

$$h_{12} = \mu_{ec}\alpha \tag{13-119}$$

Thus, h_{12} has the same frequency variation as does α. However, this is of no significance because firstly, μ_{ec} is quite small at very low frequencies and secondly, the base resistance r'_B introduces a considerable effect. For the latter, consider the new h-parameter Eq. (13-92), which is

$$h_{rb} = \frac{h_{12} + r'_B h_{22}}{1 + r'_B h_{22}} \tag{13-120}$$

We will assume that $h_{22} = g_{22} + j\omega C_{22}$, that is, h_{22} has real and reactive admittances. At the lower frequencies, where C_{22} is negligible, (13-120) becomes

$$h_{rb} \approx \mu_{ec} + r'_B g_{22} \tag{13-121}$$

which is identical to the low-frequency results obtained in Chap. 12, where $g_{22} = g_c$. At the higher frequencies, the admittance associated with C_{22} becomes dominant, so that

$$h_{rb} = \mu_{ec} + r'_B g_{22} + j\omega r'_B C_{22} \tag{13-122}$$

or

$$h_{rb} \approx j\omega r'_B C_{22} \tag{13-123}$$

where C_{22} is the combined transition and diffusion capacitance at the collector junction; it will be defined in the next section. Thus, the base resistance r'_B contributes just about all the high-frequency voltage feedback.

13-10. Frequency Variation of h_{ob}. The last, but one of the most important, h parameters concerns the output admittance of the junction transistor. For good high-frequency power gain, it is essential that this be small at high frequencies. Repeating Eq. (13-89), we see that

$$h_{22} = j\omega C_{Tc} + \frac{\alpha_o I_E}{W} \frac{\partial W}{\partial V_C} \frac{W^2(1 + j\omega\tau_{pb})}{L_{pb}^2} \frac{\tanh (j2.43\omega/\omega_b)^{1/2}}{(j2.43\omega/\omega_b)^{1/2}} + \frac{\alpha_o I_E}{W} \frac{\partial W}{\partial V_C} \frac{\gamma}{\gamma_o} (\beta^*)^2(1 - \gamma_o)(1 + j\omega\tau_{ne})^{1/2} \tag{13-124}$$

which, at this point, is too complicated to analyze conveniently. It would be desirable to transform (13-124) into a form such as

$$h_{22} = g_{22} + j\omega C_{22} \tag{13-125}$$

wherein the frequency variation of the real and reactive components can be examined independently.

In the derivation of the output conductance g_c of Chap. 11, we saw that

$$g_c \approx \frac{\alpha_o I_E}{W} \frac{\partial W}{\partial V_C} [2(1 - \beta_o) + (1 - \gamma_o)] \tag{13-126}$$

Actually, α_o had been set equal to unity in the initial derivation, but it will be retained here. Suppose that the low-frequency emitter efficiency γ_o is equal to unity; (13-126) becomes, after substituting W^2/L_{pb}^2 for $2(1 - \beta_o)$,

$$g_c \approx \frac{\alpha_o I_E}{W} \frac{\partial W}{\partial V_C} \frac{W^2}{L_{pb}^2} \tag{13-127}$$

If we assume in Eq. (13-124) that $\gamma_o = 1$, the third term drops out. After substituting (13-127) into the second term of (13-124), we have, simply,

$$h_{22} \approx j\omega C_{Tc} + \left(g_c + j\omega \frac{\alpha_o I_E W \tau_{pb}}{L_{pb}^2} \frac{\partial W}{\partial V_C}\right) \frac{\tanh (j2.43\omega/\omega_b)^{1/2}}{(j2.43\omega/\omega_b)^{1/2}} \tag{13-128}$$

Substituting $1/D_{pb}$ for τ_{pb}/L_{pb}^2 in the reactive term above, we see that the result is equivalent to the collector diffusion capacitance determined in Chap. 11 from base-region widening effects.

$$C_{Dc} = \frac{\alpha_o I_E W}{D_{pb}} \frac{\partial W}{\partial V_C} \tag{13-129}$$

Therefore, we have

$$h_{22} = j\omega C_{Tc} + (g_c + j\omega C_{Dc}) \frac{\tanh (j2.43\omega/\omega_b)^{1/2}}{(j2.43\omega/\omega_b)^{1/2}} \tag{13-130}$$

We see that as a result of the frequency analysis of the diffusion process, both the output conductance and diffusion capacitance are related to frequency by the hyperbolic tangent function. This function may be expanded into its power series as was done for h_{ib}, yielding

$$h_{22} \approx j\omega C_{Tc} + (g_c + j\omega C_{Dc})\left(1 - j\frac{\omega}{\omega_b}\right) \tag{13-131}$$

Multiplying the complex product of Eq. (13-131) and neglecting the term $-jg_c(\omega/\omega_b)$ which would normally be insignificant, we get, finally,

$$h_{22} \approx g_c + \omega_b C_{Dc}\left(\frac{\omega}{\omega_b}\right)^2 + j\omega(C_{Dc} + C_{Tc}) \tag{13-132}$$

Strictly speaking, the next term of the series expansion should be included to show that the diffusion capacitance $j\omega C_{Dc}$ is not independent of fre-

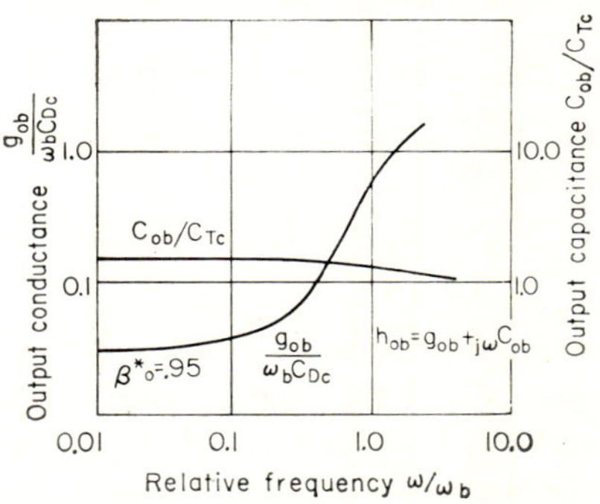

FIG. 13-7. Frequency variation of h_{ob} for $\gamma/\gamma_0 = 1$. (*After R. L. Pritcnard.*)

quency, but does decrease in magnitude at about the square root of frequency near ω_b. However, we recall that C_{Dc} is dependent upon I_E or, more explicitly, the injection level, and since the theoretical treatment is based on small-signal injection, C_{Dc} becomes small compared to C_{Tc}.

From Eqs. (13-125) and (13-132), there result, finally,

$$g_{22} \approx g_c + \omega_b C_{Dc}\left(\frac{\omega}{\omega_b}\right)^2 \tag{13-133}$$

$$C_{22} \approx C_{Dc} + C_{Tc} \tag{13-134}$$

The important result to note is the conductive term in (13-133) which is introduced from the complex diffusion admittance. This term causes the output conductance g_{22} to increase as the square of the frequency. The relative-frequency variations of g_{22} and C_{22} are plotted in Fig. 13-7, where g_{22} and C_{22} are normalized by $\omega_b C_{22}$ and C_{Te} respectively.[2] Thus, at high frequencies, the effect of g_c becomes negligible compared to the diffusion conductance. In Fig. 13-7, it is also shown that the effective output capacitance decreases from its low-frequency value $C_{Dc} + C_{Tc}$ to its ultimate value of C_{Tc}. One should also note that the higher the alpha, the smaller g_c becomes, thereby making the frequency-dependent term in (13-133) more predominant at lower frequencies.

To include the effect of base resistance on h_{22}, we refer to Eq. (13-94), which is

$$h_{ob} = \frac{h_{22}}{1 + r'_B h_{22}} \tag{13-135}$$

Let us presume that $h_{22} \approx j\omega C_{Tc}$; Eq. (13-135) becomes

$$h_{ob} = \frac{j\omega C_{Tc}}{1 + j\omega r'_B C_{Tc}} \tag{13-136}$$

Conjugate multiplying (13-136) and letting $\omega^2 r'_B C_{Tc} \ll 1$, we get

$$h_{ob} \approx r'_B \omega^2 C_{Tc}{}^2 + j\omega C_{Tc} \tag{13-137}$$

where we see that the effect of r'_B is to increase the high-frequency output conductance of the transistor by a very slight amount. If we neglect this effect, we have $h_{ob} \approx h_{22}$, or

$$h_{ob} \approx g_c + \omega_b C_{Dc} \left(\frac{\omega}{\omega_b}\right)^2 + j\omega(C_{Tc} + C_{Dc}) \tag{13-138}$$

13-11. High-frequency Equivalent Circuit. To properly conclude this chapter, it is desirable to review briefly what has been accomplished. By including small-signal a-c variations in the one-dimensional analysis of our simple junction-transistor model, we have been able to come up with the basic y-parameter or admittance equations which relate the currents to the applied potentials. Since it was assumed that the injection level was low, so that the carriers move principally by diffusion, the solutions obtained were the so-called diffusion admittances, which are characterized by hyperbolic functions of complex arguments. After some basic simplifications of the admittance equations, the results were converted to the equivalent h parameters. Finally, the effect of base resistance was included to yield (approximately) the complete frequency-dependent h parameters for the grounded-base transistor, valid for frequencies up to ω_b. These are summarized below as:

$$h_{fb} \approx \frac{-\alpha_o}{(1 + j\omega r_e C_{Te})(1 + j\omega/\omega_b)} \tag{13-139}$$

$$h_{ib} \approx \frac{1}{1/r_e + j(\omega/r_e\omega_b + \omega C_{Te})} + r'_B(1 + h_{fb}) \tag{13-140}$$

$$h_{rb} \approx j\omega r'_B(C_{Tc} + C_{Dc}) \tag{13-141}$$

$$h_{ob} \approx g_c + \omega_b C_{Dc} \left(\frac{\omega}{\omega_b}\right)^2 + j\omega(C_{Tc} + C_{Dc}) \tag{13-142}$$

The reader will note that considerable emphasis was placed on making suitable approximations in order to obtain relatively simple expressions. In spite of this, the errors involved may be no more than 20 per cent in the frequency range up to the base-cutoff frequency. Beyond ω_b, other more exact solutions can be used.[8] Nevertheless, Eqs. (13-139) to (13-142) are adequate to enable one to predict the frequency variations of grounded-base h parameters fairly well, as evidenced by Pritchard's analysis[2] presented in this chapter. Actually, an exact analysis would involve the study of transmission lines, since the complex diffusion admittances are similar to those obtained for an R-C transmission line.

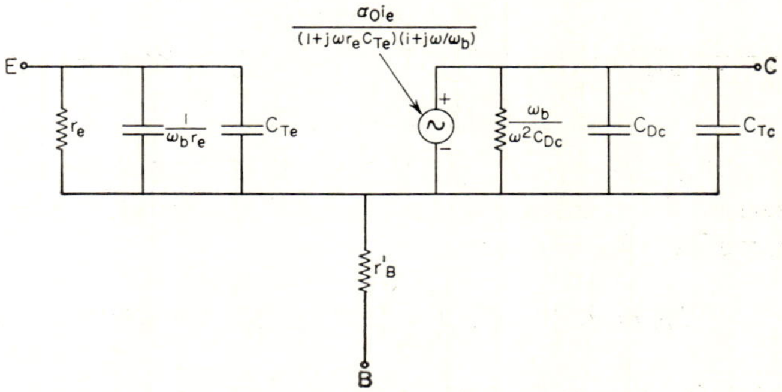

FIG. 13-8. Approximate high-frequency equivalent circuit for the grounded-base transistor.

The employment of such a complicated approach would be of more academic interest than practical value in understanding the transistor.

It is quite useful to represent the characteristics of the junction transistor at high frequencies by an equivalent circuit. Unfortunately, this is not a simple matter at high frequencies, since a circuit which fully describes the device physics is usually too unwieldy for the circuit designer. Because of this, many approximate equivalent circuits have been proposed. Some writers have utilized transmission lines[6,7,9] and others have used lumped hybrid[10] and admittance forms.[11] Rather than present all possible ramifications,[12] this writer shall describe an approximate high-frequency equivalent circuit which is based on the relations developed thus far. This is drawn in Fig. 13-8. For this circuit, which is applicable up to ω_b, it is assumed that the emitter efficiency is limited primarily by r_eC_{Te}. At the input, the emitter junction capacitance is shown in parallel with the a-c emitter resistance and the effective emitter diffusion capacitance. At the output, the collector junction capacitance is shown in parallel with the collector diffusion admittance. Both low-frequency parameters, μ_{ec} and g_c, have been omitted from the circuit.

PROBLEMS

13-1. Calculate the alpha-cutoff frequency $f_{\alpha b}$ for a diffused-base n-p-n germanium planar transistor at $V_{CB} = 5$ volts and $I_E = -2$ ma. At these conditions, $W = 0.9\ \mu$ and $\alpha_o = 0.98$. The collector resistivity is 0.18 ohm-cm. The emitter is a diffused circle 2 mils in diameter and may be approximated by a uniform concentration of 10^{20} atoms/cm^3. The impurity concentration in the base side of the emitter junction is 10^{17} atoms/cm^3.

13-2. For the p-n-p epitaxial mesa transistor described in Prob. 9-1, determine the magnitudes of the resistances (real terms only) of the short-circuit input impedance h_{ib} and the reciprocal of the open-circuit output admittance $1/h_{ob}$ at $V_{CB} = -5$ volts, $I_E = 2$ ma, and $f = 160$ megacycles.

REFERENCES

1. Early, J. M.: Design Theory of Junction Transistors, *Bell System Tech. J.*, vol. 32' pp. 1271–1312, November, 1953.
2. Pritchard, R. L.: Frequency Variations of Junction Transistor Parameters, *Proc. IRE*, vol. 42, May, 1954.
3. Guillemin, E. A.: "Communication Networks," vol. II, John Wiley & Sons, Inc., New York, 1935.
4. Pritchard, R. L.: Frequency Variations of Current-amplification Factor for Junction Transistors, *Proc. IRE*, vol. 40, pp. 1476–1481, November, 1952.
5. Pritchard, R. L.: Frequency Response of Grounded-base and Grounded-emitter Junction Transistors, presented at AIEE Winter Meeting, New York, January, 1954.
6. Pritchard, R. L.: Electric-network Representation of Transistors—A Survey, *IRE, Trans.*, vol. CT-3, no. 1, March, 1956.
7. Chow, W. F., and Suran, J. J.: Transient Analysis of Junction Transistor Amplifiers, *Proc. IRE*, vol. 4, pp. 1126–1127, September, 1953.
8. Shea, R. F., et al., "Transistor Circuit Engineering," chap. 2, John Wiley & Sons, Inc., New York, 1957.
9. Zawels, J.: The Natural Equivalent Circuit of Junction Transistors, *RCA Rev.*, vol. 16, p. 366, September, 1955.
10. Giacoletto, L. J.: Study of PNP Alloy Junction Transistor from DC through Medium Frequencies, *RCA Rev.*, vol. 15, pp. 506–562, December, 1954.
11. Middlebrook, R. D.: A Junction Transistor High-frequency Equivalent Circuit, ERL Technical Report 83, Stanford University, May, 1955.
12. Pritchard, R. L., J. B. Angell, R. B. Adler, J. M. Early, and W. M. Webster: Transistor Internal Parameters for Small-signal Representation, *Proc. IRE*, vol. 49, pp. 725–738, April, 1961.

14

Frequency Response of Transistors

14-1. Frequency Variation of Grounded-base Current Gain. One of the important high-frequency parameters of the transistor is the frequency response of the current-gain characteristic. In the previous chapter, it was shown that the grounded-base current gain is given by the expression

$$h_{fb} = -\gamma \operatorname{sech} \left(\frac{j2.43\omega}{\omega_b} \right)^{1/2} \tag{14-1}$$

By carrying out the necessary series expansions and approximations, it was also shown that

$$h_{fb} \approx \frac{-\alpha_o}{(1 + j\omega r_e C_{Te})(1 + j\omega/\omega_b)} \tag{14-2}$$

where we see that the frequency response is limited by the emitter capacitance and the base-cutoff frequency. Furthermore, the alpha-cutoff frequency was approximately defined as

$$\frac{1}{\omega_{\alpha b}} \approx r_e C_{Te} + \frac{1}{\omega_b} \tag{14-3}$$

The reader will observe that in Eq. (14-3), the total frequency response $\omega_{\alpha b}$ is simply the reciprocal of the sum of particular time factors. These factors may be referred to as either time delays or time constants. Regardless of nomenclature, a very useful approach for approximating the frequency response of a complex network is to properly add up all the time constants and/or transit times to obtain the total signal delay. In this chapter, this approach will be utilized to show that the actual alpha-cutoff frequency of a transistor is lower than that predicted by Eq. (14-3), due to additional time delays which have not been previously considered. These are the transit time through the collector depletion layer and the collector time constant. Another correction factor will be incorporated to obtain the correct phase angle for current gain at the cutoff frequency.

Once $f_{\alpha b}$ is properly corrected for both magnitude and phase, we can examine the behavior of the grounded-emitter current gain, beta, as a function of frequency. This analysis yields two additional frequency parameters, viz., the beta-cutoff frequency $f_{\alpha e}$ and the current-gain-bandwidth frequency f_T. The latter is probably the most significant parameter that characterizes the frequency performance of the transistor, since it represents the upper limit of frequency with regard to maximum switching speed or power gain where practical circuit applications are concerned.

In the next four sections, we shall proceed to examine each of the critical time delays that affect the frequency response of the transistor. Each of these will be related to the physical characteristics of the transistor structure, and its variation with emitter current and collector voltage will be explained.

14-2. Emitter Delay-time Constant. If one is to consider each of the time delays in order as the high-frequency signal travels from emitter to collector, the first to be encountered is the emitter delay-time constant. This is defined as

$$\frac{1}{\omega_e} = r_e C_{Te} \tag{14-4}$$

the product of the emitter resistance r_e and the emitter transition capacitance C_{Te}. We have already seen that r_e is the resistance of the emitter junction and is given by

$$r_e = \frac{kT}{qI_E} \tag{14-5}$$

The emitter transition capacitance is generally expressed by the step-junction formula, which uses the impurity concentration just at the base side of the emitter junction for N and the net junction potential under actual forward bias conditions for V. Referring to Eq. (5-82), we have

$$C_{Te} = A_e \sqrt{\frac{q\kappa\epsilon_o N_B}{V_T}} \tag{14-6}$$

where A_e is the emitter area. In this result, it has been assumed that the forward voltage is approximately $V_T/2$. Examining Eqs. (14-5) and (14-6), we see that the time constant $1/\omega_e$ is virtually independent of collector voltage but decreases with increasing emitter current. Thus, $1/\omega_e$ becomes significantly large at low emitter currents, particularly for $I_E < 1$ ma. It is apparent that in order to minimize $1/\omega_e$, the emitter junction capacitance must be made as small as possible by designing for small emitter areas and high base sheet resistance.

14-3. Base-transit Time. The second signal delay that is encountered is related to the time required for the injection carriers to travel across the base to the collector. This, of course, is the base-transit time which has already been thoroughly discussed in Chap. 8. For purposes of review, we recall from Secs. 8-6 and 13-6 that in the equation describing the frequency variation of the base-transport factor,

$$\beta^* = \operatorname{sech}\left(\frac{j2.43\omega}{\omega_b}\right)^{1/2} \tag{14-7}$$

we related the base signal delay to the transit time by the equation

$$\frac{1}{\omega_b} = \frac{W^2}{2.43D_{pb}} \tag{14-8}$$

where W^2/D_{pb} is the base-transit time. In Eq. (14-8), $\omega_b = 2\pi f_b$, where f_b is the base-cutoff frequency. Furthermore, this result applies only to a uniform base. In a graded base, the aiding built-in electric field reduces the transit time, so that we have

$$\frac{1}{\omega_b} = \frac{W^2}{2.43D_{pb} \ln N'_B/N_{BC}} \tag{14-9}$$

where D_{pb} is the diffusion constant associated with N'_B. Thus we see that $1/\omega_b$ or the base time delay is directly proportional to the square of the base width and is therefore quite sensitive to collector voltage variations, particularly if the widening factor $\partial W/\partial V_C$ is large.

The frequency variation of the base-transport factor was given in the last chapter as

$$\beta^* \approx \frac{\beta_o^*}{1 + j\omega/\omega_b} \tag{14-10}$$

where $1/\omega_b$ is given by either Eq. (14-8) or Eq. (14-9). Examining this result, we see that at the cutoff frequency the magnitude of β_o^* is down by $1/\sqrt{2}$ and the phase angle is 45°. Actual measurements of junction transistors show that at the cutoff frequency the phase shift of β^* is not 45° but about 57°.[1,*] This applies to uniform-base transistors. For graded-base transistors, it is observed that the *excess phase* is even greater than 12°. In fact, it was shown by Thomas and Moll[2] that the excess phase is directly dependent on the magnitude of the base built-in field, which, we recall, is related to the steepness of the base impurity gradient, i.e., the N'_B/N_{BC} ratio. Although the theoretical analysis will not be carried out here, it can be shown that Eq. (14-10) must be corrected for the observed excess phase shift as follows:[2]

$$\beta^* \approx \frac{\beta_o^* \epsilon^{j[(K_\theta - 1)/K_\theta]\omega/\omega_b}}{1 + j\omega/\omega_b} \tag{14-11}$$

* References, indicated in the text by superscript figures, are listed at the end of the chapter.

where K_θ is a phase-correction constant which can vary from 0.5 to 1.0. For a uniform base transistor, $K_\theta = 0.82$; for most graded-base transistors having error-function or gaussian-type impurity distributions, $K_\theta \approx 0.7 \pm 0.05$; for those graded-base transistors having steep exponential impurity distributions such as are found in MADT transistors, $K_\theta \approx 0.6 \pm 0.05$. It is interesting to note that a $K_\theta \approx 1$ can be achieved by introducing an impurity gradient (using outdiffusion techniques) which produces a *retarding* field in the base layer. Examining Eq. (14-11), we see that when $K_\theta = 1$ we have no excess phase and the phase shift is 45°. The significance of K_θ in the frequency response of the transistor will be better appreciated in a later section of this chapter.

14-4. Collector Depletion-layer Transit Time.[3] When the injected signal carriers have completed their transit across the base width, they reach the edge of the depletion layer of the reverse-biased collector junction of the transistor. At this point, the carriers experience the strong electric field of the depletion layer and are swept across this region at a drift rate determined by their mobility and the magnitude of $\mathcal{E}$. Thus, we have now to consider a third signal delay which arises from the finite time required for the carriers to travel across the collector depletion layer. This is the collector depletion-layer transit time, denoted by t_d.

In Chap. 4 it was shown that the drift velocity of carriers in the presence of an electric field increases with same until a limiting velocity is reached which is independent of $\mathcal{E}$. Reference to Fig. 4-2 indicates that for germanium the limiting velocity of 6×10^6 cm/sec is reached at field intensities of a little less than 10,000 volts/cm. For silicon, the limiting velocity is approximately 8.5×10^6 cm/sec and is achieved at field intensities slightly higher than 10,000 volts/cm. In most practical junction transistors, the collector and base resistivities are such that the electric field in the collector depletion layer is either on the order of 10,000 volts/cm or considerably more, even at very low collector voltages. Therefore, it is reasonable to assume that in the normal operation of a transistor, the signal carriers travel through the collector depletion layer at their limiting velocities. If the thickness of the depletion layer, regardless of whether the junction is graded or is a step, is x_m, then the total carrier transit time is

$$t_d = \frac{x_m}{v_{SC}} \tag{14-12}$$

where v_{SC} is the scattering-limited velocity of the carriers. The signal-delay time corresponding to this transit time was shown by Early[3] to be equal to $t_d/2$, so that

$$\frac{1}{\omega_d} = \frac{x_m}{2v_{SC}} \tag{14-13}$$

For a given collector voltage, $1/\omega_d$ is minimized by maintaining x_m small. This is accomplished by lowering the resistivity of that side of the collector junction into which most of the depletion layer extends. It should be apparent that $1/\omega_d$ is quite sensitive to the applied collector voltage, since x_m varies anywhere between the square root and cube root of the voltage, depending on the nature of the junction.

14-5. Collector Delay-time Constant. As in the case of the emitter junction, we must also consider the signal delay-time constant associated with the capacitance of the collector junction. The collector capacitance C_{Tc} must be charged through the combined series resistance from emitter to collector. This consists of three major terms, viz., the series resistance of the emitter region r_{SE}, the a-c resistance of the emitter junction r_e, and the series resistance of the collector region r_{SC}. Therefore, the collector delay-time constant is written as[4]

$$\frac{1}{\omega_c} = (r_{SE} + r_e + r_{SC})C_{Tc} \tag{14-14}$$

However, in most well-designed transistors, the emitter series resistance r_{SE} is negligibly small. Further, r_e is readily bypassed by C_{Te}. Therefore, as a reasonable approximation, we have

$$\frac{1}{\omega_c} \approx r_{SC}C_{Tc} \tag{14-15}$$

which is a significant term only for transistors made by the single-ended impurity-contact processes.

It is interesting to note the resistivity dependence of Eq. (14-15). Let us assume that we have a step collector junction whose characteristics are completely determined by the collector-region resistivity. In this case, $r_{SC} = \rho_c l/A$ and C_{Tc} is proportional to $A/\sqrt{\rho_c}$. Thus the $r_{SC}C_{Tc}$ product is independent of area and directly proportional to $\rho_c^{1/2}$. We see then that increasing collector resistivity increases the magnitude of $1/\omega_c$. This can be offset by increasing the collector voltage such that C_{Tc} is reduced.

14-6. Alpha-cutoff Frequency $f_{\alpha b}$. The results of the preceding four sections may now be combined to obtain the total signal-delay time from emitter to collector.[4,5] For grounded-base operation, this yields the true cutoff frequency for the current-gain characteristic (alpha). This is known as the alpha-cutoff frequency and is given as

$$\frac{1}{\omega_{\alpha b}} \approx \frac{1}{\omega_e} + \frac{1}{\omega_b} + \frac{1}{\omega_d} + \frac{1}{\omega_c} \tag{14-16}$$

Because of the additional terms, it is seen that $\omega_{\alpha b}$ is less than the base-cutoff frequency ω_b. Substituting the appropriate equations for each of the terms in Eq. (14-16), we obtain

$$\frac{1}{\omega_{\alpha b}} \approx r_e C_{Te} + \frac{W^2}{2.43 D_{pb}} + \frac{x_m}{2 v_{SC}} + r_{SC} C_{Tc} \tag{14-17}$$

This is the result for a uniform base; for a graded base, the second term is divided by ln (N'_B/N_{BC}). For an n-p-n structure, D_{nb} is substituted for D_{pb}. In most transistor structures, the first two terms are usually the

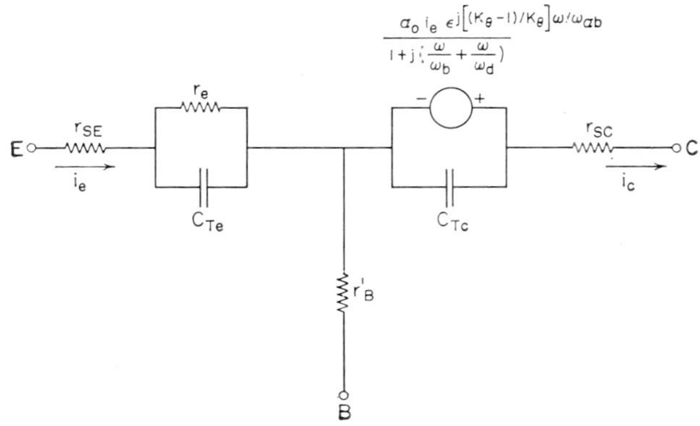

FIG. 14-1. Approximate high-frequency equivalent circuit, illustrating various signal delays making up total alpha-cutoff frequency.

more dominant terms affecting $\omega_{\alpha b}$. Using complex notation, we can therefore approximate the frequency variation of alpha as follows:

$$\alpha \approx \frac{\alpha_o \epsilon^{j[(K_\theta - 1)/K_\theta]\omega/\omega_{\alpha b}}}{1 + j\omega/\omega_{\alpha b}} \tag{14-18}$$

The reader should compare this result with Eq. 13-106, which did not take into account the effects of excess phase and the additional signal delays.

Figure 14-1 best summarizes the analysis of the frequency response of the grounded-base transistor in that it illustrates the approximate high-frequency equivalent circuit that is the inherent model for Eq. (14-17).

14-7. Frequency Variation of Grounded-emitter Current Gain. At this point in the high-frequency analysis of the transistor we are in a position to turn our attention to the frequency response of the grounded-emitter configuration, widely used in circuit applications of transistors. In this case, we are concerned with the frequency variation of the grounded-emitter current gain, beta. From Chaps. 7 to 9 we recall that

$$\beta = \frac{\alpha}{1 - \alpha} \tag{14-19}$$

To obtain the frequency-dependent beta, we substitute Eq. (14-18) into

(14-19), first with $K_\theta = 1$ (no excess phase). Thus we have

$$\beta = \frac{\alpha_o}{1 - \alpha_o + j\omega/\omega_{\alpha b}} \tag{14-20}$$

Factoring $(1 - \alpha_o)$ from the denominator and letting $\beta_o = \alpha_o/(1 - \alpha_o)$, we have[6]

$$\beta = \frac{\beta_o}{1 + j\omega/(1 - \alpha_o)\omega_{\alpha b}} \tag{14-21}$$

Here we see that the variation of β with frequency is identical to that of α, except that $\beta_o \gg \alpha_o$ (higher current gain); the cutoff frequency is reduced considerably by the factor $(1 - \alpha_o)$. This is quite understandable from a qualitative point of view. Because of the phase shift of the base current, the magnitude of $(1 - \alpha)$ increases rapidly with frequency, thereby causing beta to fall off at a very rapid rate. The relative dependence of β on α_o is shown in Fig. 14-2, where the absolute magnitude of beta is plotted in decibels as a function of $\omega/\omega_{\alpha b}$. For the ideal case, where $\alpha_o = 1$, beta is infinite and decreases at a rate of 6 db per octave to zero db at the alpha-cutoff frequency. For $\alpha_o = 0.99$, beta equals 99 and is somewhat constant with frequency, dropping 3 db at $\omega = 0.01\omega_{\alpha b}$. By lowering α_o to 0.9, however, we decrease beta about 20 db; beta cutoff increases by a factor of 10 to a frequency of $0.1\omega_{\alpha b}$. Examination of these results will verify that a low alpha is necessary for high grounded-emitter frequency response. Thus, the current-gain-bandwidth concept is preserved for the junction transistor.

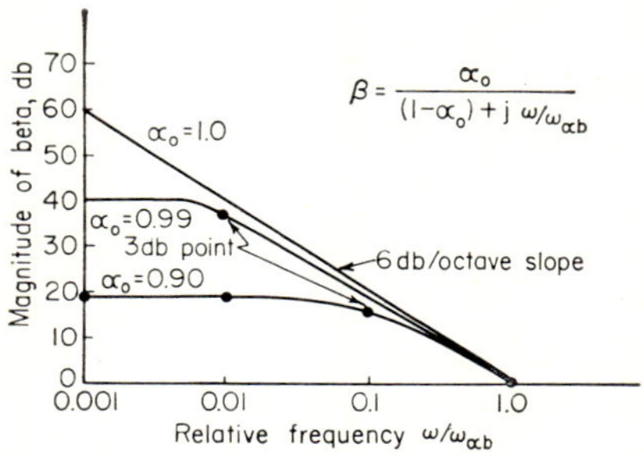

FIG. 14-2. Variation of beta with frequency.

To be completely correct, we cannot assume that $K_\theta = 1$ is the general case for all transistors. Thomas and Moll have shown that for values of K_θ less than unity, Eq. (14-21) must be corrected as follows:[2]

$$\beta = \frac{\beta_o \epsilon^{j[(K_\theta - 1)/\sqrt{K_\theta}]\omega/\omega_{\alpha b}}}{1 + j\omega/K_\theta(1 - \alpha_o)\omega_{\alpha b}} \tag{14-22}$$

Thus, because of the excess phase, K_θ becomes smaller and therefore the cutoff frequency for beta becomes even lower. Nevertheless, the fundamental point to recognize is the fact that the frequency response of the grounded-emitter current gain is largely dependent on the alpha-cutoff frequency.

14-8. Beta-cutoff Frequency $f_{\alpha e}$ and Current-gain-bandwidth Frequency f_T. By definition, the frequency at which beta β_o is reduced in magnitude by $1/\sqrt{2}$ or 3 db is called the beta-cutoff frequency $\omega_{\alpha e} = 2\pi f_{\alpha e}$. Using the general expression of Eq. (14-22), we see that

$$\omega_{\alpha e} = K_\theta(1 - \alpha_o)\omega_{\alpha b} \tag{14-23}$$

Beta cutoff is maximized by the combination of high $\omega_{\alpha b}$ and low α_o. If we consider frequencies $\omega > \omega_{\alpha e}$, then the imaginary term in Eq. (14-22) becomes greater than 1 and we have

$$\beta \approx \frac{\alpha_o K_\theta \omega_{\alpha b}}{j\omega} \qquad \omega > \omega_{\alpha e} \tag{14-24}$$

We infer from this result that β is inversely proportional to frequency in this range, and, therefore, decreases at a rate of 6 db per octave. Of extreme interest is that frequency for which the magnitude of beta is equal to 1. This is known as the *current-gain-bandwidth frequency* $\omega_T(\omega_T = 2\pi f_T)$, and is undoubtedly the most important of all the high-frequency transistor parameters. Referring to Eq. (14-24), we have

$$\omega_T \approx \alpha_o K_\theta \omega_{\alpha b} \tag{14-25}$$

or[2]

$$f_T \approx \alpha_o K_\theta f_{\alpha b} \tag{14-26}$$

In the grounded-emitter configuration, f_T represents the uppermost limit of useful current gain. Since the current gain is unity at this frequency, it represents the grounded-emitter bandwidth of the transistor. If Eq. (14-25) is substituted into (14-24), we have

$$f_T \approx \beta f \qquad \omega_{\alpha e} < \omega < \omega_T \tag{14-27}$$

which denotes a constant current-gain-bandwidth product, applicable only in the 6 db/octave region of the beta characteristic. This makes the measurement of f_T quite simple. For example, a beta of 4 at a frequency of 100 megacycles corresponds to an f_T of 400 megacycles, provided that the 100-megacycle measurement falls on the 6 db/octave slope. The beta versus frequency characteristic shown in Fig. 14-3 clearly illustrates the relationships of $f_{\alpha e}$ and f_T to $f_{\alpha b}$ for a representative transistor. Note that the magnitude of the current gain is expressed in decibels (i.e., 20 log β).

14-9. Variation of f_T with Voltage and Current. Since f_T is directly related to $f_{\alpha b}$ by Eq. (14-26), f_T can be related to the individual signal delays by the substitution of Eq. (14-17). In this section, a typical p-n-p diffused-base germanium mesa transistor will be referred to in order to illustrate the manner in which f_T varies with collector voltage and current. This is a high-frequency power device designed to deliver over 0.5 watt of r-f output at 160 megacycles. For this p-n-p model having a graded base, we can write

$$\frac{1}{f_T} \approx \frac{2\pi}{\alpha_o K_\theta}\left(r_e C_{Te} + \frac{W^2}{2.43 D_{pb} \ln (N'_B/N_{BC})} + \frac{x_m}{2v_{SC}} + r_{SC}C_{Tc}\right) \tag{14-28}$$

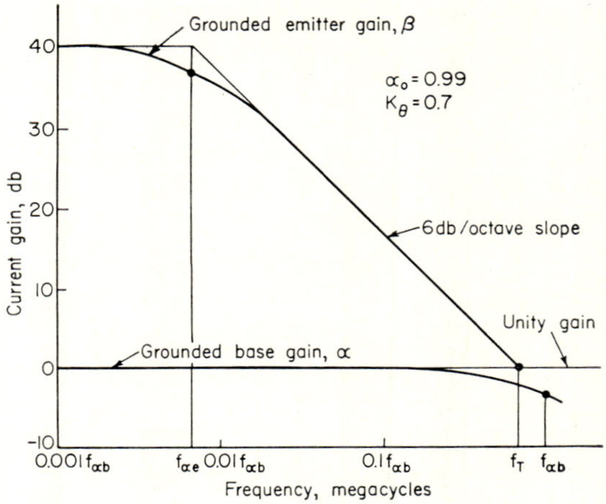

FIG. 14-3. Typical current gain versus frequency curves, illustrating relationship of f_T and $f_{\alpha e}$ to $f_{\alpha b}$.

This equation will be referred to in the discussion to follow. Since this is a reciprocal relationship, it should be borne in mind that each of the terms in Eq. (14-28) should be small in order to maximize f_T.

Figure 14-4 illustrates the typical behavior of f_T with voltage and current for the 2N1692 p-n-p germanium power mesa transistor. The reference operating conditions are $V_{CE} = -10$ volts d-c, $I_C = -50$ ma, $T_A = 25°C$, and $f_T = 500$ megacycles. Referring to the voltage varia-

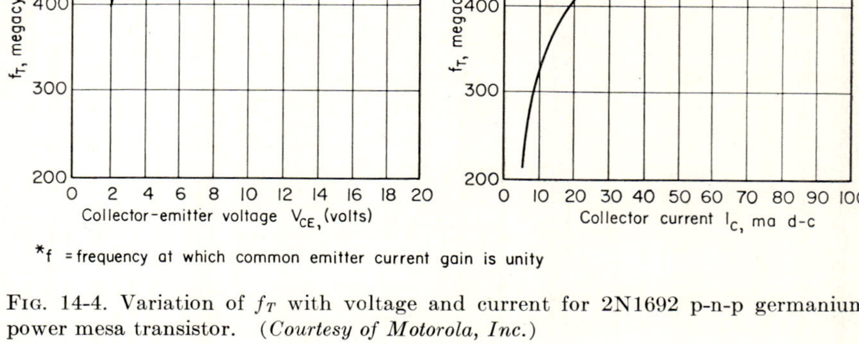

FIG. 14-4. Variation of f_T with voltage and current for 2N1692 p-n-p germanium power mesa transistor. (*Courtesy of Motorola, Inc.*)

tion, we see that at very low voltages on the order of 2 volts, f_T is less than 400 megacycles and is decreasing rapidly. Under these conditions, a significant portion of the applied voltage is dropped across the collector series resistance. Consequently, the effective base width is at its maximum, the drift velocity of the carriers in the depletion layer is probably less than the limiting value, and C_{Tc} is at its maximum. This tends to increase the magnitudes of the last three terms of Eq. (14-28), causing f_T to be low. As voltage is increased toward 10 volts, these three terms decrease in value, allowing f_T to increase. However, the most predominant effect is attributed to the decrease of base width W in the second term. At voltages greater than 10 volts, f_T decreases slightly, due primarily to the increase in x_m.

The variation of f_T with current is also shown in Fig. 14-4. At very low currents, it is seen that f_T is very low and is falling off at a very rapid rate. This is the effect of the first term, r_eC_{Te}, wherein we recall that $r_e = kT/qI_E$. At low currents, r_e is quite large and completely dominates the frequency response of the transistor. As current is increased toward 50 ma, the r_eC_{Te} time constant becomes progressively smaller, until the second term takes over and becomes the dominant signal delay affecting f_T. As the collector current is increased further, f_T starts to decrease slightly. This is attributed again to the increasing voltage drop across the collector series resistance; its effect is to make W slightly larger.

In summary, it should be understood that in most well-designed high-frequency transistors, the emitter time constant and the base-transit time are by far the most important factors affecting the magnitude of the current-gain-bandwidth frequency f_T.

14-10. Design Theory for Optimum f_T. In designing transistors for maximum frequency response, the primary objective is to minimize the magnitude of each of the terms of Eq. (14-28). Complete optimization cannot be achieved, however, without some sacrifice of other parameters, particularly such characteristics as BV_{CBO}, $V_{CE(SAT)}$, r'_B, BV_{EBO}, V_{PT}, and C_C. It is not possible in this section to cover all the ramifications of optimum f_T design. Each design situation must be analyzed on its own merits, based on the transistor process and geometry that is selected and the specifications for the intended circuit application. Nevertheless, certain design concepts can be extracted from Eq. (14-28) which are universally applicable for optimizing f_T, namely:

1. The emitter junction area should be as small as possible to achieve the smallest emitter junction capacitance consistent with the allowable current rating.
2. The base should be as thin as possible to achieve the lowest base-transit time consistent with the allowable voltage rating.

In Chap. 15, it will be shown that optimum f_T alone is not adequate for maximum power gain at high frequencies; in this case, the magnitude of $r'_B C_C$ becomes important.

PROBLEMS

14-1. Measurements of the parameters of a high-frequency transistor indicate that $\alpha_o = 0.97$ and $h_{fe} = 15.2$ db at 100 megacycles. For this structure it is estimated that $K_\theta = 0.7$. Determine the magnitudes of $f_{\alpha e}$, f_T, and $f_{\alpha b}$.

14-2. For the simple case of a transistor having no excess phase shift ($K_\theta = 1$) and alpha close to unity, show that the expression for beta in the frequency range beyond beta cutoff is $h_{fe} \approx -j\omega_{\alpha'}/\omega$.

14-3. Calculate the value of the current-gain-bandwidth frequency f_T for the n-p-n silicon planar-epitaxial transistor described below at conditions of $V_{CE} =$ 5 volts, $I_E = -10$ ma, and $T_A = 25°C$. β_o at these conditions is equal to 50.

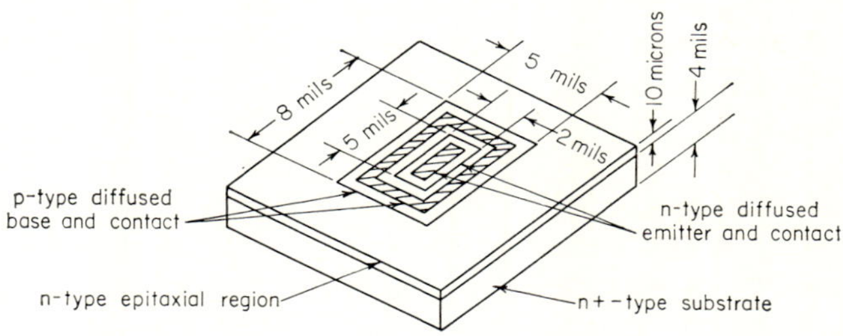

The epitaxial region is 10 μ thick and has a resistivity of 1.5 ohm-cm. The substrate resistivity is 0.001 ohm-cm and is 4 mils thick. The base is diffused in a 5 $\times$ 8 mil rectangle to a junction depth of 2.8 μ from a surface concentration of 10^{19} atoms/cm^3 and the resulting impurity distribution is an error-function distribution. The emitter is diffused in a 2- by 5-mil rectangle to a junction depth of 1.7 μ and may be approximated by a uniform impurity distribution of 10^{21} atoms/cm^3. The impurity concentration N'_B at the base side of the emitter junction is 3.5×10^{16} atoms/cm^3.

REFERENCES

1. Pritchard, R. L.: Frequency Variations of Current-amplification Factor for Junction Transistors, *Proc. IRE*, vol. 40, pp. 1476–1481, November, 1952.
2. Thomas, D. E., and J. L. Moll: Junction Transistor Short-circuit Current Gain and Phase Determination, *Proc. IRE*, vol. 46, pp. 1177–1184, June, 1958.
3. Early, J. M.: PNIP and NPIN Junction Transistor Triodes, *Bell System Tech. J.*, vol. 33, pp. 517–533, May, 1954.
4. Buie, J. L.: High Frequency Silicon NPIN Oscillator Transistor, Professional Group on Electron Devices, Washington, D.C., October, 1958.
5. Early, J. M.: Structure-determined Gain-band Products of Junction Triode Transistors, *Proc. IRE*, vol. 46, pp. 1924–1927, December, 1958.
6. Thomas, D. E.: Transistor Amplifier Cutoff Frequency, *Proc. IRE*, vol. 40, pp. 1481–1483, November, 1952.

15

Junction-transistor Amplifiers

15-1. Introduction. Having reduced the over-all physical aspects of the junction transistor from the physics of semiconductor junctions to the equivalent electric circuit for both low and high frequencies, we are now in a position to consider the problem of the transistor as an amplifier. In this chapter, all of the theory developed thus far is applied to the general power-gain equation for a four-terminal network. This results in expressions in terms of the transistor h parameters, which are then analyzed at both low and high frequencies. In the case of the latter, the emphasis is on matched power gain for the grounded-emitter configuration, since this arrangement yields results which clearly point out to the device designer the requirements for optimum transistor design. It is not the intent here to cover all the ramifications of the manifold circuit applications of transistors. Rather, the objective of this chapter is to determine what fundamental parameters (base width, impurity distributions, etc.) contribute to maximum amplifier performance, thereby establishing basic device-design criteria. Furthermore, both the transistor designer and transistor user are accorded an insight into what inherent limitations there are for certain transistor structures (or processes), from the amplifier point of view.

15-2. Power Gain for General Four-terminal Network. Any four-terminal network such as the junction transistor may be represented as in Fig. 15-1. In this simple and basic arrangement, a load represented by the conductance g_L is placed across the output terminals. At the input terminals, the signal to be amplified is depicted by the voltage generator e_g in series with its internal impedance r_g. For this network, the power gain is defined simply as the

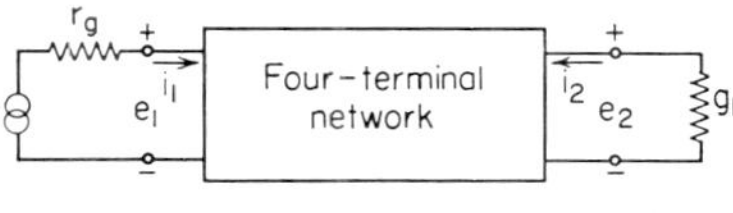

FIG. 15-1. General four-terminal network.

ratio of the power delivered to the load to the power delivered by the generator. This may be written as

$$G = \frac{P_{\text{OUT}}}{P_{\text{IN}}} = \frac{-e_2 i_2}{e_g i_1} = -\frac{e_2}{e_g}\frac{i_2}{i_1} \tag{15-1}$$

or in other words, the power gain is the product of the voltage gain and the current gain. Since the terminal voltages and currents of Fig. 15-1 may be related by the general h parameters of the network, we may proceed to obtain the equation for the voltage gain e_2/e_g. In Chap. 12 we saw that

$$e_1 = h_{11}i_1 + h_{12}e_2 \tag{15-2}$$
$$i_2 = h_{21}i_1 + h_{22}e_2 \tag{15-3}$$

We can make the substitutions $e_1 = e_g - i_1r_g$ and $i_2 = -e_2g_L$, so that Eqs. (15-2) and (15-3) become, after simple manipulation,

$$e_g = (h_{11} + r_g)i_1 + h_{12}e_2 \tag{15-4}$$
$$0 = h_{21}i_1 + (h_{22} + g_L)e_2 \tag{15-5}$$

Solving (15-5) for i_1, which is

$$i_1 = \frac{-(h_{22} + g_L)e_2}{h_{21}} \tag{15-6}$$

and substituting (15-6) into (15-4), we obtain

$$\frac{e_g}{e_2} = \frac{-(h_{11} + r_g)(h_{22} + g_L) + h_{12}h_{21}}{h_{21}} \tag{15-7}$$

Factoring out a minus sign and taking the reciprocal of (15-7), we get

$$\frac{e_2}{e_g} = \frac{-h_{21}}{(h_{11} + r_g)(h_{22} + g_L) - h_{12}h_{21}} \tag{15-8}$$

This is the expression for the voltage gain of the general four-terminal network.

Additionally, if in Eq. (15-3) we set $e_2 = -i_2/g_L$, we have

$$i_2 = h_{21}i_1 - h_{22}i_2/g_L \tag{15-9}$$

Solving for the current gain i_2/i_1, we obtain

$$\frac{i_2}{i_1} = \frac{h_{21}g_L}{h_{22} + g_L} \tag{15-10}$$

As in most amplifier circuits, if the load admittance g_L is large compared to the output admittance h_{22}, Eq. (15-10) simply becomes

$$\frac{i_2}{i_1} \approx h_{21} \qquad \text{for } g_L > h_{22} \tag{15-11}$$

Applying the same approximation to the voltage gain Eq. (15-8) and substituting both (15-8) and (15-11) into Eq. (15-1) for power gain, we have, finally,

$$G = \frac{h_{21}{}^2}{(h_{11} + r_g)g_L - h_{12}h_{21}} \tag{15-12}$$

This relation will be used in the next section in examining the power gain of junction transistors as low-frequency amplifiers.

15-3. Low-frequency Amplifier Power Gain. At very low frequencies within the audio range, the h parameters for the junction transistor are purely resistive; that is, reactive effects are completely negligible. Consequently, in analyzing low-frequency performance, the h-parameter relations developed in Chap. 12 may be used with Eq. (15-12).

As an audio-signal amplifier, the transistor is generally operated in either the grounded-base or grounded-emitter circuit configuration. Let us consider the grounded-base connection first. The general power-gain expression given by (15-12) becomes, specifically,

$$G_b = \frac{h_{fb}{}^2}{(h_{ib} + r_g)g_L - h_{rb}h_{fb}} \tag{15-13}$$

At low frequencies, the product $h_{rb}h_{fb}$ is negligibly small since $h_{fb} \approx 1$ and $h_{rb} \approx 10^{-4}$ for most transistors. Since $h_{fb} = -\alpha$, (15-13) becomes

$$G_b \approx \frac{\alpha^2}{(h_{ib} + r_g)g_L} \tag{15-14}$$

If the source resistance r_g is much greater than h_{ib}, it is seen that the power gain is directly proportional to the ratio r_L/r_g. For a typical device where $h_{ib} = 40$ ohms, $h_{fb} = 0.981$, and $h_{ob} = 0.25$ μmhos, let us assume that $r_g \approx 0$ and $r_L = 8{,}000$ ohms. Thus, the power gain is

$$G_b \approx \frac{\alpha^2 R_L}{h_{ib}} = (0.981)^2 \frac{8{,}000}{40} = 192$$

or a gain of 23 db.

For the grounded-emitter configuration, the power-gain relation is written as

$$G_e = \frac{h_{fe}{}^2}{(h_{ie} + r_g)g_L - h_{re}h_{fe}} \tag{15-15}$$

Since the same approximations are applicable here, we have

$$G_e \approx \left(\frac{\alpha}{1 - \alpha}\right)^2 \frac{1}{(h_{ie} + r_g)g_L} \tag{15-16}$$

Using the equivalent grounded-emitter magnitudes of the previous example, we have $h_{ie} = 2{,}000$ ohms, $h_{fe} = 50$, and $h_{oe} = 12.5$ μmhos.

The power gain is calculated as

$$G_e \approx \left(\frac{\alpha}{1-\alpha}\right)^2 \frac{r_L}{h_{ie}} = (50)^2 \frac{8{,}000}{2{,}000} = 10{,}000$$

which is a power gain of 40 db.

Thus we see that at low frequencies the grounded-emitter stage gives considerably higher power gain than the grounded base. In fact, since $h_{ib} = (1 - \alpha)h_{ie}$, it is apparent that G_e is greater than G_b by the factor $1/(1 - \alpha)$. For a beta of 50, this would be an increase of 17 db, which checks with the previous calculations. As a rule of thumb, for Eq. (15-16) to be applicable, g_L must be at least 10 times h_{oe} or $h_{ob}/(1 - \alpha)$. This points out the importance of having a very low output admittance h_{ob} in addition to high alpha for maximum low-frequency amplification.

Generally, because of the current gain as well as the high power gain, the grounded-emitter configuration is more widely used in audio circuits.

TABLE 15-1. LOW-FREQUENCY AMPLIFIER CHARACTERISTICS

Grounded emitter	Grounded base	Grounded collector
Large current gain	Approximate unity current gain	Large current gain
Large voltage gain	Large voltage gain	Approximate unity voltage gain
Highest power gain	Intermediate power gain	Lowest power gain
Low input resistance	Very low input resistance	High input resistance
High output resistance	Very high output resistance	Low output resistance
Analogous to grounded cathode	Analogous to grounded grid	Analogous to grounded plate

However, the junction transistor is used in either grounded-base or grounded-collector circuits whenever the specific circuit requirements warrant it. Although no attempt will be made here to cover all the circuit aspects, a general comparison of the features of the configurations is made in Table 15-1.[1] One will note that the essential differences rest in the magnitudes of the input and output resistances. Also included in the table are the analogies to the equivalent vacuum-tube configurations.

It must be emphasized that in the foregoing discussion we were concerned only with small-signal variations about the d-c operating points. Therefore, in determining the power gain, the appropriate small-signal *h* parameter must be determined for the particular combination of emitter current and collector voltage used. There are, however, many applications where transistors are used as large-signal amplifiers. One example is the output stage of an audio amplifier. There, the input signal is large, so that the collector current swings over the full range of rated current, delivering a fair amount of a-c power to the load. In that case, the small-signal *h* parameters are not applicable, especially if the transistor characteristics deviate from linearity.

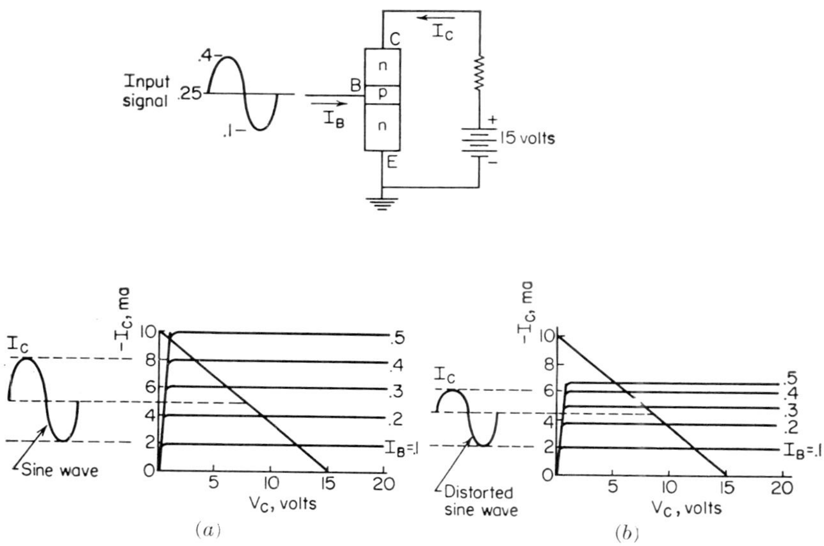

FIG. 15-2. Large-signal transistor amplifier characteristics, illustrating effect of alpha nonlinearity on output gain and distortion. (*a*) Linear V_c-I_c characteristic; (*b*) nonlinear V_c-I_c characteristic due to decrease of α with I_c.

Let us consider the case of a grounded-emitter large-signal n-p-n amplifier as an illustrative example. If the electrical characteristics are completely linear over the allowable range of collector current and voltage, the collector characteristic appears as shown in Fig. 15-2*a*. Here the direct current gain (beta) is equal to 20 and is constant over the range of 10 ma collector current. For the load line shown, the transistor is biased at $I_B = 0.25$ ma. If the input base current signal swings sinusoidally about this bias point from 0.1 ma to 0.4 ma, the output waveform is a true undistorted replica of the input wave, as shown in (*a*).

However, if the grounded-emitter $V_c - I_c$ characteristics are nonlinear (as shown in Fig. 15-2*b*) due to the falloff of alpha with collector current, the output waveform becomes severely distorted. Note that for the characteristics shown in (*b*), beta is 20 at $I_B = 0.1$ ma, but falls off to 15 at $I_B = 0.4$. Consequently, for the same bias conditions and input-signal swing, the output-signal amplitude is reduced and waveform distortion is introduced. Thus, for large-signal applications, the power-gain expression given by Eq. (15-12) is not too meaningful if the characteristics are highly nonlinear. However, for linear conditions the large-signal parameters are somewhat equivalent to the small-signal parameters, making Eq. (15-12) applicable.

On this basis, the primary problem in designing junction-transistor large-signal amplifiers (power transistors) is to maintain beta $[\alpha/(1 - \alpha)]$ as high and as constant as possible over the widest possible range of current consistent with the ratings of the device. From the design standpoint, this reduces to maintaining the emitter efficiency close to unity by means of heavy emitter doping and small base width as discussed in Chap. 10.

15-4. Maximum Available Power Gain. From fundamental network theory, it is known that maximum power transfer is achieved by matching the load to the source resistance of the generator. Consequently, in order to determine the maximum power gain that is realizable for a given active four-terminal network, it is necessary to derive the general gain equation, based on the conditions that the load g_L is matched to the output admittance of the network and the generator resistance r_g is matched to the input resistance. Referring to Fig. 15-1, let us assume that the generator e_g sees a load equal to r_g, such that the maximum available power input to the network is

$$P_{\text{in}} = \frac{e_g^2}{4r_g} \tag{15-17}$$

Also, let us assume that the output is not necessarily matched, so that g_L may have any value. Therefore, the output power is

$$P_{\text{out}} = e_2^2 g_L \tag{15-18}$$

The power gain for matched input only is therefore

$$G = \frac{P_{\text{out}}}{P_{\text{in}}} = 4r_g g_L \left(\frac{e_2}{e_g}\right)^2 \tag{15-19}$$

which we see is proportional to the square of the voltage gain. Substituting Eq. (15-8) into (15-19), we obtain

$$G = \frac{4r_g g_L h_{21}^2}{[(h_{11} + r_g)(h_{22} + g_L) - h_{12}h_{21}]^2} \tag{15-20}$$

This is the general power-gain equation in terms of the h parameters of the four-terminal network, for the condition that the generator resistance is matched to the input resistance of the network with the load g_L.

Further, this represents the power gain of a junction transistor with specified h parameters at a particular frequency. Keeping the h parameters constant, we can now attempt to maximize G by selecting appropriate values for r_g and g_L. It is apparent that we can achieve infinite gain by merely setting the denominator of Eq. (15-20) equal to zero. This, of course, corresponds to the transistor acting as an oscillator, as opposed to being a stable amplifier. For the right combination of conditions, the transistor may very well oscillate, particularly if the feedback parameter h_{12} is of the correct magnitude and sign; however, we shall not concern ourselves with this problem here.

As was shown by Pritchard,[2,*] the maximum available power gain is determined by differentiating G with respect to r_g and g_L, setting the results equal to zero, and solving the simultaneous equations to obtain the values for optimum gain. Letting the denominator of Eq. (15-20) equal H^2,

$$H = (h_{11} + r_g)(h_{22} + g_L) - h_{12}h_{21} \tag{15-21}$$

It follows that

$$\frac{dG}{dg_L} = \frac{4r_g h_{21}^2}{H^2} - \frac{8r_g g_L h_{21}^2(h_{11} + r_g)}{H^3} \tag{15-22}$$

$$\frac{dG}{dr_g} = \frac{4g_L h_{21}^2}{H^2} - \frac{8r_g g_L h_{21}^2(h_{22} + g_L)}{H^3} \tag{15-23}$$

Setting these equations equal to zero and simplifying terms,

$$1 - \frac{2g_L(h_{11} + r_g)}{H} = 0$$

$$1 - \frac{2r_g(h_{22} + g_L)}{H} = 0$$

Reinserting Eq. (15-21) and rearranging terms, we get

$$2g_L(h_{11} + r_g) = (h_{11} + r_g)(h_{22} + g_L) - h_{12}h_{21} \tag{15-24}$$

$$2r_g(h_{22} + g_L) = (h_{11} + r_g)(h_{22} + g_L) - h_{12}h_{21} \tag{15-25}$$

or

$$g_L = h_{22} - \frac{h_{12}h_{21}}{(h_{11} + r_g)} \qquad \text{mhos} \tag{15-26}$$

$$r_g = h_{11} - \frac{h_{12}h_{21}}{(h_{22} + g_L)} \qquad \text{ohms} \tag{15-27}$$

Note that in these results g_L and r_g are mutually dependent. Furthermore, for maximum gain conditions, g_L must be equal to h_{22} less the reflected admittance of the input circuit. Similarly, r_g must equal h_{11} less the reflected impedance of the output circuit. In both cases, $h_{12}h_{21}$ is a mutual transfer product. The essential fact to note here is that Eq. (15-26) represents the output admittance of a four-terminal network with

* References, indicated in the text by superscript figures, are listed at the end of the chapter.

the resistance r_g connected across the input terminals. Similarly, Eq. (15-27) represents the input impedance of a four-terminal network with a load g_L connected across the output terminals.

To obtain a result in terms of h parameters only, we can substitute (15-27) into (15-26) and solve for g_L or vice versa and solve for r_g. Carrying out the algebra, we obtain the following concise results:

$$g_L = h_{22}\sqrt{1 - \frac{h_{12}h_{21}}{h_{11}h_{22}}} \tag{15-28}$$

$$r_g = h_{11}\sqrt{1 - \frac{h_{12}h_{21}}{h_{11}h_{22}}} \tag{15-29}$$

Letting the term within the radical equal p^2, we get

$$\frac{g_L}{h_{22}} = \frac{r_g}{h_{11}} = p \tag{15-30}$$

which are the general criteria for maximum power gain. Substituting the relations of (15-30) into Eq. (15-20), we get

$$G_{\text{max}} = \frac{4p^2h_{11}h_{22}h_{21}^2}{[h_{11}h_{22}(1 + p)^2 - h_{12}h_{21}]^2} \tag{15-31}$$

Furthermore, since

$$p = \sqrt{1 - \frac{h_{12}h_{21}}{h_{11}h_{22}}} \tag{15-32}$$

then

$$h_{12}h_{21} = h_{11}h_{22}(1 - p^2) \tag{15-33}$$

Inserting this identity into (15-31) and simplifying the result, we get, finally,

$$G_{\text{max}} = \frac{h_{21}^2}{h_{11}h_{22}(1 + p)^2} \tag{15-34}$$

where p is defined by Eq. (15-32). Equation (15-34) is the general expression for the maximum power gain available from a transistor (or any four-terminal network), provided that the conditions specified by Eq. (15-30) are met.

We saw in the previous chapter that the h parameters are complex at the higher frequencies, so that

$$h_{11} = r_{11} + jx_{11} \tag{15-35}$$

$$h_{22} = g_{22} + jb_{22} \tag{15-36}$$

Usually, b_{22} is capacitive; at the input, however, x_{11} may be inductive. Nevertheless, it is possible to tune out the input reactance and output susceptance by adding like elements of equal magnitude and of opposite sign at the generator and load, respectively. As a result, we obtain a conjugate match at both input and output, provided that

$$\frac{g_L}{g_{22}} = \frac{r_g}{r_{11}} = p \tag{15-37}$$

where g_{22} and r_{11} are the real parts of h_{11} and h_{22}. Consequently, for these conditions, Eq. (15-34) for the maximum available power gain specifically becomes the equation for the real power gain if the real parts of all terms are taken.

$$G_{\max} = \frac{h_{21}^2}{r_{11}g_{22}(1 + p)^2} \qquad (15\text{-}38)$$

where r_{11} and g_{22} are the real parts of h_{11} and h_{22}, respectively, and $p = \sqrt{1 - Re\,(h_{12}h_{21})/r_{11}g_{22}}$. $Re\,(h_{12}h_{21})$ denotes the real part of this product.

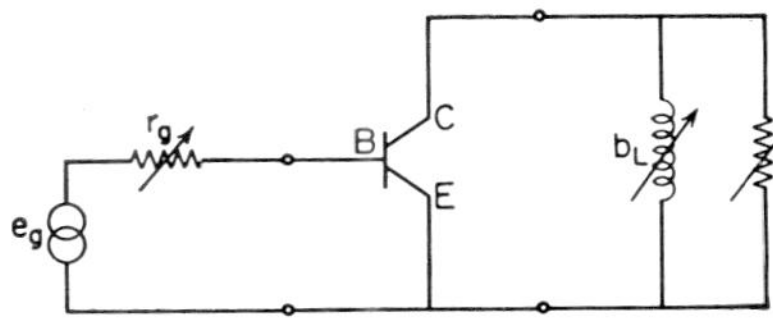

FIG. 15-3. Simple power-gain circuit for conjugate-matched output and resistive-matched input.

The foregoing matching conditions are referred to as *conjugate-matching* of both the input and output characteristics. In actual circuit or measurement practice, this is very difficult to achieve. The transistor is apt to oscillate when both input and output are tuned. Therefore, it is quite common to measure power gain by conjugate-matching the output admittance, but only resistive-matching the input. In other words, x_{11} is not tuned out. Under these conditions, we would expect that the resulting power-gain expression would be different from that of (15-38). Pritchard has shown that the difference between the two is so slight that,

TABLE 15-2. FOUR-TERMINAL-NETWORK RELATIONS

Voltage gain..................	$\dfrac{e_2}{e_g} = \dfrac{-h_{21}}{(h_{11} + r_g)(h_{22} + g_L) - h_{12}h_{21}}$
Current gain.................	$\dfrac{i_2}{i_1} = \dfrac{h_{21}g_L}{h_{22} + g_L}$
Input impedance.............	$Z_{\text{in}} = h_{11} - \dfrac{h_{12}h_{21}}{(h_{22} + g_L)}$
Output admittance...........	$Y_{\text{out}} = h_{22} - \dfrac{h_{12}h_{21}}{(h_{11} + r_g)}$
Matched power gain..........	$G = \dfrac{h_{21}^2}{h_{11}h_{22}(1 + \sqrt{1 - h_{12}h_{21}/h_{11}h_{22}})^2}$

for all practical purposes, the simpler Eq. (15-38) serves as a very good criterion for transistor power gain for the condition of resistive-matched input.[2] This measurement procedure is shown symbolically in Fig. 15-3 for a grounded-emitter transistor. An inductance is shown in the output, since the output susceptance of the transistor is capacitive.

It would be appropriate at this point to summarize the important relations that have been developed thus far regarding the general four-terminal network of Fig. 15-1. These are included in Table 15-2 for reference. All the equations are expressed in terms of the h parameters and are applicable to both low and high frequencies. They apply to any of the three junction-transistor circuit configurations, viz., grounded base, grounded emitter, and grounded collector, provided that the correct h parameters are used.

15-5. High-frequency Grounded-emitter h Parameters.[2] In most circuit applications where junction transistors are used as tuned high-frequency amplifiers, the grounded-emitter configuration is widely used, since it is known that for frequencies less than alpha cutoff it will give more gain than the grounded-base configuration. Additionally, from the point of view of the device designer, the grounded-emitter power-gain equation offers a far simpler relationship to the basic physical parameters of the transistor than does the grounded-base equation, making it easier to evaluate a particular design through matched gain measurements. As we shall see, the high-frequency power-gain equation for the grounded-emitter transistor is extremely useful to the device and circuit designer in determining high-frequency performance. Therefore, in this section the appropriate expressions for the high-frequency grounded-emitter h parameters will be obtained for application to the general power-gain equation given by Eq. (15-38).

The general expressions for the grounded-base h parameters as a function of relative frequency ω/ω_b were derived in Chap. 13. If we include the additional signal delays which were discussed in Chap. 14 so as to obtain equations expressed in terms of $\omega/\omega_{\alpha b}$, we get

$$h_{ib} \approx \frac{1}{1/r_e + j\omega/r_e\omega_{\alpha b}} + r'_B(1 + h_{fb}) \tag{15-39}$$

$$h_{rb} \approx j\omega r'_B(C_{Tc} + C_{Dc}) \tag{15-40}$$

$$h_{fb} \approx \frac{-\alpha_o\epsilon^{j[(K_\theta - 1)/K_\theta]\omega/\omega_{\alpha b}}}{1 + j\omega/\omega_{\alpha b}} \tag{15-41}$$

$$h_{ob} \approx \omega_{\alpha b}C_{Dc}(\omega/\omega_{\alpha b})^2 + j\omega(C_{Tc} + C_{Dc}) \tag{15-42}$$

Several assumptions are involved here which need clarification. Firstly' in Eq. (15-40) it is assumed that the contribution of $(\mu_{ec} + r'_Bg_c)$ to h_{rb} is negligible at high frequencies. Secondly, in Eq. (15-41), it is assumed that emitter efficiency γ is unity for all frequencies. Finally, in Eq. (15-42), it is assumed that the effect of r'_B on h_{ob} is negligible and we neglect the term $\omega^2 r'_B C_{Tc}{}^2$. Actually, more exact expressions may be obtained by picking up additional terms in the series expansions of the hyperbolic functions (see Chap. 13); however, for frequencies up to the vicinity of $\omega_{\alpha b}$, Eqs. (15-39) to (15-42) are adequate.

To obtain the equivalent grounded-emitter h parameters, we make use of the fact that the parameters for the two configurations, except for h_{rb}

or h_{re}, are simply related by the factor $(1 - \alpha)$. This was first shown in the study of h parameters in Chap. 12. Thus, Eqs. (15-39), (15-41), and (15-42) may be divided term by term by $(1 - \alpha)$ to yield the grounded-emitter results. It should be pointed out that $(1 - \alpha)$ is the same as $(1 + h_{fb})$.

Referring to Eq. (15-39), if we divide through by $(1 + h_{fb})$ and evaluate the result, we obtain

$$h_{ie} \approx r'_B \tag{15-43}$$

where the assumption is made that at high frequencies r'_B is the dominant part of the final expression for h_{ie}. Although reactive terms should be included in Eq. (15-43), in the next section these will not be used, since we will assume that all reactive terms will be tuned out by conjugate matchings for power gain.

The transformation of h_{fb} to h_{fe} was obtained in Chap. 14 in a similar manner; the result is repeated here as follows:

$$h_{fe} \approx \frac{-\beta_o \epsilon^{j[(K_\theta - 1)/\sqrt{K_\theta}]\omega/\omega_{\alpha b}}}{1 + j\omega/K_\theta(1 - \alpha_o)\omega_{\alpha b}} \tag{15-44}$$

In the frequency range beyond $\omega_{\alpha e}$, i.e., $K_\theta(1 - \alpha_o)\omega_{\alpha b}$, Eq. (15-44) becomes, in magnitude only,

$$h_{fe} \approx \frac{-\alpha_o K_\theta \omega_{\alpha b}}{j\omega} \approx \frac{-\omega_T}{j\omega} \tag{15-45}$$

Since $h_{fe} = -\alpha/(1 - \alpha)$, we can write, for α close to unity,

$$\frac{1}{1 - \alpha} \approx \frac{\omega_T}{j\omega} \tag{15-46}$$

This result can be applied to Eq. (15-42) to obtain the grounded-emitter output admittance h_{oe}. Again, considering only the real terms in the result, we have

$$h_{oe} \approx \omega_T C_c \tag{15-47}$$

where we have defined $C_c = C_{Tc} + C_{Dc}$. Thus, the output impedance of the grounded-emitter transistor at high frequencies is approximately inversely proportional to the product of the current-gain-bandwidth frequency and the collector capacitance. For example, if a transistor has $f_T = 400$ mcps and $C_c = 10\ \mu\mu f$, then $h_{oe} = 25\ \mu$mhos and we have an output impedance on the order of 40 ohms.

It should not be inferred from Eqs. (15-43) or (15-47) that the real parts of h_{ie} and h_{oe} are independent of frequency. These are only approximations which were established for the purpose of evaluating transistor designs and to aid in deriving the power-gain equation in the next section. Variations of h_{ie} and h_{oe} with frequency, voltage, and current for a typical high-frequency p-n-p germanium mesa power transistor are given in Fig. 15-4.

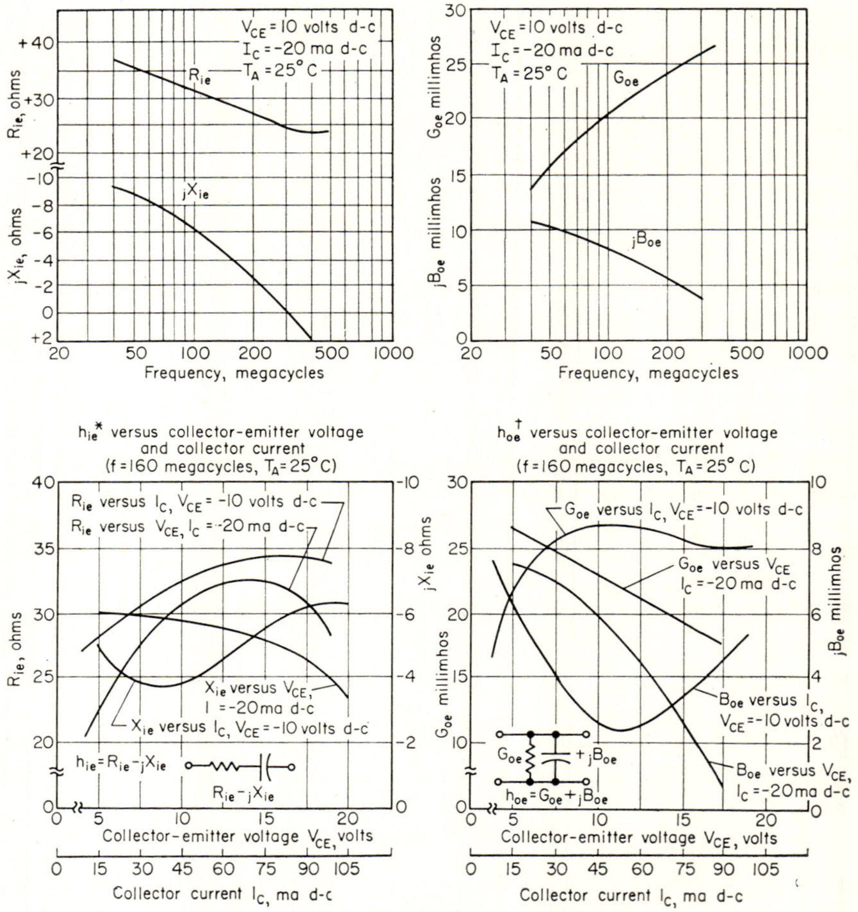

* Common emitter short-circuit series equivalent input impedance
† Open-circuit parallel equivalent output admittance

FIG. 15-4. Typical variations of high-frequency grounded-emitter h parameters of the 2N1692 p-n-p germanium power mesa transistor. (*Courtesy of Motorola, Inc.*)

15-6. High-frequency Amplifier Power Gain. In earlier sections we obtained the general expression for the maximum available power gain for a four-terminal device, subject to the conditions of conjugate matching at both input and output. The result used only the real terms of the device h parameters. Repeating this equation using the grounded-emitter transistor subscripts, we have

$$G_e = \frac{h_{fe}^2}{r_{ie}g_{oe}(1 + p)^2} \tag{15-48}$$

where

$$p = \sqrt{\frac{1 - Re\,(h_{re}h_{fe})}{r_{ie}g_{oe}}} \tag{15-49}$$

and where r_{ie} and g_{oe} are the real parts of h_{ie} and h_{oe}, respectively.

In the previous section we saw that

$$r_{ie} \approx r'_B \tag{15-50}$$

$$g_{oe} \approx \omega_T C_c \tag{15-51}$$

and that the magnitude of h_{fe}^2 is

$$h_{fe}^2 \approx \frac{\omega_T^2}{\omega^2} \tag{15-52}$$

To determine p, it is necessary to evaluate the real part of the mutual-transfer product, $Re\,(h_{re}h_{fe})$. It has been shown by Pritchard[3] that in most junction transistors, p may be approximated by unity with small error. Physically, this means that for the prescribed matching conditions, the resistance reflected into the input circuit from the output load, and vice versa, is negligible. In other words, we are saying that the real part of the transfer product is very close to zero. Therefore, inserting the final results given by Eqs. (15-50) to (15-52) into Eq. (15-48) and letting $p = 1$, we obtain the expression for the grounded-emitter power gain.[3]

$$G_e \approx \frac{\omega_T}{4\omega^2 r'_B C_c} \tag{15-53}$$

Expressing this result in terms of frequency, we have, finally,[4]

$$G_e \approx \frac{f_T}{8\pi f^2 r'_B C_c} \qquad 2f_{\alpha e} < f < 2f_{\alpha b} \tag{15-54}$$

where f = frequency of operation
f_T = current-gain-bandwidth frequency
r'_B = base resistance
$C_c = C_{Dc} + C_{Tc}$

C_{Tc} usually refers to the active portion of the collector, which is approximately equal to the active emitter area and the area bounded by the base contact. A more general definition of C_{Tc} would consider the area of the transistor which contains each of the elements that make up r'_B.

For the approximations involved, Eq. (15-54) holds very well for transistors that are conjugate-matched at the output and resistive-matched at the input. It is seen that the power gain is directly proportional to the current-gain-bandwidth frequency and inversely proportional to the product of base resistance and collector capacitance. Further, G_e will decrease as the square of the frequency, thereby decreasing at a rate of 6 db/octave. The variation of G_e with frequency for a typical high-frequency mesa amplifier transistor is shown in Fig. 15-5.

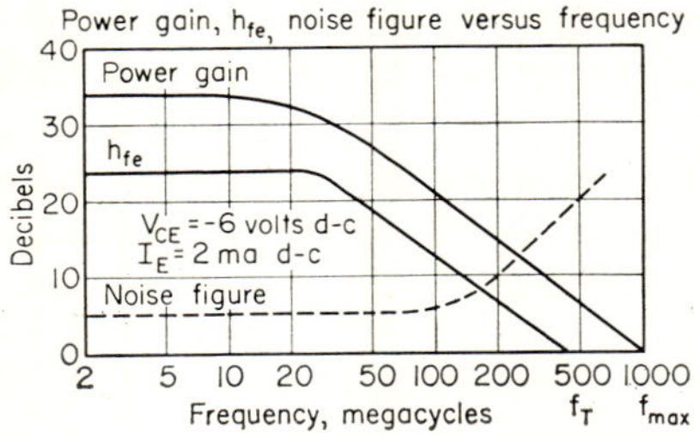

FIG. 15-5. Typical power gain versus frequency for the 2N700 p-n-p germanium amplifier mesa transistor. (*Courtesy of Motorola, Inc.*)

15-7. Maximum Frequency of Oscillation. It is interesting to note that for the frequency characteristics of the transistor given in Fig. 15-5, there is considerable power gain at the frequency f_T, for which the grounded-emitter current gain is equal to 1. At this frequency, Eq. (15-54) becomes

$$G_e \approx \frac{1}{4\omega_T r'_B C_c} \qquad \text{for } f = f_T \tag{15-55}$$

Even though the current gain is unity, we still have power gain because the ratio of the output impedance to the input impedance is greater than 1. In Eq. (15-55), the output resistance is $1/\omega_T C_c$ and the input resistance is r'_B. Therefore, f_T is not the uppermost limit of useful operation for the transistor.

We must now consider that particular frequency at which the grounded-emitter power gain is equal to 1. This frequency is denoted as f_{max} and is obtained by letting $G_e = 1$ in Eq. (15-54) and solving for f. Thus,

$$f_{max} \approx \sqrt{\frac{f_T}{8\pi r'_B C_c}} \tag{15-56}$$

Since the power gain is unity at this frequency, it becomes apparent that f_{max} represents the maximum frequency of oscillation of the transistor. This introduces the second important gain-bandwidth product for the transistor, namely,[5]

$$(\text{Power gain})^{1/2}\,(\text{bandwidth}) \approx \text{maximum oscillation frequency} \tag{15-57}$$

At this time, there are many transistors available having gain-bandwidth products in excess of 1 kilomegacycle.

15-8. Variation of Power Gain with Voltage and Current. Because of the individual dependence of such device parameters as f_T, r'_B, and C_c

on emitter current and collector voltage, we would expect the matched power gain of a junction transistor at high frequencies to display a distinct variation with d-c operating point. As a typical example, let us consider the grounded-emitter power-gain performance of the 2N700 transistor for which $G_e(f)$ was given in Fig. 15-5. For this p-n-p germanium mesa transistor, the typical power gain at 70 mc, $V_{CB} = -6$ volts d-c and $I_E = 2$ ma d-c, is 23 db, and the variation with voltage and current is shown in Fig. 15-6.

With Eq. (15-54) in mind, we may examine the collector-voltage variation first. As V_C is increased, we know that the collector depletion-layer widens, causing the collector junction capacitance C_{Tc} to decrease.

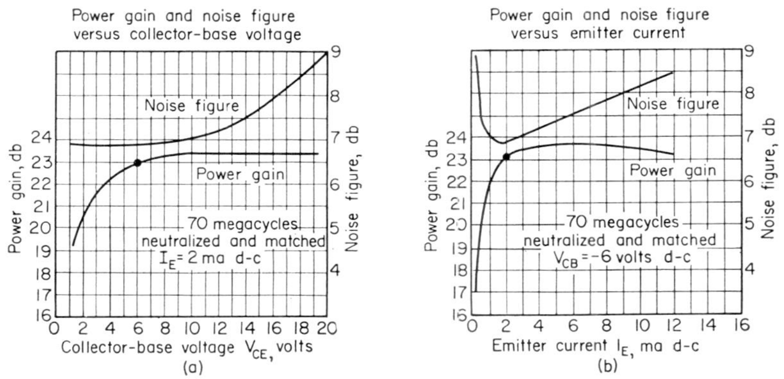

FIG. 15-6. Variations of power gain with voltage and current for the 2N700 p-n-p germanium mesa amplifier transistor. (*Courtesy of Motorola, Inc.*)

Additionally, the resulting decrease in base width W causes both f_T and base resistance to increase. However, because f_T varies as $1/W^2$ whereas r'_B varies as $1/W$, the total effect is an increase in the power gain. This is offset by the decrease in capacitance, and the power gain levels off with increasing voltage.

In a similar manner we can explain the variation with emitter current at a constant collector voltage. Firstly, we know that the collector diffusion capacitance varies directly with I_E; therefore, C_c increases with current. Secondly, because of high-level injection effects, r'_B decreases with emitter current. Thirdly, the emitter resistance r_e varies inversely with I_E, which causes f_T to increase because of the reduced bypassing effect of the emitter junction capacitance at the higher emitter currents. Combining these effects, we see that at low currents power gain is reduced because of the emitter capacitance and the large value of r_e. As I_E increases, the increase in C_c is offset by the simultaneous increase of f_T and decrease of r'_B, so that G_e reaches a maximum. At the higher currents, the power gain falls off because of the increase in diffusion capacitance and decrease of current gain.

15-9. Design Theory for Optimum Power-gain Bandwidth. The gain-bandwidth relationships depicted by Eqs. (15-56) and (15-57) are unquestionably the most important design equations for achieving optimum high-frequency performance from the transistor.[5] It is seen that the gain-band product is proportional to the ratio of f_T to the $r'_B C_c$ product.

$$f_{\max} \propto \frac{f_T}{r'_B C_c} \tag{15-58}$$

This ratio might be regarded as a high-frequency figure of merit for any particular transistor design. In this section, the relationship of Eq. (15-58) to the physical characteristics of the transistor such as base width, area, base impurity concentration, and collector resistivity will be evaluated in order to arrive at a design theory for maximizing the gain-band product.

In order to simplify the discussion, a generalized model of the transistor will be considered. This is necessary in order to arrive at a set of theoretically universal design criteria applicable to all transistor structures regardless of the fabrication process. For this model, it is assumed that the active collector area is equal to the emitter area A and that the base has a uniform sheet resistance based on an average mobility $\bar{\mu}$, an average impurity concentration $\bar{N}_B$, and a thickness W. Further, it is assumed that the collector capacitance and depletion-layer thickness are primarily functions of the average collector resistivity ρ_c and are virtually independent of the base and emitter characteristics. Finally, all series resistances are neglected so that $1/\omega_c$ becomes relatively insignificant in its effect on f_T. With this assumed model, we can write

$$f_{\max} \propto \frac{1}{r'_B C_c(1/\omega_e + 1/\omega_b + 1/\omega_d)} \tag{15-59}$$

If we take the reciprocal of the right-hand side of Eq. (15-59), we can examine each of the terms independently in order to establish minima which in turn maximize $f_{\max}$.

The first term of interest concerns the emitter time constant $1/\omega_e = r_e C_{Te}$, or the total product $r'_B C_c r_e C_{Te}$. Disregarding any effects of geometry on the magnitude of r'_B, we recognize that r'_B is proportional to the base sheet resistance or to $1/\bar{\mu}\bar{N}_B W$. The mobility term is also proportional to impurity concentration; although Figs. 4-3 and 4-4 indicate that electron and hole mobilities vary differently with N and with the parent semiconductor, it is reasonable to fit the following approximation for μ in the range of typical base dopings:

$$\mu \propto N^{-0.28} \tag{15-60}$$

Therefore, we establish that r'_B varies as $1/\bar{N}_B{}^{0.72}W$. The collector capacitance is proportional to $A/\sqrt{\rho_c}$. The emitter capacitance is proportional to $A\sqrt{\bar{N}_B}$. We have, finally,

$$r'_B C_c r_e C_{Te} \propto \frac{A^2 N_B{}^{0.22}}{W\sqrt{\rho_c} I_E} \tag{15-61}$$

It is seen that this product is minimized primarily by designing for small active junction areas, because of the A^2 dependence. Light base doping is also indicated by Eq. (15-61), but this is not a primary factor, because of the small exponent. D-c operation for large values of I_E has a very pronounced effect because of the reduction of the magnitude of r_e.

The second term of Eq. (15-59) is related to the base-transit time factor; we have the product $r'_B C_c/\omega_b$. We saw that $1/\omega_b$ is proportional to $W^2/\bar{\mu}$ where $\bar{\mu}$ is the minority-carrier mobility. Thus we have

$$\frac{r'_B C_c}{\omega_b} \propto \frac{AW}{\bar{N}_B{}^{0.44}\sqrt{\rho_c}} \tag{15-62}$$

In this proportion the minimum is dominated by the requirement for small base width W and heavy base doping.

The third term in Eq. (15-59) has to do with the transit time through the collector depletion layer and yields the product $r'_B C_c/\omega_d$, where $1/\omega_d$ is proportional to the depletion-layer thickness x_m. Since $x_m \propto \sqrt{\rho_c}$, we have, finally

$$\frac{r'_B C_c}{\omega_d} \propto \frac{A}{\bar{N}_B{}^{0.72}W} \tag{15-63}$$

The significant points here are the independence of collector resistivity and the fact that $\bar{N}_B{}^{0.72}W$ should be large. The latter is simply a direct argument for low sheet resistance or low r'_B.

From the results given by Eqs. (15-61) to (15-63), we can establish the criteria for optimizing $f_{\max}$ on the power-gain-bandwidth product. These are:

1. Design for very small base widths.
2. Design for very small emitter and collector areas.
3. Design for very low base sheet resistances or heavy base doping.
4. Design for high collector resistivities.
5. Design for low r'_B by selecting optimum emitter and base-contact geometry.

Each of the above design steps must be modified to some extent by giving consideration to the desired d-c characteristics for the transistor. For example, the limitation on (1) would be voltage punchthrough; the limitation on (2) would be the current rating; the limitation on (3) would be emitter efficiency or BV_{EBO}; the limitation on (4) may be collector series resistance.

Of particular interest is the transistor geometry problem implied by (5) for minimizing r'_B. As was discussed previously in Chap. 9, minimum r'_B can be obtained from a geometrical point of view by completely enclosing the emitter area in a very closely spaced base contact. The distance between the outside edge of the emitter and the inside edge of the base contact becomes critically important when one also considers the base-region self-crowding effect discussed in Chap. 10. Because of the small junction areas found in most high-frequency transistors, this current-density effect causes the injected emitter current to crowd to the periphery of the emitter (see Fig. 10-8). Consequently, the current-density capability of an emitter is proportional, not to the area, but to the peripheral length.[6] Therefore, most of the base region under the emitter does not significantly add to r'_B, and that portion of the base region between the emitter and base contact becomes the major factor. In view of this and the fact that a small emitter area is required for optimum f_{max}, the optimum emitter geometry is one that offers the maximum perimeter-to-area ratio. This leads to the choice of a thin rectangular stripe as shown in Fig. 15-7, with the base contact having the same form and spaced as close as possible.

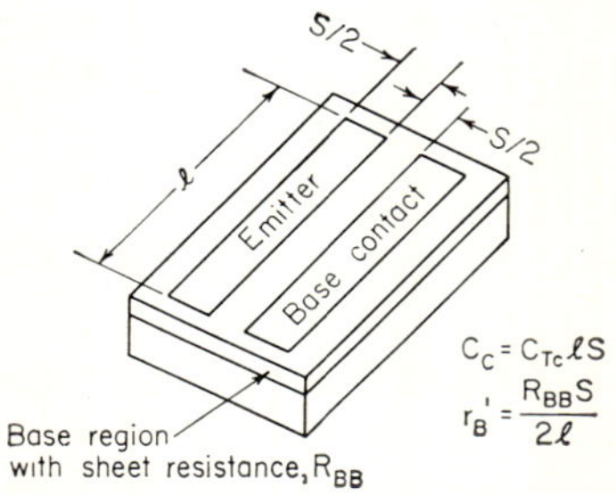

FIG. 15-7. Basic idealized transistor structure having stripe geometry for maximum emitter perimeter-to-area ratio. (*After J. M. Early.*)

It is interesting to note that the idealized structure given by Fig. 15-7 has an $r'_B C_c$ product which is independent of the stripe length l. This results from the fact that C_c is directly related to the area lS, r'_B is directly related to the ratio S/l, and in the product of the two, the l's cancel out. Thus, with regard to the product $r'_B C_c r_e C_{Te}$, if the stripe length is increased by a factor of three, for example, then the product is increased by the same amount due to the threefold increase in C_{Te}. However, this can be offset simply by increasing I_E by the same factor. Therefore, the fundamental structure of Fig. 15-7 can be scaled up in length without deteriorating the frequency response, provided that the current is increased by the same amount. (This, of course, assumes that $1/\omega_e$ is greater than $1/\omega_b$.) By so doing, high-frequency high-power transistors (high current ratings) can be readily designed. One approach[6] which achieves a long emitter edge length utilizes a "comb" emitter interleaved between the teeth of a "comb" base contact, as shown in Fig. 15-8. This structure is a 10-amp unit having an f_T of about 100 megacycles.

The structure of Fig. 15-7 has been analyzed by Early[5] for optimum gain-band product. With regard to stripe dimension S, it was shown that

Germanium: $$f_{\max} \approx \frac{7.5 \times 10^6}{S} \quad \text{cps} \tag{15-64}$$

Silicon: $$f_{\max} \approx \frac{4 \times 10^6}{S} \quad \text{cps} \tag{15-65}$$

where S is in centimeters. Furthermore, in optimum transistor designs in which the frequency response is limited primarily by the base width W,

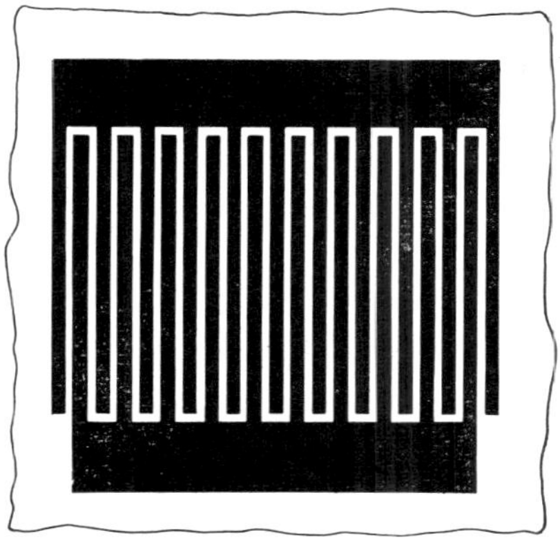

FIG. 15-8. n-p-n silicon power-transistor structure having a maximum collector current rating of 10 amp with an emitter edge length of 2.8 in. (*Courtesy of Pacific Semiconductors, Inc.*)

there is no difference in power gain between p-n-p and n-p-n devices. This is understandable if one considers that

$$\frac{r'_B C_c}{\omega_b} \propto \frac{W C_c}{\mu_p \mu_n \bar{N}_B} \tag{15-66}$$

It is seen that this factor is inversely related to the mobility product $\mu_p\mu_n$, which is constant for any particular level of $\bar{N}_B$.[7] Whether the device is n-p-n or p-n-p, one of the mobilities is a minority type associated with base-transit time and the other is a majority mobility associated with r'_B. However, since f_T alone is not dependent on r'_B, f_T for an n-p-n device is higher than for the equivalent p-n-p.

PROBLEMS

15-1. For the general four-terminal network of Fig. 15-1, show that the power gain for $r_g = 0$ is given by

$$G = \frac{h_{21}{}^2 r_L}{(1 + h_{22}r_L)(h_{11} + \Delta^h r_L)}$$

where $r_L = 1/g_L$ and Δ^h is the h-parameter determinant, equal to $(h_{11}h_{22} - h_{12}h_{21})$.

15-2. The grounded-base h parameters for a 2N404 p-n-p germanium-alloy transistor are as follows ($I_E = 1$ ma, $V_C = -5$ volts, $T_A = 25°C$):

$$h_{ib} = 28 \text{ ohms} \qquad h_{rb} = 8 \times 10^{-4}$$
$$h_{fb} = 0.98 \qquad h_{ob} = 0.9\ \mu\text{mho}$$

For the given bias and temperature conditions, calculate the low-frequency power gain for the grounded-emitter configuration, where $r_L = 20{,}000$ ohms and $r_g = 100$ ohms. Express the answer in decibels.

15-3. For the junction transistor described in Prob. 15-2, what should the values of r_L and r_g be for maximum grounded-emitter power gain? What is the matched power gain in decibels under these conditions?

15-4. Determine the maximum frequency of oscillation of an n-p-n germanium double-doped transistor (grown-junction process) at conditions of $I_E = -2$ ma, $V_C = 5$ volts, and $T_A = 25°C$. For this bar structure, the effective base width is 1 mil, the cross-sectional area is 100 mil^2, the base is uniform, having a resistivity of 0.2 ohm-cm, the collector is a step junction, and the collector resistivity is 4 ohm-cm. The contact to the base region is a single-line contact along one side of the bar. For this problem, assume that $\alpha \approx 1$ and that the only significant signal delay is due to base transit, such that $1/\omega_{\alpha b} \approx 1/\omega_b$.

15-5. Calculate G_e at 70 megacycles for the n-p-n silicon planar-epitaxial transistor described in Prob. 14-3 at the operating conditions of $V_{CE} = 20$ volts, $I_E = -10$ ma, and $T_A = 25°C$. Assume that at this level of current density, the active base resistance is between the edge of the diffused emitter and the inside edge of the base contact.

15-6. A p-n-p silicon microalloy transistor is to be designed for high power gain at a particular frequency. What base resistivity (uniform impurity profile) is required for maximum G_e? Assume that the majority-carrier mobility is the same as the corresponding minority value and that the alpha-cutoff frequency is determined only by the base-cutoff frequency ω_b.

15-7. A p-n-p germanium mesa transistor consists of a pair of 1×2 mil stripes (emitter and base) separated by 0.5 mil. f_T for this transistor is 500 megacycles and is assumed to be limited only by the $r_e C_{Te}$ time constant. The effective C_c is 0.4 $\mu\mu f$. It is desired to scale up this structure for high-power operation. What is the maximum emitter edge length of the final design if the tuned output impedance of the final power transistor is not to be less than 10 ohms?

REFERENCES

1. Hunter, L. P.: Handbook of Semiconductor Electronics, chap. 11, p. 20, McGraw-Hill Book Company, Inc., New York, 1956.
2. Pritchard, R. L.: Unpublished memo reports, General Electric Research Laboratory, Schenectady, New York.
3. Pritchard, R. L.: High-frequency Power Gain of Junction Transistors, *Proc. IRE*, vol. 43, pp. 1075–1085, September, 1955.
4. Pritchard, R. L.: Frequency Response of Grounded-base and Grounded-emitter Transistors, AIEE Winter Meetings, New York, January, 1954.
5. Early, J. M.: Structure-determined Gain-band Product of Junction Triode Transistors, *Proc. IRE*, vol. 46, pp. 1924–1927, December, 1958.
6. Roach, William E.: Transistor Scaling Theory, Pacific Semiconductor, Inc., 1960.
7. Giacoletto, L. J.: Comparative High-frequency Operation of Junction Transistors Made of Different Semiconductor Materials, *RCA Rev.*, vol. 16, pp. 34–42, March, 1955.

16

Junction-transistor Switches

16-1. Introduction to Switching Transistors. In addition to its use as an amplifier, the junction transistor serves as an excellent switching device for use in electronic computers or other forms of switching circuitry. Although switching circuits are many and varied, one can readily establish the fundamental requirements for the active element of the circuit for optimum switching performance. Let us examine these requirements in the light of an ideal switching transistor.

In general, a switching transistor has two states of conduction; it is either on or off. For the switch to be ideal, the off state must have a very high resistance, approaching that of an open circuit. In the on state, on the other hand, the ideal transistor switch should have a very low resistance, approaching that of a short circuit. In either case, the objective is to minimize the load power dissipated within the switch, whether it is on or off. The amount of power required to drive the ideal switch from one state to the other should be as small as possible.* Furthermore, the ideal device should be capable of switching large amounts of power; i.e., it should be able to withstand large voltages during the off condition and large currents during the on condition. Finally, in changing from one state to the other, the transistor should be able to switch as rapidly as possible. This requires that the transistor respond to the driving signal instantaneously. In other words, the transistor should have a very high gain-bandwidth response. It becomes apparent from the foregoing that the criteria for a junction-transistor switch have been established. In this chapter we shall discuss, in particular, the large-signal transient response of the transistor in order to relate the significant switching characteristics to the basic electrical parameters of the device.

* There is a lower practical limit to driving power, since it would not be desirable to have a device so sensitive that random noise pulses would cause it to switch.

In transistor switching circuits, the grounded-emitter configuration is by far the most widely used. For this arrangement, we shall be concerned with the collector output characteristic, since this will enable us to clearly understand the mechanism of switching from one state to another. A typical $V_C - I_C$ characteristic for a p-n-p junction transistor is given in Fig. 16-1. Examining this figure, we see that the output characteristic is divided into three defined regions of operation,[1,2,*] namely, the off region, the active region, and the on region. Each of these regions will be detailed separately.

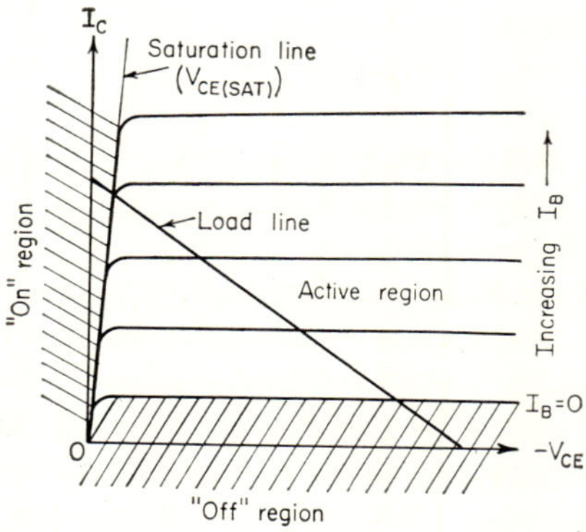

Region	Emitter	Collector
Off.	Reverse	Reverse
Active.	Forward	Reverse
On.	Forward	Forward

FIG. 16-1. Collector output characteristic for a p-n-p grounded-emitter transistor, illustrating three operating regions.

In the off region there exists the situation in which both the emitter and collector junctions are reverse-biased. Under these conditions, the collector current is very small, resulting in a large output impedance. As was shown in Chap. 9, the magnitude of this current is on the order of I_{CES}. It is seen that the off region is bounded by the curve for $I_B = 0$. This corresponds to the point of zero applied voltage to the base-to-emitter junction and also represents the boundary for the active region of the transistor. In Fig. 16-2 is a sketch of the minority-carrier concentration (holes) in the base layer for each of the three regions.[2] For the off condition, since both junctions are reverse-biased, the hole con-

* References, indicated in the text by superscript figures, are listed at the end of the chapter.

centration is zero at the junctions, as shown. At the approximate center of the base layer, the hole concentration is equal to the equilibrium value.

In the active region we have the normal mode of operation for the junction transistor; that is, the emitter is forward-biased and the collector is reverse-biased. In switching from off to on, the active region is traversed along the load line, as indicated in Fig. 16-1. As we shall see later, the speed of transition through the active region (transient response) is a function of the gain and frequency response in that region. The minority-carrier concentration for the active region, as shown in Fig. 16-2, is the familiar form discussed in previous chapters.

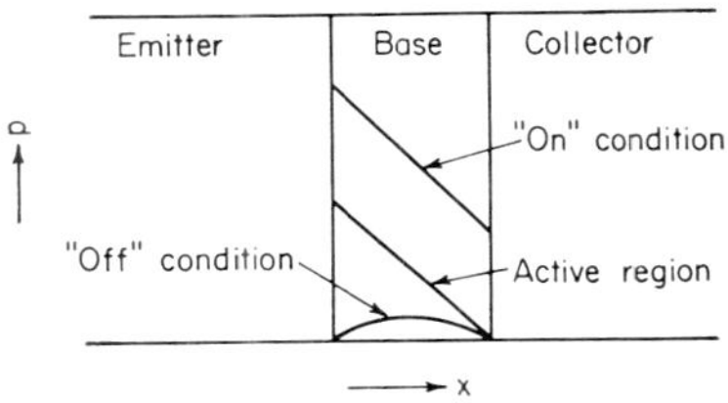

FIG. 16-2. Minority-carrier concentrations in the base layer as related to the region of operation.

The last of the three regions of operation is the on region, which is often referred to as a *saturation* condition. In this mode, the collector bias reverses polarity such that both emitter and collector junctions are forward-biased. As was shown in Sec. 9-8, the forward-bias potential of the emitter junction is slightly greater than that of the collector, resulting in a net negative potential from collector to emitter. This potential is represented by the saturation line drawn in Fig. 16-1, the slope of which corresponds to the saturation resistance of the transistor in the on condition. As is evident from Eq. (9-80), $V_{CE(\text{SAT})}$ also includes the effect of any series resistance in the collector region. The hole concentration in the base layer for the on condition is also included in Fig. 16-2. At the collector junction, the hole concentration increases significantly because of the forward-bias condition at the collector. However, the slope of the hole concentration in the on condition remains the same as that at the edge of the active region, in order to maintain the same collector current.

The next section will present a physical description of the mechanisms that take place as the transistor is switched through the three defined regions of operation.

16-2. Qualitative Description of Switching Process. Before we establish a specific mathematical analysis of the switching transistor, it would be appropriate to present a qualitative description of the physical events that occur when a transistor is switched from off to on and then to off again. Let us refer to the basic switching circuit shown in Fig. 16-3, in which the transistor is driven by a constant base-current source. When the transistor is turned on, the collector current I_C flows for an input base current I_{B1}. The potential $V_{BE(\text{OFF})}$ holds the transistor in the off state because of the reverse-biased emitter. The generator in the base circuit is a pulse generator and is assumed to generate an ideal step-function input pulse. For this circuit, the waveforms that are observed for the base and collector current pulses, respectively, are as shown in Fig. 16-4. One observes that the output collector pulse is far from being an exact replica of the input base pulse. The reasons for these particular wave-shape discrepancies will be explained in the paragraphs to follow.

At time t_0, the pulse generator delivers a step base current to the transistor. At this instant, the transistor is sitting in the off condition because of the emitter reverse-bias voltage $V_{BE(\text{OFF})}$. The collector current that is flowing is extremely low (on the order of I_{CES}); since this is negligible, the voltage on the collector junction is equal to the sum of the supply voltage V_{CC} and the turnoff voltage $V_{BE(\text{OFF})}$. After t_0, the base current rises immediately to I_{B1}, but it is observed that the collector current does not begin to increase until t_1. The time between t_0 and t_1 is called the *delay time* t_d and is defined as the time required to bring the transistor from the initial off condition to the edge of conduction, i.e., to the beginning of the active region. This may be defined as that instant of time, t_1, at which the applied base-to-emitter voltage is zero. Physically, the finite time required for t_d comes about because of the reverse bias on both the emitter and collector junctions. As the effective base-to-emitter voltage goes from $V_{BE(\text{OFF})}$ to zero, the depletion layers on both junctions must reduce in thickness accordingly. This corresponds to an increase in the junction capacitances; the delay time is the time required to charge these capacitances to the new voltage level. It should be apparent that if $V_{BE(\text{OFF})} = 0$, then $t_d = 0$.

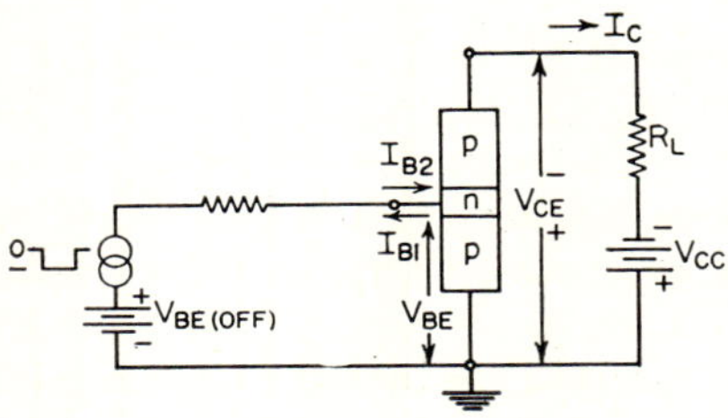

FIG. 16-3. Basic grounded-emitter transistor switching circuit. The base is driven by a constant-current source.

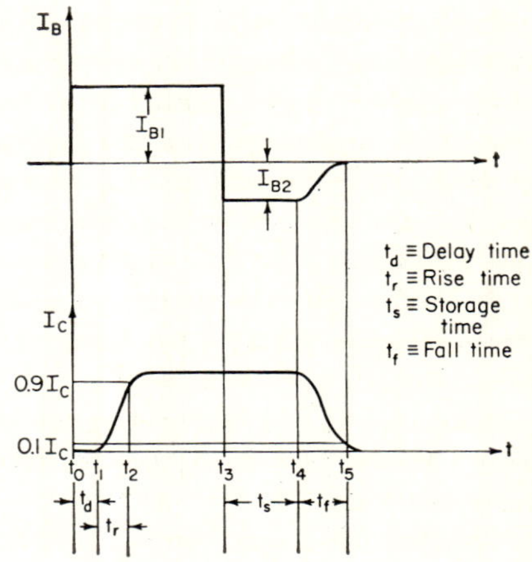

FIG. 16-4. Input base-current waveform and output collector-current waveform for a transistor, illustrating definition of switching times.

At time t_1, the operating point of the transistor is at the beginning of the active region, in which the emitter starts to become forward-biased and begins to inject current into the base. Now the collector current begins to increase toward its saturation value, corresponding approximately to V_{CC}/R_L. However, rather than increasing instantaneously at t_1, it requires a finite time to reach 90 per cent of the final value. This occurs at time t_2; the time interval $t_2 - t_1$ is defined as the *rise time* t_r of the collector current pulse. Rise time is attributed to the fact that there exists a finite transit delay between the base and collector currents and is a manifestation of the frequency response and current gain (beta) of the active region of the transistor. It should be recognized that neither the gain nor the frequency response remains constant through the active region. D-c beta (h_{FE}) will vary with current for the reasons already discussed in Chap. 10 and may pass through a maximum value depending on the $h_{FE}(I_C)$ characteristic and the magnitude of current to be switched. Also, the effective base width increases as the collector voltage decreases from the supply value to the saturation value. This, in turn, causes the base-transit time to increase so that the frequency response falls off.

During the same active switching interval, both the emitter and collector transition capacitances must be charged to account for the depletion-layer changes with voltage. In the emitter, the capacitance increases because the contact potential is reduced by the forward-bias portion of V_{BE}; in the collector, the capacitance also increases and must be charged through the collector series resistance, which is predominantly the external load resistor R_L. All these effects take place simultaneously and will be analyzed quantitatively to determine the collective effect on rise time.

The transistor will remain in the on state as long as the input base current I_{B1} is maintained. With reference to Fig. 16-4 again, at time t_3 the base input pulse steps off immediately; however, it is observed that the collector pulse does not respond until time t_4. The time interval between t_3 and t_4 is referred to as the *storage time* t_s and is attributed to the same basic storage phenomenon described in Chap. 6 for the p-n junction. The storage time is a measure of the time required for the minority carriers in the base and collector to recombine back to the level corresponding to the boundary between the active and saturation regions. These excess carriers arise because the collector junction becomes forward-biased when the base current I_{B1} is greater than the I_B necessary to produce I_C; i.e., $I_{B1} > I_C/h_{FE}$. Thus, storage time is related to a carrier-recombination process and is a measure of the minority-carrier lifetime in the base and collector regions.

Finally, at time t_4, the transistor comes out of saturation and the operating point traverses the load line again through the active region into the off state. This is the turnoff portion of the collector waveform; the time interval between t_4 and t_5 is defined as the *fall time* t_f. At t_5, the collector current has reduced to $0.1I_C$. The description of the switching process for fall time is similar to that for rise time, except that the active region is traversed in the reverse direction.

In summary, we see that in response to a step input of base current, the collector current requires a total *turnon time*, which is

$$t_{\mathrm{ON}} = t_d + t_r \tag{16-1}$$

Also, when the base current is removed as a step, the collector current requires a total *turnoff time*, which is

$$t_{\mathrm{OFF}} = t_s + t_f \tag{16-2}$$

16-3. Stored-base-charge Transistor Analysis. In all the analyses of the transistor that have appeared in the preceding chapters, the theory has been based on the argument that the transistor is a current-controlled device; i.e., the collector current is affected by changes in either the emitter current or base current. Thus, the basic equations derived for d-c characteristics, h parameters, frequency response, etc., are all a function of the currents of the transistor. Without a doubt, this is a very appropriate and worthwhile approach to analyzing the transistor theoretically from either the steady-state or small-signal point of view, because one can readily approximate linear operation for the transistor. However, when one has to apply a large-signal analysis to the transistor, as is the case in describing switching characteristics, the current approach becomes relatively complex because of the high degree of nonlinearity displayed by the transistor for large changes in current and/or voltage.

To aid in the large-signal analysis of the switching transistor, a powerful new concept has been introduced by Beaufoy and Sparkes, in which they regard the transistor as a *charge-controlled* device.[3] In this concept, the collector current is related to the charge in the base; in order to effect a change in collector current there must be a change in the total charge in the base. For example, in the grounded-emitter transistor, the base current provides the necessary total base charge which controls the collector current.

The fundamental relationship in charge-control transistor theory is the obvious fact that current is charge per unit time, or

$$\Delta I = \frac{\Delta Q}{\tau} \tag{16-3}$$

where τ is defined as a time constant which relates the current to the charge. Therefore, if we are to relate the collector current I_C to the total base charge Q_B, the collector time constant τ_C is Q_B/I_C. In a similar manner, the emitter and base currents can be related to Q_B, so that we have

Emitter time constant: $$\tau_E = \frac{Q_B}{I_E} \tag{16-4}$$

Base time constant: $$\tau_B = \frac{Q_B}{I_B} \tag{16-5}$$

Collector time constant: $$\tau_C = \frac{Q_B}{I_C} \tag{16-6}$$

We shall now proceed to relate these time constants to the characteristics of the transistor. Let us consider the active region of the p-n-p transistor in which the emitter is forward-biased and the collector is reverse-biased. Under these conditions, the emitter is injecting holes into the base, and if $W \ll L_{pb}$, then a linear hole gradient is established in the base which decreases to zero at the collector. As was shown in Chap. 8, the emitter current is given by

$$I_E = -qAD_{pb}\frac{dp}{dx} \tag{16-7}$$

In the base region,

$$\frac{dp}{dx} \approx -\frac{p_{nb}\epsilon^{qV_E/kT}}{W} \tag{16-8}$$

so that Eq. (16-7) becomes

$$I_E \approx \frac{qAD_{pb}p_{nb}\epsilon^{qV_E/kT}}{W} \tag{16-9}$$

But $qAp_{nb}\epsilon^{qV_E/kT}$ is the charge per unit length just at the base side of the emitter junction ($x = 0$); it decreases linearly to zero at $x = W$. Therefore, the total charge in the base is the area of the charge gradient (triangle), or

$$Q_B = \frac{qAp_{nb}\epsilon^{qV_E/kT}W}{2} \tag{16-10}$$

Substituting Eq. (16-10) into (16-9), we obtain

$$I_E \approx \frac{2D_{pb}Q_B}{W^2} \tag{16-11}$$

From Eq. (9-13) for $\gamma \approx 1$ we see that $W^2/2D_{pb}\tau_{\text{eff}} \approx I_B/I_E$, so that Eq. (16-11) becomes, finally,

$$I_B = \frac{Q_B}{\tau_{\text{eff}}} \tag{16-12}$$

Thus we see that the base time constant τ_B is equal to the effective minority-carrier lifetime in the base; i.e., surface recombination is included. In other words, base current flows to maintain the total base charge constant by making up for the amount of charge being lost by recombination.

To determine the emitter time constant, we refer to Eq. (16-11), where we saw that

$$\tau_E = \frac{Q_B}{I_E} = \frac{W^2}{2D_{pb}} = \tau_B(1 - \alpha) \tag{16-13}$$

From Eq. (14-8), we see that for the uniform base, the base-cutoff frequency is given as

$$\frac{1}{\omega_b} = \frac{W^2}{2.43D_{pb}} \tag{16-14}$$

Substituting this result into Eq. (16-13), we have[3]

$$\tau_E = \frac{1.2}{\omega_b} \qquad \text{uniform base} \tag{16-15}$$

For a graded base, it was shown in Sec. 8-7 that $\tau_E \approx W^2/4D_{pb}$; therefore,

$$\tau_E = \frac{0.6}{\omega_b} \qquad \text{graded base} \tag{16-16}$$

where ω_b is already corrected by the ln (N'_B/N_{BC}) term.

Finally, for the collector time constant, since $I_C = \alpha I_E$, we get, simply,

$$\tau_C = \frac{\tau_E}{\alpha} \tag{16-17}$$

With each of the time constants defined, we are now in a position to apply them to the large-signal transient response of the transistor. The basic equation is the equation of charge continuity, which is [3,4]

$$I = \frac{dQ}{dt} + \frac{Q}{\tau} \tag{16-18}$$

If this result is integrated over a given time interval, then

$$\int_0^t I\,dt = \int_{Q_1}^{Q_2} dQ + \int_0^t \frac{Q\,dt}{\tau} \tag{16-19}$$

This says that the total charge supplied during the interval t is equal to the amount of charge necessary to change the current in the volume to a new level plus the amount necessary to replenish that lost by recombination. Herein lies the beauty of the charge concept for large-signal analysis. In solving Eq. (16-19), one is concerned only with the absolute total change of charge during the time interval as determined by the limits of integration, thereby eliminating any concern for the variations of the time constants during the same interval.

In order to attach physical meaning to all of the foregoing, we shall now present a qualitative description[3] of how the total base charge of a grounded-emitter transistor changes as the base current shifts suddenly from I_{B1} to a new level I_{B2}. In this particular situation, we shall assume that the switching all takes place in the active region. Reference is made to the switching circuit of Fig. 16-3 and the base-charge diagram in Fig. 16-5. At the initial current level, the charge distribution in the base appears as the unshaded portion of Fig. 16-5. The charge level below

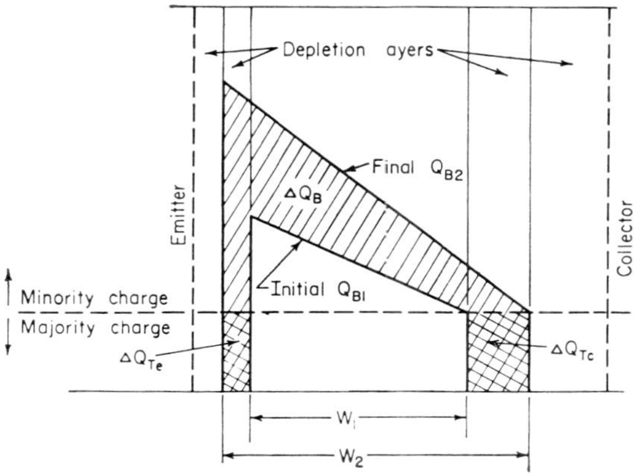

FIG. 16-5. Base-charge distributions illustrating switching in the active region. The shaded area denotes additional stored charge.

the dashed line corresponds to the equilibrium majority-carrier electrons in the n-type base. The charge above the dashed line represents the minority-carrier hole concentration injected by the emitter. This hole gradient is matched by an equal charge of electrons injected by the base in order to maintain neutrality.[3] Also shown in Fig. 16-5 are the depletion layers of the emitter and collector junctions, which have spread to establish the effective base width W_1. Thus, we have described the nature of the base charge at the initial active condition where the base current I_{B1} produces a collector current I_{C1}.

When I_{B1} is switched rapidly to I_{B2}, several changes occur simultaneously. Firstly, the increased forward-bias potential at the emitter causes the depletion layer to narrow, corresponding to an increase in the emitter transition capacitance. Secondly, the level of injected holes increases and the gradient steepens, corresponding to the new level of collector current I_{C2}. Thirdly, the increase in collector current causes the collector junction voltage to decrease due to the larger voltage drop across the load resistor R_L. This causes the collector depletion layer to narrow, establishing an increase in the collector transition capacitance.

Finally, because of both depletion-layer variations, the base width increases from its initial value W_1 to its final value W_2.

To account for all the changes described, an additional charge, equal to all the shaded area shown in Fig. 16-5, must be injected into the base region. This charge is composed of several parts. Below the dashed line, the shaded regions represent the amount of majority-electron charges necessary to neutralize the impurity ions exposed by the reduced depletion regions. These are denoted by ΔQ_{Te} and ΔQ_{Tc}, the additional charges in the transition capacitances of the junctions. The shaded area above the dashed line is the charge required to set up the new current level and maintain it against the recombination rate. This area also includes the extra base charge required to compensate for the increased base width or, more specifically, to charge up the emitter and collector diffusion capacitances. It should be recalled that diffusion capacitance is the amount of charge that must be added or removed from the base when a voltage variation causes the base width to change; i.e., $C_D = dQ/dV$. To express the total base-charge variation mathematically, we use the form of Eq. (16-19).[3,5,6]

$$\int_0^t I_B\,dt = \int_{Q_{Te1}}^{Q_{Te2}} dQ_{Te} + \int_{Q_{Tc1}}^{Q_{Tc2}} dQ_{Tc} + \int_{Q_{B1}}^{Q_{B2}} dQ_B + \int_0^t \frac{Q_B\,dt}{\tau_B} \tag{16-20}$$

Since the first two right-hand terms represent the transition-region charges, we can change their limits of integration, and we have

$$\int_0^t I_B\,dt = \int_{V_{BE1}}^{V_{BE2}} C_{Te}\,dV_{BE} + \int_{V_{CB1}}^{V_{CB2}} C_{Tc}\,dV_{CB} + \int_{Q_{B1}}^{Q_{B2}} dQ_B + \int_0^t \frac{Q_B\,dt}{\tau_B} \tag{16-21}$$

It was the purpose of this section to illustrate the general application of the base-charge theory to the solution of large-signal transistor problems. In the sections to follow, the concept will be utilized to derive the specific equations for the delay, rise, storage, and fall times of the transistor in the switching circuit of Fig. 16-3.

16-4. Delay Time. In Sec. 16-2, delay time t_d was defined as the time required to bring the transistor from the initial off condition to the edge of the active region. To do this, the base requires a total charge equal to that necessary to charge the transition capacitances of the emitter and collector junctions as the voltage changes by the amount $\Delta V = V_{BE(\text{OFF})}$. Since we are at the edge of conduction, we can assume that $I_C = 0$ (neglecting I_{CES}), and therefore $Q_B \approx 0$. The charge equation is written as[7]

$$I_B = \frac{dQ_{Te}}{dt} + \frac{dQ_{Tc}}{dt} \tag{16-22}$$

Integrating this result with respect to time and changing the limits, we have

$$\int_0^{t_d} I_B\,dt = \int_{-V_{BE(\text{OFF})}}^{0} C_{Te}\,dV_{BE} + \int_{-(V_{CC}+V_{BE(\text{OFF})})}^{-V_{CC}} C_{Tc}\,dV_{CB} \tag{16-23}$$

where V_{CC} is the collector supply voltage. In this equation we cannot assume that the transition capacitances are constants, since we know that they vary as $V^{-1/n}$. In most transistors, the emitter approximates a step junction, and therefore C_{Te} varies as $V_{BE}^{-1/2}$, or

$$C_{Te} = C'_{Te}V_{BE}^{-1/2} \tag{16-24}$$

where C'_{Te} is the emitter capacitance measured at a total voltage $(V_{BE} + V_T)$ equal to 1 volt. Further, for the collector capacitance it may be assumed that C_{Tc} varies as $V_{CB}^{-1/2}$, regardless of whether the junction is a step or is graded. This is so for the graded case as long as the magnitude of V_{CC} is high enough to establish square-root behavior. At lower values of V_{CC}, C_{Tc} will vary as $V_{CB}^{-1/2.5}$ or $V_{CB}^{-1/3}$. Nevertheless,

$$C_{Tc} \approx C'_{Tc}V_{CB}^{-1/2} \tag{16-25}$$

where C'_{Tc} is the collector capacitance for $(V_{CB} + V_T) = 1$ volt. Substituting Eqs. (16-24) and (16-25) into Eq. (16-23) and integrating, we have

$$I_{B1}t_d = 2C'_{Te}V_{BE(\text{OFF})}^{1/2} + 2C'_{Tc}[(V_{CC} + V_{BE(\text{OFF})})^{1/2} - V_{CC}^{1/2}] \tag{16-26}$$

or, finally,

$$t_d = \frac{2}{I_{B1}}\{C'_{Te}V_{BE(\text{OFF})}^{1/2} + C'_{Tc}[(V_{CC} + V_{BE(\text{OFF})})^{1/2} - V_{CC}^{1/2}]\} \tag{16-27}$$

Thus we see that the delay time increases with the magnitude of the turnoff voltage $V_{BE(\text{OFF})}$ and decreases as the turnon base current I_{B1} becomes larger. From a transistor-design point of view, since the second term of Eq. (16-27) is usually negligible when $V_{BE(\text{OFF})} \ll V_{CC}$, we see that delay time is minimized when the emitter transition capacitance is small. This means that one should design for small emitter areas primarily.

16-5. Rise Time. Rise time t_r is defined as the time required for the collector current to increase to 90 per cent of its final value in the switching circuit. Essentially, this transition occurs completely in the active region of the transistor, starting at the edge of conduction and ending just below the edge of saturation. Based on the discussions in the earlier sections, we can write the charge equation as[3,6]

$$I_B = \frac{dQ_{Te}}{dt} + \frac{dQ_{Tc}}{dt} + \frac{dQ_B}{dt} + \frac{Q_B}{\tau_B} \tag{16-28}$$

where τ_B is the base time constant. The first right-hand term corresponds to the charging of the emitter junction capacitance as the forward-bias voltage reduces the junction contact potential; the second term corresponds to the charging of the collector capacitance as the collector voltage decreases with increasing current; the third and fourth terms represent the charging of the base region to attain the new collector-current level and maintain it against the recombination rate.

Eq. (16-28) is a first-order differential equation and may be readily solved by separation of variables. It may be arranged as

$$\frac{1}{I_B - Q_B/\tau_B} = \frac{dt}{dQ_{Te} + dQ_{Tc} + dQ_B} \tag{16-29}$$

Since we are interested in a solution in terms of collector current I_C, we can multiply both sides of Eq. (16-29) by dI_C and integrate the result. Since the collector current increases from 0 to $0.9I_C$ during the time interval from 0 to t_r, we have, finally,

$$\int_0^{0.9I_C} \frac{dI_C}{I_{B1} - Q_B/\tau_B} = \int_0^{t_r} \frac{dt}{dQ_{Te}/dI_C + dQ_{Tc}/dI_C + dQ_B/dI_C} \tag{16-30}$$

Note that $I_B = I_{B1}$, since we are assuming that the base current steps up instantaneously at the start of turnon. The dQ/dI_C terms in the denominator of Eq. (16-30) are time constants as defined by Eq. (16-3). The emitter time constant is given by

$$\frac{dQ_{Te}}{dI_C} = \frac{1}{I_C}\int_{Q_{Te1}}^{Q_{Te2}} dQ_{Te} = \frac{1}{I_C}\int_{V_{BE1}}^{V_{BE2}} C_{Te}\, dV_{BE} \tag{16-31}$$

During the rise-time interval, the applied emitter voltage changes from 0 to approximately $V_T/2$, where V_T is the junction contact potential. Assuming a step junction,

$$\frac{dQ_{Te}}{dI_C} \approx \frac{1}{I_C}\int_0^{V_T/2} \frac{C'_{Te}\, dV_{BE}}{V_{BE}^{1/2}} = \frac{2C'_{Te}(V_T/2)^{1/2}}{I_C} \tag{16-32}$$

where C'_{Te} is the capacitance at 1 volt. However, Eq. (16-32) can be modified such that

$$\frac{dQ_{Te}}{dI_C} \approx \frac{C_{Te}V_T}{I_C} \tag{16-33}$$

where C_{Te} is the emitter capacitance at the final current level. Since V_T/I_C is the d-c junction resistance, we can approximate the time constant as

$$\frac{dQ_{Te}}{dI_C} \approx R_E C_{Te} \approx \frac{1}{\omega_e} \tag{16-34}$$

In a similar manner, the collector-junction time constant is given by

$$\frac{dQ_{Tc}}{dI_C} = \frac{1}{I_C} \int_{Q_{Tc1}}^{Q_{Tc2}} dQ_{Tc} = \frac{1}{I_C} \int_{V_{CB1}}^{V_{CB2}} C_{Tc}\, dV_{CB} \tag{16-35}$$

Assuming that the collector depletion layer varies as $V_{CB}{}^{1/2}$, we have

$$\frac{dQ_{Tc}}{dI_C} \approx \frac{1}{I_C} \int_{-V_{CC}}^{-V_{CE(\text{SAT})}} \frac{C'_{Tc}\, dV_{CB}}{V_{CB}{}^{1/2}} \approx \frac{2C'_{Tc} V_{CC}{}^{1/2}}{I_C} \tag{16-36}$$

In this result, the $V_{CE(\text{SAT})}$ term is neglected because it is usually quite small compared to V_{CC}. We can express Eq. (16-36) in terms of capacitance at the initial voltage V_{CC}. Also, we recognize that[7,8]

$$R_L = \frac{V_{CC} - V_{CE(\text{SAT})}}{I_C} \approx \frac{V_{CC}}{I_C} \tag{16-37}$$

Therefore, Eq. (16-36) becomes

$$\frac{dQ_{Tc}}{dI_C} \approx 2R_L C_{Tc} \qquad \text{step junction} \tag{16-38}$$

For a graded junction, where $C_{Tc} \propto V_{CB}{}^{-1/3}$, it is easily shown that

$$\frac{dQ_{Tc}}{dI_C} \approx 1.5 R_L C_{Tc} \qquad \text{graded junction} \tag{16-39}$$

Thus we see that the charging of the collector capacitance through the load resistor can be a significant factor.[8] It should be understood that the collector resistance r_{SC} is in series with R_L but is neglected because we assume $V_{CE(\text{SAT})}$ is small.

The final time constant in Eq. (16-30) is the dQ_B/dI_C term.

$$\frac{dQ_B}{dI_C} = \frac{1}{I_C} \int_{Q_{B1}}^{Q_{B2}} dQ_B = \frac{Q_{B2} - Q_{B1}}{I_C} \tag{16-40}$$

Since the initial total charge Q_{B1} is zero ($I_C \approx 0$),

$$\frac{dQ_B}{dI_C} \approx \frac{Q_{B2}}{I_C} = \tau_C \tag{16-41}$$

From Eqs. (16-15) to (16-17), we have

$$\frac{dQ_B}{dI_C} \approx \frac{1.2}{\alpha\omega_b} \qquad \text{uniform base} \tag{16-42}$$

$$\frac{dQ_B}{dI_C} \approx \frac{0.6}{\alpha\omega_b} \qquad \text{graded base} \tag{16-43}$$

where ω_b is the base-cutoff frequency at the edge of saturation.

Examining Eqs. (16-34), (16-38), (16-42), and (16-43), we see that they are in the form of

$$\frac{dQ_{Te}}{dI_C} + \frac{dQ_{Tc}}{dI_C} + \frac{dQ_B}{dI_C} \approx \frac{1}{\omega_e} + \frac{K}{\alpha\omega_b} + 2R_LC_{Tc} \tag{16-44}$$

However, the first two terms may be approximated by $1/\omega_T$, the current-gain bandwidth frequency (see Eq. 14-28), wherein we neglect the depletion-layer transit time since this is small at the edge of saturation. Therefore we obtain, finally,

$$\frac{dQ_{Te}}{dI_C} + \frac{dQ_{Tc}}{dI_C} + \frac{dQ_B}{dI_C} \approx \frac{1}{\omega_T} + 1.7R_LC_{Tc} \tag{16-45}$$

The constant multiplying R_LC_{Tc} has been selected as a nominal value to generalize the result for all transistor types. Thus Eq. (16-45) may be substituted into Eq. (16-30) to solve for t_r.

To integrate the left-hand part of Eq. (16-30), we recognize from Eqs. (16-13) and (16-17) that

$$\frac{Q_B}{\tau_B} = \frac{\tau_C I_C}{\tau_B} = \frac{I_C(1-\alpha)}{\alpha} = \frac{I_C}{h_{FE}} \tag{16-46}$$

Substituting this result into Eq. (16-30) and completing the integration, we have

$$t_r \approx h_{FE}\left(\frac{1}{\omega_T} + 1.7R_LC_{Tc}\right)\ln\frac{h_{FE}I_{B1}}{h_{FE}I_{B1} - 0.9I_C} \tag{16-47}$$

which is the final expression for rise time.* All terms except C_{Tc} are measured at the edge of saturation. One observes that the logarithm function is multiplied by both h_{FE} and a time constant which is the effective cutoff frequency of the transistor having R_L as a load resistor. The logarithm function times h_{FE} is called the *circuit-drive function* and is plotted in Fig. 16-6 as a function of alpha $[\alpha = h_{FE}/(1 + h_{FE})]$ for different values of the circuit current ratio I_C/I_{B1}. It is apparent that this function (or rise time) is reduced by the choice of a high-h_{FE} transistor and high circuit drive (low I_C/I_{B1} ratio).

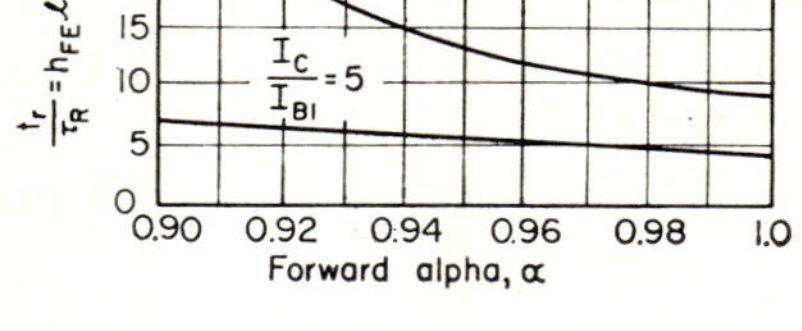

FIG. 16-6. Magnitude of circuit-drive function for rise time t_r.

From a transistor-design point of view, Eq. (16-47) indicates that low rise time is achieved with high f_T. Thus, the requirement for very

* This result is very similar to that given by Moll.[2]

small emitter and collector junction areas becomes extremely important if very high switching speeds are needed.

16-6. Storage Time. Once the transistor is fully turned on, it remains in saturation because the driving base current I_{B1} is greater than that necessary to establish the collector current I_C, that is, $I_{B1} > I_C/h_{FE}$. In saturation, the collector junction becomes forward-biased and the charge diagram[3] appears as shown in Fig. 16-7. In the base region, the total stored charge in saturation is denoted by Q_{BS}. It is also seen that the carrier concentration at the collector is not zero, but has a finite value depending on the magnitude of the collector forward bias in saturation. Because the collector is injecting holes into the base (for p-n-p), one also has to consider the injection of electrons into the collector from the same junction potential. If the minority-carrier concentration in the collector is at all appreciable (high collector resistivity), the charge stored in the collector cannot be neglected. In Fig. 16-7, the collector charge is denoted by Q_{CS}. Thus the total stored transistor charge is $Q_{BS} + Q_{CS}$. If Q_B is the amount of base charge required to bring the transistor to the boundary between active and saturated operations, then the excess stored charge is

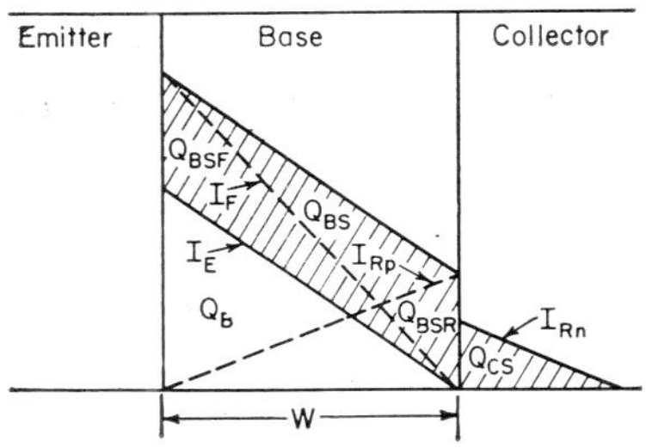

FIG. 16-7. Transistor charge distribution in saturation region. The shaded area denotes excess stored charge. Illustrated also is the stored charge in the collector region.

$$Q_X = Q_{BS} + Q_{CS} - Q_B \tag{16-48}$$

The crosshatched portions of Fig. 16-7 represent the excess charge Q_X.

Thus, storage time t_s is the time required for the transistor to come out of saturation, or specifically, the time required to reduce the excess charge to zero. The latter occurs when the collector forward potential is reduced to zero and starts to reverse polarity. The storage interval begins when the base current I_{B1} is reversed to a turnoff value I_{B2}. Since the excess base current may be defined as[3,9]

$$I_{BX} = I_{B1} - I_B \tag{16-49}$$

the storage time constant is

$$\tau_S = \frac{dQ_X}{dI_B} = \frac{Q_X}{I_{BX}} \tag{16-50}$$

To aid in the analysis of the saturation region, we can make use of the concept that the total base charge is made up of two components,[2]

$$Q_{BS} = Q_{BSF} + Q_{BSR} \tag{16-51}$$

where these total charges are defined by the regions shown in Fig. 16-7. The first term corresponds to a forward emitter-to-collector current I_F; the second term is a reverse collector-to-emitter current I_R. The latter current is

$$I_R = I_{Rp} + I_{Rn} \tag{16-52}$$

where I_{Rp} is the hole current injected into the base and I_{Rn} is the electron current injected into the collector. Substituting Eq. (16-51) into Eq. (16-48) and the latter into Eq. (16-50), we have

$$\tau_S = \frac{dQ_X}{dI_B} = \frac{Q_{BSF} + Q_{BSR} + Q_{CS} - Q_B}{I_{BX}} \tag{16-53}$$

or, in terms of time constants,

$$\tau_S = \frac{dQ_X}{dI_B} = \frac{\tau_E I_F + \tau_{ER} I_{Rp} + \tau_{CS} I_{Rn} - \tau_E I_E}{I_{BX}} \tag{16-54}$$

where τ_{ER} is the emitter time constant in the reverse direction and τ_{CS} is the time constant of the collector region. We can now proceed to simplify Eq. (16-54).

Firstly, if base recombination is sufficiently low so that base transport is unity in either direction, we can establish a reverse alpha given by the reverse collector efficiency as

$$\alpha_R \approx \gamma_R = \frac{I_{Rp}}{I_{Rp} + I_{Rn}} = \frac{I_{Rp}}{I_R} \tag{16-55}$$

From this relation, we see that $I_{Rp} = \alpha_R I_R$ and $I_{Rn} = (1 - \alpha_R) I_R$.

Secondly, to determine the expression for I_{BX} in terms of α_R and I_R, we refer to Eq. (16-49), in which[3]

$$I_{B1} = (1 - \alpha) I_F + (1 - \alpha_R) I_R \tag{16-56}$$

and

$$I_B = (1 - \alpha) I_E \tag{16-57}$$

Therefore

$$I_{BX} = (1 - \alpha)(I_F - I_E) + (1 - \alpha_R) I_R \tag{16-58}$$

Since $I_C = \alpha I_E = \alpha I_F - I_R$,

$$I_F - I_E = \frac{I_R}{\alpha} \tag{16-59}$$

Substituting Eq. (16-59) into Eq. (16-58) and simplifying, we obtain[3]

$$I_{BX} = \frac{I_R}{\alpha} (1 - \alpha \alpha_R) \tag{16-60}$$

Inserting Eq. (16-60) and the relations from Eq. (16-55) into Eq. (16-54) and rearranging, we obtain the final expression for the storage time constant,

$$\tau_S = \frac{dQ_X}{dI_B} = \frac{Q_X}{I_{BX}} = \frac{\tau_E + \alpha \alpha_R \tau_{ER}}{1 - \alpha \alpha_R} + \frac{\alpha (1 - \alpha_R) \tau_{CS}}{1 - \alpha \alpha_R} \tag{16-61}$$

This result will be discussed in the latter part of this section.

The equation for storage time t_s is obtained by solving the differential charge equation, which is[3]

$$I_B = \frac{Q_B}{\tau_B} + \frac{Q_{BS}}{\tau_S} + \frac{dQ_B}{dt} + \frac{dQ_{BS}}{dt} \tag{16-62}$$

Since the storage-time interval ends when $Q_{BS} = 0$, during that interval $dQ_B/dt = 0$; Q_B is constant during that time and equal to $\tau_C I_C$. Furthermore, during the same interval, I_B was switched to I_{B2}. Thus, Eq. (16-62) becomes

$$I_{B2} - \frac{\tau_C I_C}{\tau_B} - \frac{Q_{BS}}{\tau_S} = \frac{dQ_{BS}}{dt} \tag{11-63}$$

Introducing dI_B and separating the variables for integration, we obtain

$$\int_{I_{BX}}^{0} \frac{dI_B}{I_{B2} - \tau_C I_C/\tau_B - Q_{BS}/\tau_S} = \int_0^{t_s} \frac{dt}{dQ_{BS}/dI_B} \tag{16-64}$$

where $I_{BX} = I_{B1} - I_B$. Recognizing that τ_C/τ_B is $1/h_{FE}$ and that dQ_{BS}/dI_B is equal to τ_S, we obtain, as a complete solution for the storage time,

$$t_s = \tau_S \ln \frac{I_{B1} - I_{B2}}{I_C/h_{FE} - I_{B2}} \tag{16-65}$$

In this result, I_{B2} is negative, so that $I_{B1} - I_{B2}|$ equals $|I_{B1}| + |I_{B2}|$. This is also true for the denominator.

The significant parameter in the storage-time equation is τ_S, as given by Eq. (16-61). For an alloy-type transistor having negligible collector stored charge, we can let $\alpha\alpha_R = 1$ and $\tau_{CS} = 0$, and Eq. (16-65) becomes

$$t_S = \frac{\tau_E + \tau_{ER}}{1 - \alpha\alpha_R} \ln \frac{I_{B1} - I_{B2}}{I_C/h_{FE} - I_{B2}} \tag{16-66}$$

Since $\tau_E = 1.22/\omega_b$ and $\tau_{ER} \approx 1.22/\omega_{bR}$, we obtain a result similar to that obtained by Ebers and Moll,[2] namely,

$$t_s = \frac{1.22(\omega_b + \omega_{bR})}{\omega_b\omega_{bR}(1 - \alpha\alpha_R)} \ln \frac{I_{B1} - I_{B2}}{I_C/h_{FE} - I_{B2}} \tag{16-67}$$

For a uniform-base device, it is reasonable to assume that $\omega_b \approx \omega_{bR}$ and therefore the storage time constant is approximately $2.4/\omega_b(1 - \alpha)$.

If the alloy-type transistor has a graded base such as that found in drift or MADT transistors, then the reverse-cutoff frequency is lowered considerably by the built-in electric field, which is retarding to the flow of carriers. Since $\tau_E = 0.6/\omega_b$ for a graded base in the normal direction, one would conclude that $\tau_{ER} \approx 2.4/\omega_b$ in the reverse direction, since the reverse base-cutoff frequency is lowered by the same factor that enhances the transit time in the normal direction. With this assumption, the storage-time equation for a diffused-base alloy transistor is, approximately,

$$t_s \approx \frac{3}{\omega_b(1-\alpha)} \ln \frac{I_{B1} - I_{B2}}{I_C/h_{FE} - I_{B2}} \tag{16-68}$$

where ω_b is the base-cutoff frequency in the normal direction and includes the effect of the graded base. Both ω_b and α are measured at the edge of saturation.

For transistors made by single-ended processes such as mesa or planar, the collector resistivity is moderately high and we cannot neglect the τ_{CS} term in Eq. (16-61), since the current I_{Rn} and the stored charge Q_{CS} are appreciable. Because of this effect, the inverse alpha is very low, $\alpha_R \approx 0$. Therefore,

$$\tau_S \approx \tau_E + \tau_{CS} \tag{16-69}$$

wherein it is assumed that the forward alpha is equal to unity. In Eq. (16-69), τ_E is given by $0.6/\omega_b$.

To determine the magnitude of τ_{CS}, we can write that in the collector

$$I_{Rn} \approx \frac{qAD_{nc}n_{pc}\epsilon^{qV_C/kT}}{L_{nc}} \tag{16-70}$$

where L_{nc} is the diffusion length of electrons and is a function of the minority-carrier lifetime τ_{nc}. Assuming a linear carrier gradient, the total stored collector charge is (area of triangle)

$$Q_{CS} = \frac{qAn_{pc}\epsilon^{qV_C/kT}L_{nc}}{2} \tag{16-71}$$

so that

$$I_{Rn} \approx \frac{2D_{nc}Q_{CS}}{L_{nc}^2} = \frac{Q_{CS}}{\tau_{nc}/2} \tag{16-72}$$

Here we see that the collector time constant is

$$\tau_{CS} = \frac{\tau_{nc}}{2} \tag{16-73}$$

or one-half the minority-carrier lifetime in the collector.

The storage-time equation for mesa- or planar-type structures is

$$t_s \approx \left(\frac{0.6}{\omega_b} + \frac{\tau_{nc}}{2}\right) \ln \frac{I_{B1} - I_{B2}}{I_C/h_{FE} - I_{B2}} \tag{16-74}$$

For low storage time, it is essential that the collector lifetime be made as small as possible. This is usually accomplished in the processing of high-speed switching transistors by the intentional introduction of certain impurities which drastically reduce the collector lifetime. The effectiveness of such lifetime treatments is determined by the measurement of the recovery time of the collector-to-base diode.[10,11]

If the thickness of the collector region is smaller than the minority-carrier diffusion length as is the case for epitaxial transistors, Eqs. (16-70) to (16-74) must be modified accordingly. For epitaxial transis-

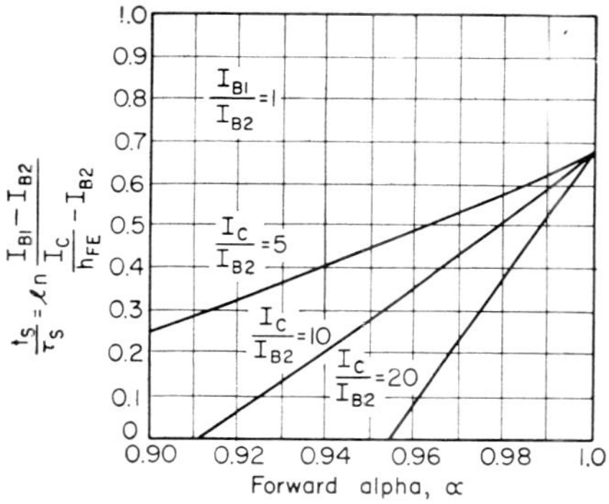

FIG. 16-8. Magnitude of circuit-drive function for storage time t_s.

tors having an epitaxial collector of thickness W_e in centimeters, it has been determined empirically that

$$\tau_{CS} \approx \frac{W_e^2}{2} \qquad \text{sec} \qquad W_e \ll L_{nc} \tag{16-75}$$

Thus, the thickness of the epitaxial layer becomes an important design factor for switching transistors.

All the foregoing expressions for t_s have in common the circuit-drive function given by the logarithmic term. A plot of this function is shown in Fig. 16-8 for different values of the turnoff ratio I_C/I_{B2}. One sees that storage time increases as the transistor h_{FE} gets larger or as the transistor is driven harder. Increasing the magnitude of I_{B2} lowers t_s. It is interesting to note that when h_{FE} is very large, or if $I_{B1} = I_{B2} = I_C$, the circuit-drive function reaches the limiting value of 0.69. This yields the useful relationship

$$t_s \approx 0.7\tau_S \tag{16-76}$$

To close this section it should be pointed out that in saturation the depletion layers of both junctions are very narrow and therefore the effective base width used in the calculation of ω_b is approximately equal to the physical base width of the transistor.

16-7. Fall Time. The last switching parameter is the fall time t_f, which is the time required for the collector current to decrease to $0.1I_C$. During this interval, the transistor is operating in the active region again, but this region is being traversed in the direction opposite to that of rise time. Therefore, during the fall-time period, the amount of charge to be removed from the base is equal to that which was added during the rise-time interval. For this reason, the charge analysis for

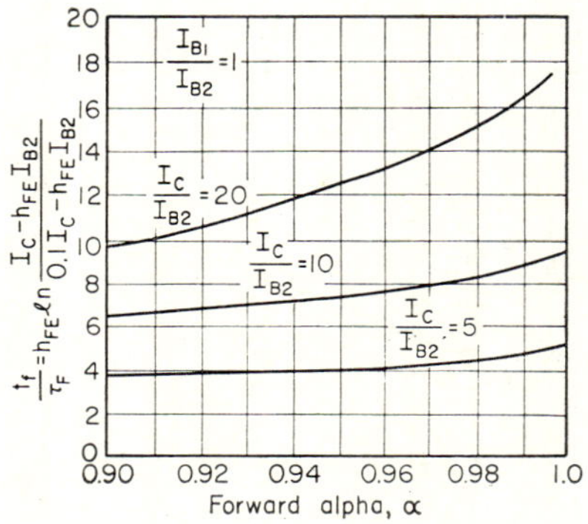

FIG. 16-9. Magnitude of circuit-drive function for fall time t_f.

fall time yields the same time constant as for rise time, given by Eq. (16-45).[3] Therefore, the integral equation (see Eq. 16-30) becomes

$$\int_{I_C}^{0.1I_C} \frac{dI_C}{I_{B2} - I_C/h_{FE}} = \int_0^{t_f} \frac{dt}{1/\omega_T + 1.7R_LC_{Tc}} \tag{16-77}$$

and the integral solution yields

$$t_f \approx h_{FE}\left(\frac{1}{\omega_T} + 1.7R_LC_{Tc}\right) \ln \frac{I_C - h_{FE}I_{B2}}{0.1I_C - h_{FE}I_{B2}} \tag{16-78}$$

The magnitude of the circuit-drive function in Eq. (16-78) is plotted in Fig. 16-9. It is seen that t_f increases with increasing h_{FE}, but primarily decreases as the magnitude of the circuit ratio I_C/I_{B2} decreases.

16-8. Charge-control Parameters of Switching Transistors. In summary, the design of high-speed switching transistors requires that the base-cutoff frequency be as large as possible and that the emitter and collector junction capacitances be as small as possible. Furthermore, to minimize charge storage in the collector, the minority-carrier concentration and/or lifetime should be made as low as possible. These requirements come as a result of the charge analysis of the grounded-emitter switching transistor which yields expressions for the delay, rise, storage, and fall times. For maximum switching speed, the magnitudes of the delay-plus-rise times (t_{ON}) and the storage-plus-fall times (t_{OFF}) must be small. Present transistor technology makes possible devices having total switching times in the range of 10 to 100 nsec (nanoseconds, 10^{-9} sec). Fig. 16-10 provides typical switching data for a high-speed n-p-n silicon epitaxial transistor. The curves give the values of t_{ON} and t_{OFF} for different levels of collector current and circuit ratio I_C/I_{B1}.

One useful way of satisfying the circuit designer's problem of predicting the switching times of a particular device for circuit-drive conditions,

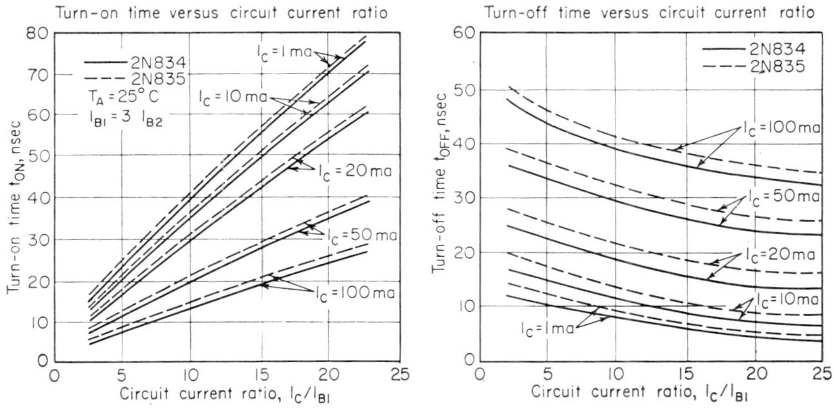

FIG. 16-10. Typical switching characteristics as functions of circuit current ratio for a 2N834 n-p-n silicon epitaxial planar transistor. (*Courtesy of Motorola, Inc.*)

different from that specified by the manufacturer, is to provide data on the so-called charge-control parameters of the transistor.[3,12] These charge-control parameters are Q_{OFF}, τ_R, τ_S, and τ_F; they manifest themselves when the various switching-time expressions are written in the form

$$t_d = \frac{Q_{OFF}}{I_{B1}} \tag{16-79}$$

$$t_r = \tau_R h_{FE} \ln \frac{h_{FE} I_{B1}}{h_{FE} I_{B1} - 0.9 I_C} \tag{16-80}$$

$$t_s = \tau_S \ln \frac{I_{B1} - I_{B2}}{I_C/h_{FE} - I_{B2}} \tag{16-81}$$

$$t_f = \tau_F h_{FE} \ln \frac{I_C - h_{FE} I_{B2}}{0.1 I_C - h_{FE} I_{B2}} \tag{16-82}$$

In each of the above equations, we see that the charge-control parameter is multiplied by a circuit-drive function. In fact, Figs. 16-6, 16-8, and 16-9 are plots of t_r/τ_R, t_s/τ_S, and t_f/τ_F, respectively.

From the theory of the previous sections, we can write

$$Q_{\text{OFF}} \approx 2C'_{Te}V_{BE(\text{OFF})}^{1/2} \qquad V_{CC} \gg V_{BE(\text{OFF})} \tag{16-83}$$

$$\tau_R \approx \tau_F \approx \frac{1}{\omega_T} + 1.7R_LC_{Tc} \tag{16-84}$$

$$\tau_S \approx \frac{2.4}{\omega_b(1-\alpha)} \qquad \begin{array}{l}\text{uniform base alloy}\\ \omega_b = \omega_{bR};\ \alpha_R = 1\end{array} \tag{16-85}$$

$$\tau_S \approx \frac{3}{\omega_b(1-\alpha)} \qquad \begin{array}{l}\text{graded base alloy}\\ \omega_b \approx 4\omega_{bR};\ \alpha_R = 1\end{array} \tag{16-86}$$

$$\tau_S \approx \frac{0.6}{\omega_b} + \frac{\tau_{nc}}{2} \qquad \begin{array}{l}\text{single-ended types}\\ \alpha_R \approx 0\end{array} \tag{16-87}$$

The above relations provide a reasonably accurate means of calculating the magnitudes of the charge-control parameters from the physical properties of the device structure. On the other hand, any attempt to determine the charge-control parameters from measurements of ω_T, ω_b, α, ω_{bR}, etc. would be exceedingly difficult, since these parameters must be measured at the edge of saturation. Therefore, a means of measuring Q_{OFF}, τ_R, τ_S, τ_F directly becomes paramount. Several methods of measurement have been proposed, the details of which are covered in the references.[3,13,14] However, the technique is based on the principle that the circuit-drive functions converge to a limiting value when I_{B1} is made at least five times greater than I_C/h_{FE}. For example, if $I_{B1} = 5I_C/h_{FE}$, the storage-time circuit-drive function becomes approximately ln 2. By setting up such circuit conditions and measuring t_s, we have

$$\tau_S \approx \frac{t_s}{\ln 2} \tag{16-88}$$

Thus, by providing measured data as to how Q_{OFF} varies with $V_{BE(\text{OFF})}$ and how τ_R, τ_S, and τ_F vary with current, the designer is equipped with a means for calculating the switching times for any set of circuit-drive conditions.[13] It should be apparent that both τ_R and τ_F are also functions of the collector supply voltage V_{CC}, since this determines the magnitude of R_LC_{Tc} at a fixed I_C.

PROBLEMS

16-1. Measurements of a 2N705 p-n-p germanium mesa transistor indicate that $C_{Te} = 3.5\ \mu\mu f$ at 2 volts, $C_{Tc} = 5\ \mu\mu f$ at 10 volts, $f_T \approx f_{\alpha b} = 250$ megacycles and $h_{FE} = 40$ at $V_{CE} = -0.3$ volt, $I_C = -10$ ma. The collector-region lifetime is determined to be 0.12 μsec. For this device, calculate t_d, t_r, t_s, and t_f in a switching circuit in which $I_C = -10$ ma, $I_{B1} = -1$ ma, $I_{B2} = 0.25$ ma, $V_{CC} = -3.5$ volts, $V_{BE(\text{OFF})} = 1.25$ volts, and $R_L = 300$ ohms.

16-2. Calculate the storage time t_s for a 2N706 n-p-n silicon mesa transistor which has a 2-ohm-cm epitaxial collector region with a thickness of 10 μ and a lifetime of

0.02 μsec. The base time constant is such that the excess base charge is negligible. Use a circuit in which $I_C = I_{B1} = -I_{B2} = 10$ ma.

16-3. An n-p-n germanium-alloy switching transistor is designed with equal emitter and collector areas of 100 mil². The base region is 1 ohm-cm p type and is 0.5 mil thick, resulting in a 50-ma beta of 100 near saturation. Calculate the rise and fall times in a circuit for which $I_C = 50$ ma, $I_{B1} = 5$ ma, $I_{B2} = -1$ ma, and $R_L = 500$ ohms.

16-4. The charge-control parameters of a 2N501 p-n-p germanium MADT transistor are measured at $V_{CC} = -3.5$ volts. They are: $Q_{\text{OFF}} = 25$ μμcoulombs at $V_{BE(\text{OFF})} = 1.25$ volts, $\tau_R = \tau_F = 3$ mμsec at $I_C = -10$ ma, and $\tau_S = 55$ mμsec at $I_{BX} = 0.6$ ma. For this device, $h_{FE} = 25$ at $I_C = -10$ ma. From this data, calculate t_d, t_r, t_s, and t_f in the switching circuit given in Prob. 16-1.

REFERENCES

1. Anderson, A. E.: Transistors in Switching Circuits, *Proc. IRE*, vol. 40, pp. 1541–1548, November, 1952.
2. Moll, J. L.: Large-signal Transient Response of Junction Transistors, *Proc. IRE*, vol. 42, pp. 1773–1784, December, 1954.
3. Beaufoy, R., and J. J. Sparkes: The Junction Transistor as a Charge-controlled Device, *ATE J.*, vol. 13, pp. 310–327, October, 1957.
4. Baker, A. N.: Charge Analysis of Transistor Operation, *Proc. IRE*, vol. 48, pp. 949–950, May, 1960.
5. Ekiss, J. A., and C. D. Simmons: Calculation of the Rise and Fall Times of an Alloy Junction Transistor Switch, *Proc. IRE*, vol. 48, pp. 1487–1488, August, 1960.
6. Cho, Y.: Calculation of the Rise and Fall Times in the Alloy Junction Transistor Switch Based on the Charge Analysis, *Proc. IRE*, vol. 49, pp. 636–637, March, 1961.
7. Severin, Ernest: Switching Time Formulae for Single-diffused Mesa Transistors, *Semiconductor Prods.*, pp. 37–42, June, 1961.
8. Easley, J. W.: The Effect of Collector Capacity on the Transient Response of Junction Transistors, *IRE Trans.*, vol. ED-4, pp. 6–14, January, 1957.
9. Simmons, C. D.: Hole Storage Delay Time and Its Prediction, *Semiconductor Prods.*, pp. 14–18, May/June, 1958.
10. Lax, Benjamin, and S. R. Neustadter: Transient Response of a p-n Junction, *J. Appl. Phys.*, vol. 25, pp. 1148–1154, September, 1954.
11. Grinich, V. H., and R. N. Noyce: Switching Time Calculations for Diffused Base Transistors, *IRE Wescon Convention Record*, part 3, Electron Devices, pp. 141–147, August, 1958.
12. Sparkes, J. J.: A Study of the Charge-control Parameters of Transistors, *Proc. IRE*, vol. 48, pp. 1696–1705, October, 1960.
13. Simmons, C. D.: High-speed Switching Transistors, *Elec. Design News*, pp. 39–47, September, 1960.
14. Hwang, Y. C., D. S. Cleverley, and D. J. Monsour: Transistor Switching Speed from Base Storage Charges and Their Lifetimes, *Electronic Design*, part I, pp. 52–55, March, 1961, part II, pp. 50–53, April, 1961.

17

Transistor Design Principles

17-1. Limitations of Manufacturing Processes on Transistor Design. The transistor theory presented in the preceding chapters has been sufficiently complete so that any device structure can be analyzed for its d-c and frequency characteristics as a function of current, voltage, and temperature. Implicit in this argument is the fact that regardless of what manufacturing process is employed to make the transistor, the analytical model reduces simply to a heavily doped emitter region, an emitter-base junction which may be a step or slightly graded junction, a base region of finite width which may be uniformly doped or graded, a collector-base junction which may be a step or graded junction, and a collector region which may be either high or low resistivity or some combination thereof. The definition of this model establishes the impurity profile of the transistor in conjunction with the fact that the device is either p-n-p or n-p-n and the material is either germanium or silicon. Further, the physical properties of the device such as the area and cross-sectional shape of the junctions are determined by the particular geometry of the emitter, the collector, and the base contact. Thus, the combination of the impurity profile and the geometry permits the determination of such important transistor properties as current and power gain, breakdown voltages, leakage currents, junction capacitances, cutoff frequencies, saturation voltages, input and output impedances, and base resistance.

The problem in transistor engineering is to design a transistor that will satisfy the requirements of the application. The applications are numerous, and include high-speed switching circuits, low-frequency and d-c amplifiers, very-high-frequency power amplifiers, and low-level tuned amplifiers, to mention but a few. In general, the specific circuit application dictates the required operating voltages, currents, frequency response, impedances, and gain for the transistor. Using the theory only, one can design an optimum device that would have ideal electrical parameters suitable for most any application. However, it would be extremely difficult to fabricate such a theoretical transistor, owing to the practical limitations of the known process and current semiconductor technology. In other words, the theoretical design of a transistor in terms of the impurity profile and geometry must be modified to make the design compatible with the limitations of the particular manufacturing process selected to fabricate the transistor.

As was pointed out in Chap. 1, transistor processes have made great strides since their inception years ago. The early point-contact devices immediately gave way to the grown- and alloy-junction techniques. Since then, extensive research and engineering made possible the advent of diffusion, mesas, oxide masking, photoresist methods, and epitaxial growth. Each of these important processes was developed with the primary purpose of enabling the transistor engineer to more readily achieve the ideal transistor design.

In the sections to follow, each of the major processes will be discussed from the viewpoint of the limitations on design, leading up to the universal features of the epitaxial-planar process. For each process the design principles will be established, illustrating how an optimum design is attained. Wherever possible, appropriate procedures will be pointed out. In general, the basic design problem is a matter of determining whether a given gain-bandwidth frequency can be obtained within the objective voltage and current ratings or vice versa, subject to the limitations of geometry on capacitances, base resistance, and saturation voltage.

17-2. Grown-junction Transistors. Transistors made by grown-junction processes such as rate growing, meltback, diffused meltback, double doping, and grown diffusion all have in common the fact that the device is in the form of a rectangular bar of semiconductor having a thin base region located in a transverse plane in the center. Because of the bar geometry, the frequency response is limited in part by the minimum cross-sectional area that can be handled practically during manufacture. Typical bar dimensions may be in the order of 0.012 to 0.020 in. on one side, although some transistors have been made as small as 0.007 in. on one side. In addition to the effect of bar size on emitter and collector capacitances, one finds that the rate-grown, meltback, and double-doped types have relatively large base widths, on the order of 0.5 mil, due to the limitations of the respective processes. The grown-junction types having diffused bases, however, may have extremely thin base widths, on the order of 0.1 mil.

In the rate-grown and meltback types, the design must be a compromise between current gain and collector breakdown voltage. This is attributed to the fact that these junctions are formed by segregation, wherein the collector impurity level is some fraction of the emitter level.

Thus, as the emitter doping is increased to increase beta, the collector resistivity decreases, yielding a lower BV_{CBO}. On the other hand, this is not a problem with double-doped and grown-diffused types, since these junctions are formed by adding impurities to form the emitter against a high-resistivity collector background.

Of all the grown-junction types mentioned, it is apparent that the grown-diffused type offers the best electrical characteristics, owing to the thin, diffused base region and the fact that the emitter and collector regions are independently doped. Nevertheless, the designer, in his

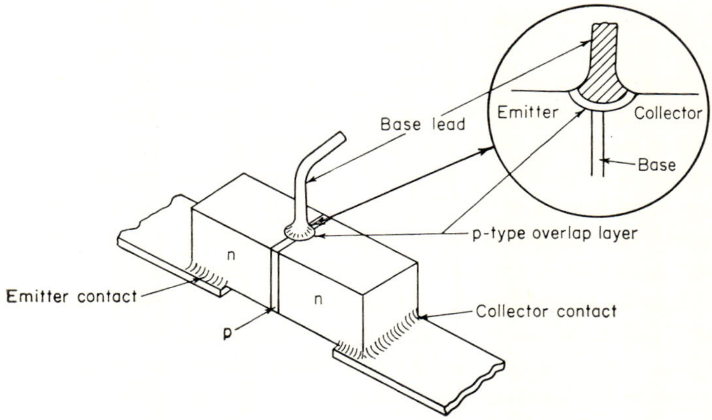

FIG. 17-1. Typical n-p-n grown-junction transistor with bar geometry, showing overlap of base-lead contact.

choice of area and collector resistivity, must make a compromise among BV_{CBO}, C_{Tc}, and collector series resistance r_{SC}. In most designs, r_{SC} usually turns out to be quite high, amounting to as much as 75 to 100 ohms in some types. If the application warrants power control at high currents or audio frequencies, then the bar may be cut quite large and r_{SC} reduced considerably.

Perhaps the most serious design limitation of grown-diffused transistors (and all other grown-junction types as well) is the ohmic connection to the base region of the bar. A sketch of a typical assembly is shown in Fig. 17-1, wherein a base lead is shown connected to the p region of an n-p-n bar. It is seen that the base contact overlaps on each side of the base into the adjacent emitter and collector regions. Such overlap is to be expected, since typical base regions vary from 0.1 to 0.5 mil, whereas practical wires are at least 1 or 2 mils in diameter. Consequently, to prevent shorting out the base region, the base wire material must be such that after bonding a p-n junction is made in the emitter and collector regions. For example, for an n-p-n device the contact overlap must be p type. A cross-sectional view of such a contact is shown in Fig. 17-1. The p-type overlap layer is created in the alloying process and serves to establish the p-n junctions in the adjacent regions. Therefore it becomes evident that the base contact simply becomes an extension of the base region. When external biases are applied to the transistor, the overlap junctions become forward or reverse biased, as the case may be.

The base-contact overlap has several deleterious effects on the transistor electrical parameters.[1,*] Firstly, because the contact is a semipoint contact on one side of the bar, r'_B is relatively high. Secondly, because of the area of the overlap junction and the very low resistivity of the overlap regrowth region, there is a considerable increase in both the emitter and collector capacitances, since the overlap capacitance adds to the intrinsic bar capacitance. Thirdly, because of the overlap into the emitter region it is necessary to establish a minimum limit on emitter resistivity if one is to meet a minimum BV_{EBO} rating; this has the effect of lowering the emitter efficiency of the bar. Finally, the emitter efficiency is further reduced by carrier injection at the overlap junction. To demonstrate the latter point analytically, let us assume that we have an n-p-n transistor of area A and an overlap junction on the emitter side of area A_O and resistivity ρ_O. Further, we assume that the emitter forward-bias potential is uniformly applied across both the overlap and emitter-base junctions. Under these conditions, the emitter injects an electron current I_{nEB} into the base and an electron current I_{nEO} into the overlap region. Similarly, the overlap and base regions inject hole currents I_{pEO} and I_{pEB}, respectively, into the emitter region. The total emitter current is the sum of all four components, but the only active term contributing to current gain is I_{nEB}. Therefore, the emitter efficiency is given by

$$\gamma = \frac{I_{nEB}}{I_{nEB} + I_{pEB} + I_{pEO} + I_{nEO}} \tag{17-1}$$

Rearranging terms, we have

$$\gamma = \frac{1}{1 + I_{pEB}/I_{nEB}(1 + I_{pEO}/I_{pEB}) + I_{nEO}/I_{nEB}} \tag{17-2}$$

In Eq. (17-2), I_{pEO}/I_{pEB} is simply the ratio of injected hole currents into a common emitter region and is therefore equal to A_O/A. Evaluating the other ratios in terms of sheet resistance for emitter efficiency, it is easily shown that for a grown-junction transistor

$$\gamma \approx \frac{1}{1 + R_{EE}/R_{BB}(1 + A_O/A) + A_O R_{OO}/A R_{EE}} \tag{17-3}$$

* References, indicated in the text by superscript figures, are listed at the end of the chapter.

where R_{OO} is the effective sheet resistance of the overlap region. If no overlap exists, i.e., $A_O = 0$, then Eq. (17-3) reduces to the familiar expression for γ. The addition of new terms in the denominator, however, decreases the magnitude of the emitter efficiency. In other words, a significant portion of the emitter current may be lost simply by injection into the overlap region, thereby decreasing the current gain. Particularly at high emitter currents, if a significant amount of base-region self-crowding occurs, then in Eq. (17-3) the value of A is decreased and γ falls off accordingly. It is for this reason, coupled with the relatively high resistivity of the emitter, that the h_{FE} of grown-junction transistors falls off quite rapidly with increasing collector current.

17-3. Alloy Transistors. The vast majority of alloy transistors in use today are p-n-p germanium types, although there are n-p-n germanium and p-n-p silicon types available also. Alloy transistors, due to their relatively simple geometry and impurity profile, lend themselves to reasonably straightforward and accurate transistor design.[2] The electrical characteristics resulting from the attributes of very-low-resistivity emitter and collector regions, planar-parallel junctions, the high-resistivity base region, and the circular base contact make these transistors very suitable for general-purpose low-frequency amplifier and switching applications and use as power transistors. The alloy process imposes a limitation on alpha-cutoff frequency because of the fact that the base width is controlled by the depth of alloying of each of two opposing junctions. The possibility of shorts occurring due to junction irregularities establishes a practical limit on the order of 0.3 mil. Consequently, the $f_{\alpha b}$ for alloy power transistors is in the audio range, for amplifier types it is on the order of 2 megacycles, and for switching types $f_{\alpha b}$ ranges anywhere from 5 to 30 megacycles.

The principal design problem in alloy transistors is one of compromising collector voltage rating with frequency response. Because of the high-resistivity base and the spreading of the collector depletion layer toward the emitter, V_{PT} decreases as $f_{\alpha b}$ increases. One can offset this by decreasing resistivity, but this increases C_{Tc} and lowers the emitter efficiency. On the other hand, the base width can be increased to the point where avalanche breakdown takes place before voltage punch-through occurs; this, of course, lowers $f_{\alpha b}$ considerably. Again, the circuit application dictates the parameter requirements, but the general procedure is to calculate and prepare design curves where BV_{CBO}, V_{PT}, C_{Tc}, and $f_{\alpha b}$ are plotted as a function of base resistivity and base width. With the required compromises displayed graphically, the necessary base-design decisions are made readily. The transistor current rating and any maximum requirements on I_{CBO} and/or C_{Tc} are the boundary conditions for determining the junction areas.

Although base width and resistivity have a pronounced effect on the magnitude of the current gain beta, it is also controlled in alloy transistors by several other variables. The designer is able to control the emitter sheet resistance directly by the choice of emitter dopant. If high betas are required out to very large currents (such as in p-n-p germanium power transistors) one can use aluminum instead of indium as the primary acceptor in the emitter, since aluminum has a considerably higher segregation coefficient than indium. A second design variable is the base surface recombination velocity, which can be directly controlled in processing by the judicious choice of final etch and encapsulation procedures. The third and final variable that affects the magnitude of the current gain is the ratio of the collector junction area to the emitter junction area. It was shown by Moore and Pankove[3] that a maximum is reached when the collector is approximately twice the area of the emitter. This is explained by the fact that if the emitter is larger than the collector, a considerable percentage of the injected carriers is lost at the surface around the collector. As the emitter is made smaller, the amount lost at the collector surface decreases and the gain increases toward a maximum value. This occurs until a point is reached when the emitter becomes so small that a greater percentage of the injected carriers is lost at the emitter surface, and the current gain begins to fall off.

Because of the aforementioned design parameters, alloy transistors have excellent beta-versus-collector-current characteristics. This feature, coupled with the extremely low $V_{CE(\mathrm{SAT})}$ afforded by the very-low-resistivity collector region, makes them excellent devices for slow-speed switching circuits. In alloy devices, $V_{CE(\mathrm{SAT})}$ is largely composed of the junction potentials whose difference is quite small due to the high inverse alpha. This can be minimized further simply by making a bilateral structure having equal emitter and collector areas. Finally, as amplifiers, alloy transistors are limited to some extent because the output impedance is affected considerably by the voltage sensitivity of the base width, namely, $\partial W/\partial V_C$.

One significant process improvement that helped increase the frequency of alloy transistors was the MAT or microalloy transistor. Because of the electrochemical technique of forming the base region and junctions, extremely thin base widths became practical to manufacture. Of course, these devices are limited by lower punchthrough voltages.

17-4. Diffused-base-alloy Transistors. The advent of the diffusion process and the theoretical understanding of the transit-time reduction afforded by the built-in electric field of graded base layers made it possible to fabricate transistors having a frequency response of 100 megacycles and several times greater and still offering adequate voltage ratings. These transistors are in the generic family of diffused-base-alloy transistors and are specifically known as drift and MADT transistors. These have in common a structure which is basically the same as that of alloy transistors except that the base region consists of a graded impurity layer diffused from the emitter side into an almost intrinsic resistivity background. Although the physical width of the base region is comparable to that of alloys, the base-cutoff frequency is increased considerably because the collector depletion layer readily fills up the intrinsic region such that the effective base width becomes equal to the depth of the diffused region only.

If these transistors are used as high-frequency amplifiers, the primary problem is to design for maximum frequency of oscillation at a particular operating voltage and current.[4] This requires low r'_B, low C_{Tc}, and high f_T, wherein the latter is controlled mainly by the emitter time constant and the base-transit time. The resistivity of the base region is usually selected high enough to provide the required collector capacitance per unit area; the width of the high-resistivity region must be adjusted so that it fills up at the lowest possible voltage below the required operating voltage. Once the resistivity and width are determined, the area of the collector junction is adjusted to obtain the required C_{Tc}.

The design of the diffused region of the base involves the choice of the values of N'_B and x_j for an exponential impurity function as a compromise between emitter efficiency and frequency response. For high low-frequency beta, it is necessary to maintain both W and N'_B low for high sheet resistance, R'_{BB}. Keeping N'_B low is advantageous to $f_{\alpha b}$ since the emitter capacitance C_{Te} is smaller at low impurity concentrations. On the other hand, a smaller ratio of N'_B/N_{BC} will yield higher r'_B and less of a base-transit-time reduction due to the lower drift-field intensity. One additional consideration is the fact that a high N'_B will lower the BV_{EBO} rating. Thus, the problem has several complex interactions of variables, and the use of design curves is necessary to obtain optimum results. One design factor that may be disregarded in dealing with diffused-base-alloy transistors is the transit time through the collector depletion layer; in most designs, the electric field is such that the limiting velocity is reached and the transit time is small compared to the emitter and base delay.

In switching applications, the higher f_T and low $V_{CE(\text{SAT})}$ of diffused-base-alloy transistors give a favorable advantage over alloy types in attaining higher speeds. The rise and fall times, however, are far from optimum because of the severe falloff of f_T as the collector voltage drops toward the saturation value. This is particularly acute in drift transistors that have relatively wide intrinsic regions. In MADT types designed for optimum switching, the rise and fall times are greatly improved by virtually eliminating the high-resistivity region, such that the tail of the diffused region is just at the collector junction.[5] This is done by electrochemically etching the base membrane thinner and adjusting the diffusions accordingly. Although the switching speed is greatly improved, the voltage rating is reduced by punchthrough. Finally, storage times are quite low in MADT transistors because there is essentially no charge storage in the low-resistivity collector region.

Probably the most serious limitation of the diffused-base-alloy transistor process in achieving very high gain bandwidths is the practical lower limit on the size of the emitter junction that is formed.

17-5. Mesa and Planar Transistors. The introduction of the mesa process made it possible to fabricate transistors in which the design could offer both extremely high frequency response and high voltage ratings. These transistors have been largely p-n-p germanium mesas with alloyed emitters and n-p-n silicon mesas with diffused emitters, although n-p-n germanium and p-n-p silicon have also been manufactured. The ease of design afforded by the mesa technique is attributed to the fact that the collector is of high resistivity, permitting the collector depletion layer to spread almost completely into the collector rather than into the base. Consequently, the base regions can be diffused to very shallow depths, resulting in widths varying from 0.5 to 2 μ, as desired. Additionally, the photoresist masking technique employed in the mesa process allows the formation of extremely small emitter and collector junction areas with photographic precision and reproducibility. The small areas coupled with the thin base widths have extended the frequency response of transistors from 300 megacycles to 3 kilomegacycles. Furthermore, the higher output impedance afforded by the relative insensitivity of base width with collector voltage (low $\partial W/\partial V_C$) makes mesa transistors excellent high-frequency amplifiers.

The basic principles involved in the design of the impurity profile for mesa devices are not significantly different from those discussed in previous sections. The design of the diffused base region still requires a compromise among emitter efficiency C_{Te}, base-cutoff frequency, r'_B, BV_{EBO}, and possibly V_{PT} if very thin base widths are required. This calls for a graphical procedure similar to that indicated for diffused-base-alloy transistors. One technique which offers a design advantage for increasing base sheet resistance for higher beta and BV_{EBO} is the use of out-diffusion, which is a method of reducing the impurity concentration just within the surface of the semiconductor.[6]

The collector region, however, does require some design modifications, especially for switching transistors. Although high resistivity is advantageous for high BV_{CBO} and low C_{Tc}, the series resistance r_{SC} becomes larger than desired. If $V_{CE(\text{SAT})}$ is to be comparable to that of alloyed junctions, the collector resistivity must be maintained as low as possible and the junction area as large as possible. Furthermore, the high minority-carrier lifetime associated with high resistivity establishes a significant amount of stored charge in the collector, resulting in high t_s in switching circuits. Storage time is lowered by keeping the collector resistivity small. Thus, the collector-region design involves a balancing of BV_{CBO} and C_{Tc} with $V_{CE(\text{SAT})}$ and t_s, wherein resistivity and area are the variables. This is noticeably so in germanium mesa devices; in silicon, collector lifetime is readily "killed" with such techniques as gold diffusion and therefore very low values of t_s are obtained without any major degradation of the other three parameters.*

In addition to the aforementioned impurity-profile advantages over other process types, mesa transistors offer considerable freedom of geometry design. As a result of the development of evaporation masks, photoresist methods, and oxide masking of diffusants, almost any geometrical configuration or pattern can be achieved for the emitter and the base contact. In germanium mesas, most designs consist of rectangular emitter stripes, with a similarly shaped base contact either on one side only or on both sides of the emitter stripe. In silicon mesas, most designs utilize rectangular or circular emitter with corresponding base contacts that go either partially or completely about the emitter. In either case, the designer has the opportunity to minimize r'_B through his choice of the base-contact geometry. Most important of all, the linear stripe geometry can be scaled up to establish a larger emitter perimeter if higher operating currents are required. Thus, the advent of the mesa process made possible the design of high-frequency high-power transistors, particularly in silicon. In the case of the latter, the junction areas increase proportionately and, because of the existence of certain imperfections in the semiconductor material, the base must be widened slightly if high collector voltages are required.

Subsequent to the mesa, the planar process was introduced as a substitute technique for fabricating silicon mesas. The design and electrical advantages of silicon planar transistors are identical to those of silicon mesas as far as impurity profile and geometry are concerned. As was discussed in Chap. 1, instead of forming the area of the collector-base junction by etching the mesa, the planar process accomplishes the same objective by using the silicon oxide as a mask to confine the base diffusion to the desired area. Nevertheless, the surface recombination velocity of a silicon planar transistor is considerably lower than that of a silicon mesa because the edges of the collector-base and emitter-base junctions are in the same surface plane, which is completely protected by passive silicon

* Actually, a severe reduction of collector lifetime will increase $V_{CE(\text{SAT})}$ slightly because of the decrease in the inverse alpha of the transistor; that is, the effective sheet resistance of the collector is increased.

oxide. This results in extremely low I_{CBO} approaching the charge-generation level and more stable h_{FE}. The growth of the silicon oxide during manufacture also serves to getter recombination centers from the emitter-base junction, which improves the emitter efficiency at very low currents through the reduction of I_{RE}. Silicon planar transistors have excellent h_{FE} versus I_C characteristics, extending from 1 μ to 1 amp.

17-6. Epitaxial Transistors. The recent development and introduction of the epitaxial process to the mesa and planar technologies represent the first opportunity since the transistor was invented to achieve the ultimate in design. The epitaxial process is a means of growing a microthin high-resistivity region on a very-low-resistivity semiconductor substrate of the same type. The pure semiconductor atoms are deposited on the semiconductor wafer with the same crystal orientation and perfection. This is done by the vapor decomposition of either $GeCl_4$ or $SiCl_4$ with hydrogen at high temperatures. By controlling such process variables as time, temperature, flow rate, and doping, one can grow uniform epitaxial layers having any desired thickness or resistivity on many substrate wafers at a time.

The significant advantages of the epitaxial semiconductor wafer as compared to the conventional uniform-resistivity wafer became apparent in its use as the collector region in mesa or planar transistors. With a thin high-resistivity epitaxial region in the collector, one does not have to compromise the design of BV_{CBO} and C_{Tc} with $V_{CE(\mathrm{SAT})}$ and t_s as indicated in the previous section. By selecting a high epitaxial resistivity for the collector, one immediately obtains a higher inherent collector breakdown voltage and very low capacitance per unit area. Then the epitaxial thickness is designed to be small enough to sustain the desired breakdown voltage. Because this is usually quite thin (on the order of microns), the total series resistance affecting $V_{CE(\mathrm{SAT})}$ is extremely small, since the resistance of the heavily doped substrate is negligible. Also, the collector stored charge is reduced considerably because the effective minority-carrier diffusion length is equal to the thickness of the epitaxial region, which is many times less than the diffusion length corresponding to the actual lifetime in the epitaxial region. This occurs because the recombination rate at the substrate is very high due to the heavy doping. Consequently, the transistor storage time t_s becomes very small; this can be reduced even further by degrading the lifetime of the epitaxial region. All of these collector parameter advantages, along with the gain-bandwidth advantages of diffused bases, are realized in epitaxial-mesa or epitaxial-planar transistors. Whether in germanium or silicon, epitaxial transistors have high voltage and current ratings, very high frequency response, and excellent d-c characteristics, making them equally applicable to amplifier and switching applications at all power levels.

Perhaps one of the most universal transistor designs that evolved from the epitaxial process is the n-p-n silicon epitaxial-planar transistor employing the star geometry. In Fig. 17-2 are a photomicrograph of the geometry and a cross-sectional sketch illustrating the various regions

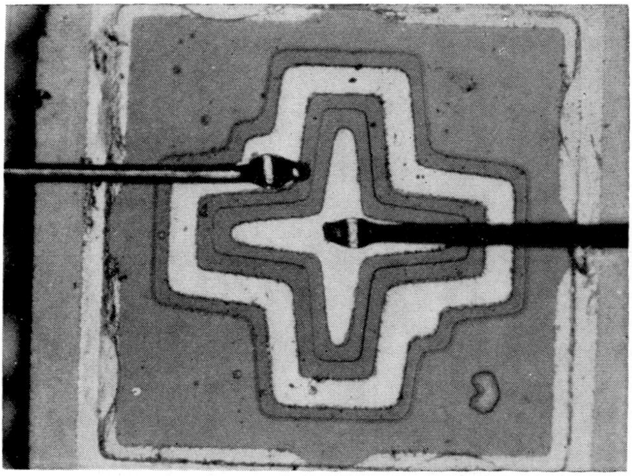

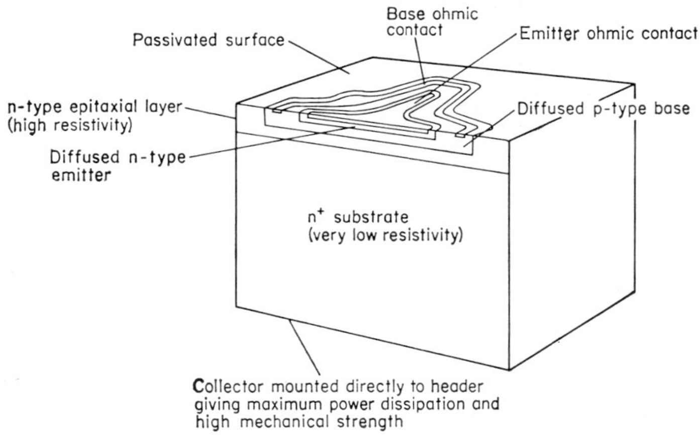

FIG 17-2. The n-p-n silicon epitaxial star planar transistor. The sequence of manufacturing processes is similar to that shown in Fig. 1-17. (*Courtesy of Motorola, Inc.*)

that make up the impurity profile. This transistor combines all the advantages of the epitaxial process with the frequency and gain characteristics of diffused structures, the temperature and power capabilities of silicon, and the surface passivation of the planar process. Furthermore, it utilizes the star geometry, which is an optimum design making use of the principle of the long thin stripe to obtain maximum perimeter-to-area ratio. The latter is essential for obtaining maximum current rating with minimum emitter capacitance area. As a result of this ideal design, an epitaxial-star-planar transistor might have the following typical characteristics: $BV_{CBO} = 120$ volts, $BV_{CEX} = 80$ volts, $BV_{EBO} = 7$ volts, $I_{CBO} = 0.001$ μa, $h_{FE} = 80$ with useful current gain from 10 μa to 1 amp, $V_{CE(\text{SAT})} = 0.2$ volt at 150 ma, $C_{Tc} = 4$ μμf, $f_T = 400$ megacycles, and a power-dissipation capability of 3 watts.

17-7. Introduction to Integrated Semiconductor Circuits. The same semiconductor technologies of diffusion, oxide passivation and masking, photoresist, and epitaxial growth that made the silicon epitaxial-planar transistor possible are now being utilized to introduce a new concept in solid-state engineering—the integrated semiconductor circuit. This is a "device" which will perform a complete circuit function by integrating all the active and passive elements of an electronic circuit into a single semiconductor substrate. Thus a solid wafer of semiconductor material, e.g., silicon, can be processed and packaged to perform virtually any amplifying or switching circuit function, thereby eliminating the need for individual circuit components such as transistors, diodes, resistors, capacitors, and inductors. Integrated circuits, which are also referred to as microcircuits, have many potential advantages for the electronics system in terms of circuit miniaturization, uniformity, reliability, and cost and have therefore established the next threshold of revolution for the electronics industry.

Integrated semiconductor circuits are made possible through the combination of many known techniques. As already indicated, all types of planar transistors may be diffused into either epitaxial or nonepitaxial silicon substrates. Likewise, all types of diodes such as high-speed switching diodes or zener diodes, can be fabricated using similar techniques. Because of the photographic nature of these processes, any multiple number of transistors and/or diodes can be formed simultaneously within a common semiconductor substrate. Further, all kinds of resistors can be fabricated within the same substrate by either diffusing regions to within a given geometry and sheet resistance, evaporating thin films of material of known resistivity, or depending directly on the distributed resistance of the substrate. Capacitors may be formed within the common material by several methods such as using the junction capacitance of a p-n junction, using silicon oxide as a dielectric between two metallic layers, or using the thin-film technology of sputtering tantalytic capacitors. Finally, inductors can be formed by utilizing spiral or other special patterns of evaporated materials to yield relatively small values of inductance or possibly by using deposited ferromagnetic materials to obtain higher values; this is the most difficult component to integrate at the time. In summary, the design of an integrated circuit involves a complex interaction of processes, photographic sequences, geometric patterns, and distributed and parasitic values, all superimposed on and contained within a single semiconductor wafer.

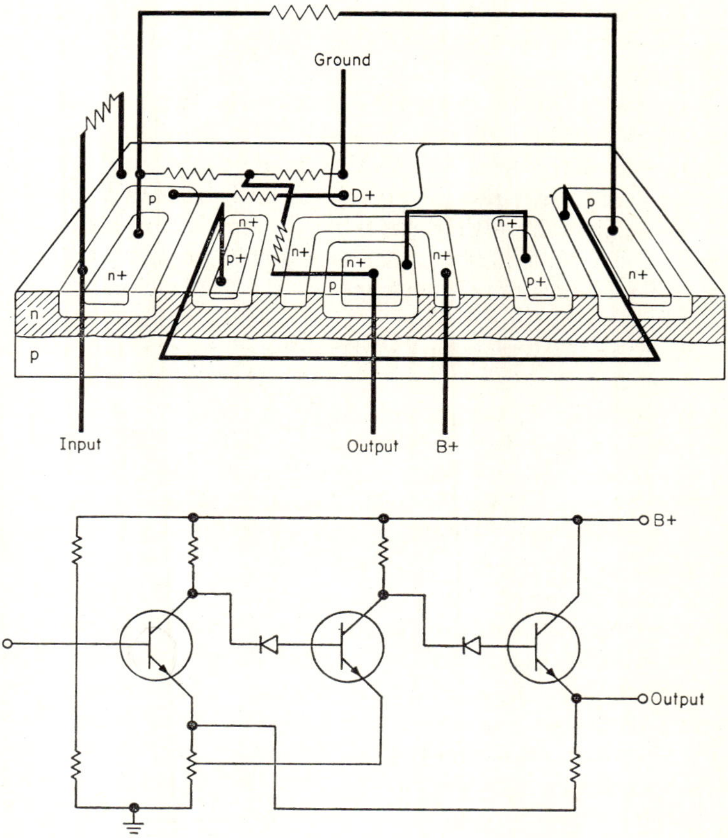

FIG. 17-3. Design layout and circuit of an integrated semiconductor video amplifier. (*Courtesy of Motorola, Inc.*)

Some typical circuit examples are shown in Figs. 17-3 and 17-4. The design layout and circuit of a three-stage video amplifier using a silicon epitaxial substrate is shown in Fig. 17-3. This integrated circuit employs three diffused n-p-n transistors of the high-frequency planar type; the stages are coupled by diffused p-n diodes which have characteristics equivalent to a parallel *R-C* network. The resistances are determined by the separation of the transistor and diode geometries, the terminal locations, and the resistivity of the epitaxial layer. Another example of an integrated circuit is the micrologic flip-flop shown in Fig. 17-4. The structure consists wholly of transistors and resistors that are diffused into a single monolithic chip of silicon. This particular switching circuit has a 50-nsec stage delay and will operate at a clock rate of 1 megacycle.

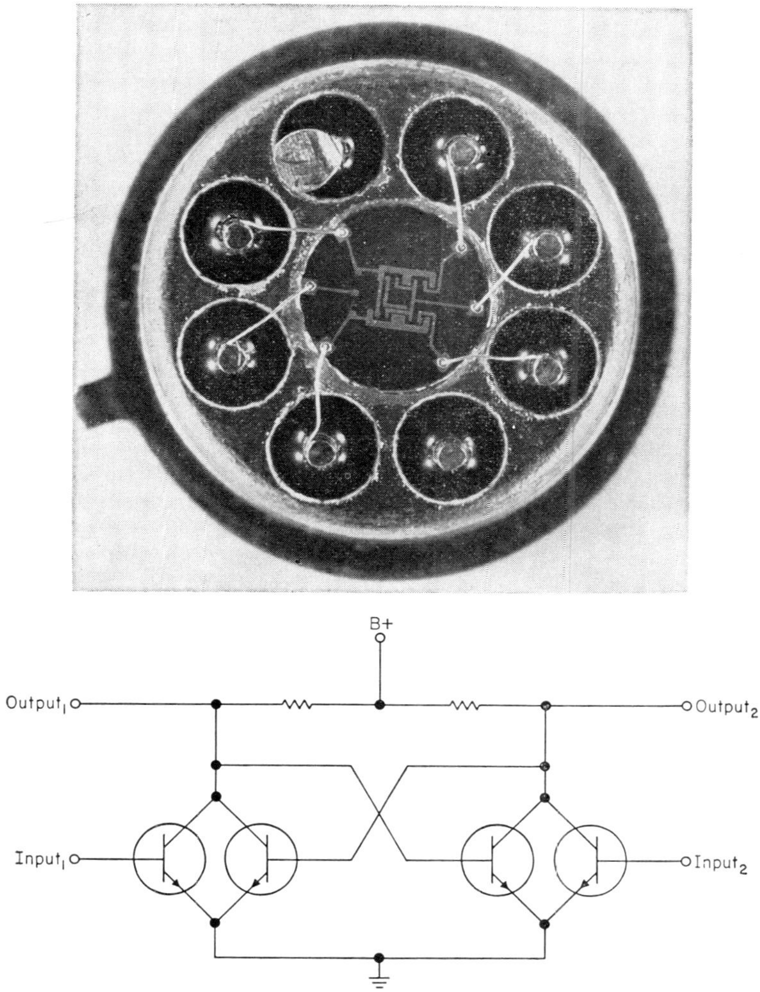

Fig. 17-4. Structure and circuit of the micrologic flip-flop, an integrated semiconductor multivibrator. (*Courtesy of Fairchild Semiconductor.*)

These two relatively simple examples are but a few of the infinite number of integrated circuit configurations that may be incorporated into solid-state systems in the years to come.

It is most appropriate that this section be used to end this textbook on transistor engineering. It should be apparent to the reader that all the theory and design principles presented herein are applicable to the engineering of integrated circuits as far as diffused transistors, diodes, resistors, oxide capacitors, and p-n junctions in silicon are concerned. These transistor engineering principles are basic to solid-state engineering and it may very well come to pass that the electronic circuit engineering function as we know it today will dissolve away between the forces of the system engineering function and the new solid-state engineering function, which will be elevated to the position of supplying the integrated semiconductor circuit as the component rather than the transistor alone. The transistor engineer and the electronic circuit engineer of today will inevitably be the solid-state engineers of tomorrow.

REFERENCES

1. Pritchard, R. L.: Effect of Base-contact Overlap and Parasitic Capacitance on Small-Signal Parameters of Junction Transistors, *Proc. IRE*, vol. 43, pp. 38–40, January, 1955.
2. Ebers, J. J., and S. L. Miller: Design of Alloyed Junction Germanium Transistors for High-speed Switching, *Bell System Tech. J.*, vol. 34, pp. 761–781, July, 1955.
3. Moore, A. R., and J. I. Pankove: The Effect of Junction Shape and Surface Recombination on Transistor Current Gain, *Proc. IRE*, vol. 42, pp. 907–913, June, 1954.
4. Kestenbaum, A. L., and N. H. Ditrick: Design, Construction, and High-frequency Performance of Drift Transistors, *RCA Rev.*, vol. 18, pp. 12–23, March, 1957.
5. Thornton, C. G., and J. B. Angell: Technology of Micro-alloy Diffused Transistors, *Proc. IRE*, vol. 46, pp. 1166–1176, June, 1958.
6. Halpern, J., and R. H. Rediker: Outdiffusion as a Technique for the Production of Diodes and Transistors, *Proc. IRE*, vol. 46, pp. 1068–1076, June, 1958.

APPENDIX:

Analysis of the Equilibrium Lifetime for Single-level Recombination Centers

According to the Shockley-Read-Hall theory, it is assumed that the density N_R of the recombination centers is small compared to the equilibrium concentration of carriers, n and p. Further, the centers all lie at a single energy value E_R in the band gap. To determine the net recombination rate R, we must consider that it is equal to the capture rate of the centers less the emission rate of same. We shall treat the case of the electron first.

It was shown in Chap. 3 that if the Fermi-Dirac function $f(E)$ represents the probability that a state at energy E is occupied, then $[1 - f(E)]$ represents the probability that the state is empty. If C_n is the capture probability of the centers for electrons, the capture rate for electrons is proportional to the electron concentration n and the number of centers that are empty or $N_R[1 - f(E_R)]$. The capture rate is therefore $nC_nN_R[1 - f(E_R)]$. The emission rate, however, is proportional to the number of occupied centers $N_Rf(E_R)$ and the thermal emission rate e_n for electrons. Therefore, the net electron-recombination rate is simply

$$R_n = nC_nN_R[1 - f(E_R)] - e_nN_Rf(E_R)$$

Similar reasoning may be applied to the case of the hole, with the exception that the capture rate is proportional to the number of filled centers. Likewise, the emission rate for holes is proportional to the number of empty centers. Thus,

$$R_p = pC_pN_Rf(E_R) - e_pN_R[1 - f(E_R)]$$

where C_p and e_p have definitions corresponding to those of C_n and e_n.

For nonequilibrium conditions, the net recombination rates R_n and R_p must be equal to each other, or

$$R = nC_nN_R[1 - f(E_R)] - e_nN_Rf(E_R) = pC_pN_Rf(E_R) - e_pN_R[1 - f(E_R)]$$

Therefore, the probability factor for filled centers is

$$f(E_R) = \frac{1}{1 + (pC_p + e_n)/(nC_n + e_p)}$$

The emission-rate parameters e_n and e_p may be eliminated by considering equilibrium conditions. At equilibrium, the net recombination rates must be zero. Therefore, if the Fermi level is adjusted such that $E_F = E_R$, the emission rates for electrons must equal the free-electron concentration n_r times their capture probability C_n. In other words,

$$e_n = n_r C_n$$

Also, for holes,

$$e_p = p_r C_p$$

Recall that n_r and p_r would be the carrier concentrations if the Fermi level coincided with the center energy level. Substituting these relations into the $f(E_R)$ expression, and inserting the result into either side of the equation for R, we get

$$R = \frac{N_R C_n C_p (np - n_r p_r)}{C_n(n + n_r) = C_p(p + p_r)}$$

If we substitute the definitions for limiting lifetimes, viz., $\tau_{pr} = 1/N_R C_p$ and $\tau_{nr} = 1/N_R C_n$ and $n_r p_r = n_i^2$, the final expression for the net rate of recombination becomes

$$R = \frac{np - n_i^2}{\tau_{pr}(n + n_r) + \tau_{nr}(p + p_r)}$$

The lifetime is defined as

$$\tau \equiv \frac{\Delta c}{R}$$

where $\Delta c = \Delta p = \Delta n$, and represents small departures from the equilibrium values. Performing the differentiation, we get, finally,

$$\tau = \frac{\tau_{pr}(n + n_r) + \tau_{nr}(p + p_r)}{n + p}$$

which is given in Chap. 4 as Eq. (4-41).

INDEX